Lecture Notes in Mechanical Engineering

Lecture Notes in Mechanical Engineering (LNME) publishes the latest developments in Mechanical Engineering - quickly, informally and with high quality. Original research reported in proceedings and post-proceedings represents the core of LNME. Volumes published in LNME embrace all aspects, subfields and new challenges of mechanical engineering. Topics in the series include:

- Engineering Design
- Machinery and Machine Elements
- Mechanical Structures and Stress Analysis
- Automotive Engineering
- Engine Technology
- Aerospace Technology and Astronautics
- Nanotechnology and Microengineering
- Control, Robotics, Mechatronics
- MEMS
- Theoretical and Applied Mechanics
- Dynamical Systems, Control
- Fluid Mechanics
- Engineering Thermodynamics, Heat and Mass Transfer
- Manufacturing
- Precision Engineering, Instrumentation, Measurement
- Materials Engineering
- Tribology and Surface Technology

To submit a proposal or request further information, please contact the Springer Editor in your country:

China: Li Shen at li.shen@springer.com
India: Dr. Akash Chakraborty at akash.chakraborty@springernature.com
Rest of Asia, Australia, New Zealand: Swati Meherishi at swati.meherishi@springer.com
All other countries: Dr. Leontina Di Cecco at Leontina.dicecco@springer.com

To submit a proposal for a monograph, please check our Springer Tracts in Mechanical Engineering at http://www.springer.com/series/11693 or contact Leontina.dicecco@springer.com

Indexed by SCOPUS. The books of the series are submitted for indexing to Web of Science.

More information about this series at http://www.springer.com/series/11236

Adam Hamrol · Marta Grabowska ·
Damjan Maletic · Ralf Woll
Editors

Advances in Manufacturing II

Volume 3 - Quality Engineering and Management

Springer

Editors
Adam Hamrol
Poznan University of Technology
Poznan, Poland

Marta Grabowska
Poznan University of Technology
Poznan, Poland

Damjan Maletic
University of Maribor
Kranj, Slovenia

Ralf Woll
Brandenburg University of Technology
Cottbus, Brandenburg, Germany

ISSN 2195-4356 ISSN 2195-4364 (electronic)
Lecture Notes in Mechanical Engineering
ISBN 978-3-030-17268-8 ISBN 978-3-030-17269-5 (eBook)
https://doi.org/10.1007/978-3-030-17269-5

This Springer imprint is published by the registered company Springer Nature Switzerland AG
The registered company address is: Gewerbestrasse 11, 6330 Cham, Switzerland

Preface

This volume of Lecture Notes in Mechanical Engineering contains selected papers presented at the 6th International Scientific-Technical Conference MANUFACTURING 2019, held in Poznan, Poland, on May 19–22, 2019. The conference was organized by the Faculty of Mechanical Engineering and Management, Poznan University of Technology, Poland, under the scientific auspices of the Committee on Machine Building and Committee on Production Engineering of the Polish Academy of Sciences.

The aim of the conference was to present the latest achievements in mechanical engineering and to provide an occasion for discussion and exchange of views and opinions. The main conference topics were:

- quality engineering and management
- production engineering and management
- mechanical engineering
- metrology and measurement systems
- solutions for Industry 4.0.

The organizers received 293 contributions from 36 countries around the world. After a thorough peer review process, the committee accepted 167 papers for conference proceedings prepared by 491 authors from 23 countries (acceptance rate around 57%). Extended versions of selected best papers will be published in the following journals: *Flexible Services and Manufacturing Journal, Research in Engineering Design, Management and Production Engineering Review and Archives of Mechanical Technology and Materials.*

The book **Advances in Manufacturing II** is organized into five volumes that correspond with the main conference topic mentioned above.

Advances in Manufacturing II - Volume 3 - Quality Engineering and Management is a collection of papers that depicts problems from various management areas, including management of the enterprise and production processes in the era of globalization. Papers included in this book contain considerations of current issues related to technical development, among other: technology transfer and cultural problems, management of the education sphere, quality control issues with statistical control and problems of continuous process improvement, including Six Sigma and lean manufacturing strategies. The book consists of 21 chapters, prepared by 55 authors from 6 countries.

We would like to thank the members of the International Program Committee for their hard work during the review process.

We acknowledge all that contributed to the staging of MANUFACTURING 2019: authors, committees, and sponsors. Their involvement and hardwork were crucial to the success of the MANUFACTURING 2019 conference.

May 2019

Adam Hamrol
Marta Grabowska
Damjan Maletic
Ralf Woll

Organization

Steering Committee

General Chair

Adam Hamrol	Poznan University of Technology, Poland

Chairs

Olaf Ciszak	Poznan University of Technology, Poland
Stanisław Legutko	Poznan University of Technology, Poland

Scientific Committee

Stanisław Adamczak, Poland
Michal Balog, Slovakia
Zbigniew Banaszak, Poland
Myriam Elena Baron, Argentina
Stefan Berczyński, Poland
Johan Berglund, Sweden
Wojciech Bonenberg, Poland
Christopher A. Brown, USA
Anna Burduk, Poland
Somnath Chattopadhyaya, India
Shin-Guang Chen, Taiwan
Danut Chira, Romania
Edward Chlebus, Poland
Damir Ciglar, Croatia
Marcela Contreras, Mexico
Nadežda Cuboňová, Slovakia
Jens J. Dahlgaard, Sweden
María de los Angeles Cervantes Rosas, Mexico
Andrzej Demenko, Poland
Magdalena Diering, Poland
Ewa Dostatni, Poland
Jan Duda, Poland
Davor Dujak, Croatia
Milan Edl, Czech Republic
Sabahudin Ekinovic, Bosnia and Herzegovina
Mosè Gallo, Italy
Bartosz Gapiński, Poland
Józef Gawlik, Poland
Hans Georg Gemuenden, Norway
Boštjan Gomiščček, UEA

Marta Grabowska, Poland
Wit Grzesik, Poland
Michal Hatala, Slovakia
Sandra Heffernan, New Zealand
Christoph Herrmann, Germany
Ivan Hudec, Slovakia
Vitalii Ivanov, Ukraine
Andrzej Jardzioch, Poland
Mieczyslaw Jurczyk, Poland
Wojciech Kacalak, Poland
Lyudmila Kalafatova, Ukraine
Anna Karwasz, Poland
Mourad Keddam, Algeria
Sławomir Kłos, Poland
Ryszard Knosala, Poland
Janusz Kowal, Poland
Drazan Kozak, Croatia
Agnieszka Kujawińska, Poland
Janos Kundrak, Hungary
Maciej Kupczyk, Poland
Ivan Kuric, Slovakia
Oleksandr Liaposhchenko, Ukraine
Piotr Łebkowski, Poland
José Mendes Machado, Portugal
Aleksandar Makedonski, Bulgaria
Ilija Mamuzic, Croatia
Krzysztof Marchelek, Poland
Tadeusz Markowski, Poland
Edison Perozo Martinez, Colombia
Thomas Mathia, France
Józef Matuszek, Poland
Adam Mazurkiewicz, Poland
Andrzej Milecki, Poland
Mirosław Pajor, Poland
Ivan Pavlenko, Ukraine
Dragan Perakovic, Croatia
Alejandro Pereira Dominguez, Spain
Marko Periša, Croatia
Emilio Picasso, Argentina
Jan Pitel, Slovakia
Alla Polyanska, Ukraine
Włodzimierz Przybylski, Poland
Luis Paulo Reis, Portugal
Álvaro Rocha, Portugal
Rajkumar Roy, UK
Iwan Samardzic, Croatia
Krzysztof Santarek, Poland
Jarosław Sęp, Poland
Bożena Skołud, Poland
Jerzy Sładek, Poland
Roman Staniek, Poland
Beata Starzyńska, Poland
Tomasz Sterzyński, Poland
Tomasz Stręk, Poland
Antun Stoić, Croatia
Manuel Francisco Suarez Barraza, Mexico
Marek Szostak, Poland
Rafał Talar, Poland
Franciszek Tomaszewski, Poland
María Estela Torres Jaquez, Mexico
Justyna Trojanowska, Poland
Stefan Trzcieliński, Poland
Maria Leonilde R. Varela, Portugal
Sachin D. Waigaonkar, India
Edmund Weiss, Poland
Michał Wieczorowski, Poland
Ralf Woll, Germany
Magdalena Wyrwicka, Poland
Jozef Zajac, Slovakia
Jan Żurek, Poland

Program Committee

Available on http://manufacturing.put.poznan.pl/en/.

Special Sessions

Collaborative Manufacturing and Management in the Context of Industry 4.0

Special Session Organizing Committee

Leonilde Varela	University of Minho, Portugal
Justyna Trojanowska	Poznan University of Technology, Poland
Vijaya Kumar Manupati	Mechanical Engineering Department, NIT Warangal
José Machado	University of Minho, Portugal
Eric Costa	Solent University, UK
Sara Bragança	Solent University, UK

Intelligent Manufacturing Systems

Special Session Organizing Committee

Ivan Pavlenko	Sumy State University, Ukraine
Sławomir Luściński	Kielce University of Technology, Poland

Tooling and Fixtures: Design, Optimization, Verification

Special Session Organizing Committee

Vitalii Ivanov	Sumy State University, Ukraine
Yiming Rong	Southern University of Science and Technology, China

Advanced Manufacturing Technologies

Special Session Organizing Committee

Jozef Jurko	TU Košice, Slovak Republic
Michal Balog	TU Košice, Slovak Republic
Tadeusz E. Zaborowski	TU Poznań, Poland

The Changing Face of Production Engineering and Management in a Contemporary Business Landscape

Special Session Organizing Committee

Damjan Maletič	University of Maribor, Faculty of Organizational Sciences, Enterprise engineering Laboratory, Slovenia
Matjaž Maletič	University of Maribor, Faculty of Organizational Sciences, Enterprise engineering Laboratory, Slovenia
Tomaž Kern	University of Maribor, Faculty of Organizational Sciences, Enterprise engineering Laboratory, Slovenia

Enabling Tools and Education for Industry 4.0

Special Session Organizing Committee

Dorota Stadnicka	Politechnika Rzeszowska, Poland
Dario Antonelli	Politecnico di Torino, Italy
Katarzyna Antosz	Politechnika Rzeszowska, Poland

Staff for the Industry of the Future

Special Session Organizing Committee

Magdalena Wyrwicka	Poznan University of Technology, Poland
Anna Vaňová	Faculty of Economics, Matej Bel University, Slovakia
Maciej Szafrański	Poznan University of Technology, Poland
Magdalena Graczyk-Kucharska	Poznan University of Technology, Poland

Advances in Manufacturing, Properties, and Surface Integrity of Construction Materials

Special Session Organizing Committee

Szymon Wojciechowski	Poznan University of Technology, Poland
Grzegorz M. Królczyk	Opole University of Technology, Poland
Sergei Hloch	Technical University of Kosice, Slovakia

Materials Engineering

Special Session Organizing Committee

Monika Dobrzyńska-Mizera	Poznan University of Technology, Poland
Monika Knitter	Institute of Materials Technology, Poznan University of Technology, Poland
Robert Sika	Institute of Materials Technology, Poznan University of Technology, Poland
Dariusz Bartkowski	Institute of Materials Technology, Poznan University of Technology, Poland
Waldemar Matysiak	Institute of Materials Technology, Poznan University of Technology, Poland
Anna Zawadzka	Institute of Materials Technology, Poznan University of Technology, Poland

Advanced Mechanics of Systems, Materials and Structures

Special Session Organizing Committee

Hubert Jopek	Institute of Applied Mechanics, Poznan University of Technology, Poland
Paweł Fritzkowski	Institute of Applied Mechanics, Poznan University of Technology, Poland
Jakub Grabski	Institute of Applied Mechanics, Poznan University of Technology, Poland
Krzysztof Sowiński	Institute of Applied Mechanics, Poznan University of Technology, Poland
Agata Matuszewska	Institute of Applied Mechanics, Poznan University of Technology, Poland

Virtual and Augmented Reality in Manufacturing

Special Session Organizing Committee

Filip Górski	Poznan University of Technology, Poland
Paweł Buń	Poznan University of Technology, Poland
Damian Grajewski	Poznan University of Technology, Poland
Jorge Martin-Gutierrez	Universidad de la Laguna, Spain
Letizia Neira	Universidad Autónoma de Nuevo León, Mexico
Eduardo Gonzalez Mendivil	Tecnologico de Monterrey, Mexico

Contents

Requirements Engineering for Production Transfer to Developing Countries

Matthias Brönner[1(✉)], Valerie Baumgartner[2], and Markus Lienkamp[1]

[1] Institute of Automotive Technology,
Technical University of Munich,
Boltzmannstr. 15, 85748 Garching b. Munich, Germany
broenner@ftm.mw.tum.de

[2] Technical University of Munich,
Boltzmannstr. 15, 85748 Garching b. Munich, Germany

Abstract. Sustainability requires localization of manufacturing sites, especially in developing countries – even when products are previously designed globally. However, the establishment of manufacturing/production locations in these countries poses crucial challenges for companies that do not yet have facilities or experience in developing countries. For this reason, production transfer planning based on verifiable facts is essential. Therefore, we present a framework to develop production transfer requirements based on the methodic proceedings of requirements engineering – discovery, classification, prioritization and specification. We focus on the identification of stakeholders within the transfer processes. In addition, the classification of the requirements, their correlation, and prioritization are included. The application is conducted exemplarily as part of the aCar mobility project, which plans the transfer of an electric vehicle to developing countries. In conclusion, we present a framework to support production transfer planning and show applicability executing the framework on a current project.

Keywords: Production transfer · Sustainability · Requirements engineering · Decision making · Developing countries

1 Production Transfer to Developing Countries

Once the decision has been taken to transfer production to developing countries, the planning phase of production transfers begins. Within these, the question ‘who poses to whom which requirements?’ must be answered. To support this process, we present a framework to develop requirements within the planning phase of production transfer to developing countries. We are convinced that domestic production of goods destined for developing countries must be part of a social and economic sustainable corporate strategy. Therefore, the factors defining production in developing countries, with an emphasis on economic and social sustainability, are presented first. These drivers of local production face challenges that companies must consider when planning production in less developed countries. To systematically develop, and overcome these challenges during the transfer planning, a framework based on the Requirements Engineering procedure is adapted for production transfer planning.

A. Hamrol et al. (Eds.): *Advances in Manufacturing II* - Volume 3, LNME, pp. 1–15, 2019.
https://doi.org/10.1007/978-3-030-17269-5_1

2 Local Production

The World Health Organization [1, p. 19] defines local production as "[f]irst, *local production* is the domestic production of […] devices by a country utilizing that device to solve a local […] need. Second, *local production* can be owned by either/both an international or national industry, though the majority of this ownership should be national". According to this definition, *local production in developing countries* is defined in this paper as the domestic production of products mainly produced for the local market.

For transnational corporations (TNC), step-by-step strategies are known from the literature to facilitate entry into developing markets. For example, Leontiade [2] presents an approach that changes local value creation based on the market situation. In the so-called pre-markets products are imported, in the take-off-markets a local assembly takes place. Large quantities are produced locally in early-mass-markets, and are adapted to local conditions. In the mature-mass-markets new technologies are developed through local R&D. Xie [3] and Karabag [4] are pursuing similar approaches and confirm these strategies for market entries in China with color TVs and in Turkey with commercial vehicles. Despite these promising approaches, local production in developing countries is still characterized by low value added [5], which is contrary to a sustainable corporate strategy that includes local value adding in developing countries [6].

2.1 Sustainability in Developing Countries

The primary corporate objectives in developing countries dealt with in this paper are social sustainability, and economic sustainability. A socially sustainable company attaches particular importance to creating jobs to combat poverty and improve local living conditions [7, 8]. This also includes the fight against child labor [9] and for equal income [7]. Thus, the World Trade Organization describes a direct link between well paid local jobs and the local demand for products [10]. Economic sustainability is enhanced by the resulting customer proximity. This allows savings in logistics due to short distribution channels [11], and a quicker response to market changes. These responses can be adjustments of the number of units produced or the product design [11–14].

One of the basic motivations to set up local production in developing countries is to enter a new market, characterized by a large number of new customers [11]. Despite their still low purchasing power, this will increase in the future, and thus the demand of the local population for higher quality products [15]. For this reason, it is essential to offer appropriately priced products for these price-sensitive markets already today. This is made possible by lower product costs, enabled by local production. On-site production not only saves logistics costs, as mentioned, it also enables price reductions through low investment costs [7, 13, 16], lower wage costs [11, 12, 17], local suppliers [11, 18], and adapted quality [13]. Furthermore, by producing local in developing countries governmental protection measures, like import duties and local content requirements [11, 18], must not be considered. On the contrary, the local government often subsidizes companies creating local value [8, 18].

2.2 Challenges of Local Production in Developing Countries

Alongside these opportunities offered by local production, possible challenges are mentioned in the literature. Below, some are described exemplarily – categorized as related to product, technology, employees, organization, sales & procurement and society & policies. These categories are in accordance with the production operational systems described by Pawellek [19].

Products in developing countries are influenced significantly by required adjustments in price, function, and quality [11, 18, 20]. These must also be profitable for the small quantities expected, which presupposes a low break-even point [13, 17, 20]. Furthermore it is essential to test the product locally [21, 22].

Next to these product challenges, appropriate technologies are required for a local production location in developing countries. That means technologies adapted to local conditions with an emphasis on labor-intensive methods [6, 11, 20]. Furthermore, they should avoid tests [23], be low in complexity [11, p. 192], and prevent high scrap rates [24]. This is particularly important as there are difficulties in finding qualified employees [11, 20, 25] with implicit product knowledge [18] or management experience [25]. Additionally, the fluctuation rate of trained employees challenges the TNCs [26, 27]. This is critical as the transfer of know-how is seen as a core challenge [11, 21]. Therefore, TNCs have to aim for long-term relationships with their employees [27, 28].

On the organizational side it is also important to aim for long-term relationships with local partners [29, 30]. These include the management staff, which should be of local birth [20]. Thus, local competency can be gained [25, 31, 32]. Additionally, the company culture should be adapted to the local culture [33, 34].

Externally, local politics as well as society influence production, e.g. by local requirements, like safety, security and recycling laws [11, 29, 30]. Another aspect is that, society and politics demand social commitment in the respective country [7, 8, 30]. Purchasing and sales are external factors influencing production in developing countries [19]. Both factors are challenged by the prevailing infrastructure, as this makes time-controlled distribution and purchasing difficult [25, 29, 32]. Therefore, it is essential to integrate suppliers into the company's own network [35]. This enables collaborative partnership [34, 36] and the development of local suppliers through training [37].

3 Theoretical Background of the Framework

Local production is the logical step in the TNC's goal of sustainability in developing countries, but the transfer of production to these regions is made difficult by the environmental challenges prevailing there, as some of them are unknown. In this chapter, the theoretical background of production transfer is presented and critically considered, followed by a presentation of requirements engineering. This prepares for the subsequent solution finding.

3.1 Production Transfer

For relocation processes, which can be adapted to production transfer, Ponton [38] proposes a stepwise approach (Fig. 1). The transfer begins with project initialization and a subsequent decision phase. Based on this, the planning phase starts, followed by the physical transfer. Finally, the process is finished by ramp-up and controlling.

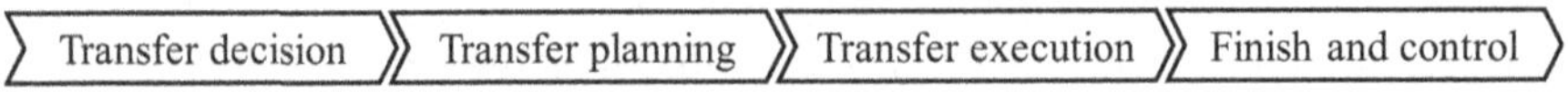

Fig. 1. Transfer process, adapted from [38].

As described above, we focus on the planning phase of the production transfer and thus on questions concerning the development of requirements. It is therefore essential to define the term production transfer and to determine the related research fields. Since there is no common understanding of the term *transfer* in the technical literature, some examples of transfers shall be described below. Among others, the terms *technology transfer*, *know-how transfer*, *product transfer* and *production transfer* will also be mentioned.

According to Simona and Axele [36] knowledge transfer takes place when process, product, technology, manufacturing techniques and management know-how are shared. Galbraith [39] defines technology and knowledge transfer as the shift of production techniques between facilities. Production transfer as described by Fredriksson [40] means the start of a transfer until a steady state is reached. The term *product transfer* describes the transfer from development to production [41]. For a common understanding of the production transfer to developing countries, we define it as follows:

Production transfer to developing countries describes the process of transferring production locations from advanced developed countries to less developed countries, including the operative production system and operational knowledge.

Related to transfer research are location decision, factory planning, and outsourcing research. Herby, location decision research is focusing on the geographical selection of the right location and thus on location criteria as well as the cost reduction due to the (re-)location [11]. In contrast, approaches to factory planning such as those made by Pawellek [19] describe the planning of the factory in various detail, depending on the planning level (e.g. system level to workplace level). Hence, focus is on classic production goals: reduction of time and costs while increasing quality. Outsourcing research deals with the fields of supplier selection, supply chain management, make-or-buy decisions and cost reduction as exemplarily described in Fredriksson [40].

3.2 Literature Review: Production Transfer Planning

Hence, the question arises, which framework supports the process of production transfer planning especially in the development of requirements for production location planning? The framework's content should include the requirements that must be met for a successful transfer as well as its sustainability in developing countries, the environment in the destination country, and the product as well as business characteristics. These should be clearly identified at the operational system levels of production (Table 1).

Table 1. Content requirements for a framework supporting production transfer.

Content requirements
Production transfer requirements
Sustainability in developing countries
Environmental requirements
Product requirements
Production requirements
Business requirements

The results of the literature research are summarized and evaluated in (Table 2).

Table 2. Literature evaluation.

Content requirements	[42]	[43]	[44]	[45]	[46]	[47]	[48]	[49]	[50]	[51]	[52]
Production transfer requirements	◐	◕	◐	◐	◔	◔	◐	◐	◔	◕	◔
Environmental requirements	◐	◔	◐	○	◐	◔	◔	◔	◔	○	◕
Product requirements	◔	○	◐	○	○	◕	◐	◔	◔	○	◔
Production requirements	◔	◔	◔	◐	◐	◐	◔	◔	◐	○	◔
Business requirements	◐	◔	◐	◔	○	○	◕	◔	◔	◔	◐

Legend: An empty Harvey Ball means "not mentioned", a quarter-filled Harvey Ball "mentioned in context", a half Harvey Ball "explicitly described", a three-quarter Harvey Ball "context systematically integrated", and a filled Harvey Ball "systematically integrated" within the source.

When reviewing the literature, it is noticeable that existing methods either describe a generic method for production transfer [42–45, 48, 49, 51] or specifically work out the product characteristics that are considered success factors in production transfer [44, 47, 48]. Economic factors against the background of production transfer are also dealt with in the literature [42, 44, 48, 52], as are the influences of environmental factors [42, 44, 46, 52]. Based on this evaluation, we found that current research lacks a systematic development of production transfer requirements relating environmental influences, product and production characteristics as well as economic influences. To fill this gap, we suggest using a known procedure. Thus, the method of requirements engineering known from software development is presented below.

3.3 Requirements and Requirements Engineering

The IEEE 830-1998 [49] focuses on a systematic definition of requirements. Hence, the development of a common understanding of requirements between customer and supplier is a core task. This reduces the development effort substantially and gives the possibility for time and cost planning. Final verification and validation of the results also takes place because of a common understanding of requirements. This enables explicitly a possible transfer of the software product to other applications [49]. According to [49], requirements should be:

- correct
- unambiguous
- complete
- consistent
- ranked for importance
- verifiable
- modifiable
- traceable.

When setting up requirements, the background is to be considered. Including environment, preparation, prototyping and project requirements [53]. In 2011, the ISO/IEC/IEEE 29148 was published explaining further the requirements development in software engineering. To detail the requirements process, various levels in business context are introduced. External environmental conditions such as market trends, laws, and regulations have an impact. Internally, the organizational environment is the highest level that defines standards & specifications. Below this is the business operation level which defines processes and constraints. The proceeding's center is the system with its elements, such as the software [54]. Sommerville [55] takes up these basic specifications to design the requirements engineering process. They include four steps (1) requirements discovery, (2) requirements classification, (3) requirements prioritization and (4) requirements specification.

4 Requirements Engineering for Production Transfer Planning

As described previously, the framework supporting production transfer planning is created to fill the lack of a process to develop requirements. This is necessary due to the challenges in less developed countries described in Sect. 2.2.

4.1 General Description and Levels of Requirements

Like the requirements engineering process of software development, production transfer planning has three major levels (I to III) as shown in (Fig. 2). For each of those levels, the four steps (a to d) of the requirements engineering approach must be conducted. Following, the process is described in detail.

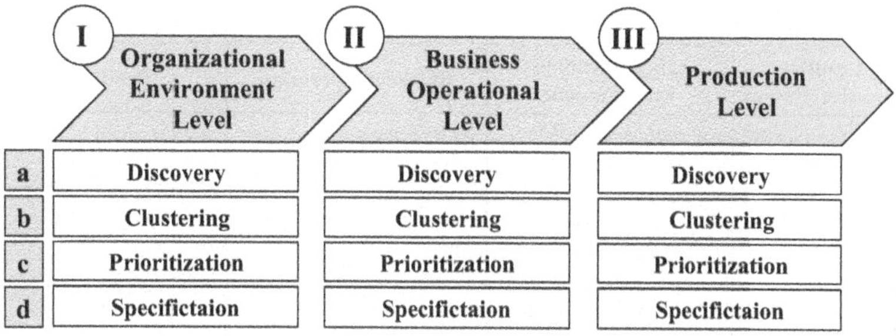

Fig. 2. Production transfer planning using requirements engineering.

The requirements' levels in the production transfer planning are organizational environment (I), business operation (II), and production (III). Further division can be done into production elements (organization, product, technology, and employees) on the respective level.

4.2 Development of Requirements

In production transfer planning, the question 'Who makes what requirements on whom?' should be answered. The framework approaches the solution by defining the included parties. Thus, requirements clustering and allocating to the executing departments is possible. Prioritization of the requirements supports efficient processing as well as the concluding specification process.

4.2.1 Requirements Discovery

The requirement search is carried out at the respective level, mentioned above, and can be further supported by querying the external and internal production operative systems: product, technology, organization, politics & society, and procurement & sales (Fig. 3). The search for requirements can be supported by specific questions, for example, at organizational environment level – *"Are restrictions on the product set by the local policy?"* – at the business operation level – *"Which features is the product supposed to fulfil?"* – and the production level – *"Has the necessary quality of components for the local conditions been defined?"*. These questions are part of a questionnaire that we have designed based on the challenges of production in developing countries to support the requirements discovery. These questions must be answered either by internal or external experts in order to determine the requirements.

Requirements Discovery

Level/ Operative System	Organizational Environment	Business Operation	Production
Product	*Are restrictions on the product set by the local policy?*	*Which features is the product supposed to fulfil?*	*Has the quality of components for the local conditions been defined?*
Technology	*What is the know-how of potential employees like?*	*What quality to sell to the customer?*	*Are the necessary safety requirements for the technology defined?*
Organization	*Are the investment costs defined?*	*Is the personnel cost situation known?*	*How is qualification of employees for the technologies ensured?*
Employee	*How are professionals recruited?*	*How is IP protected?*	*How are employees trained for work tasks?*
Politics & Society	*Have the local labor conditions been clarified?*	*Which depth of value added is the long-term goal?*	*What depth of value added is required by local politics?*
Procurement & Sales	*Is the necessary information about the local infrastructure available?*	*Is the distribution concept defined?*	*Which components can be purchased locally?*

Fig. 3. Levels and operative systems of the requirements and exemplary questions.

4.2.2 Categorization of Requirements

Once requirements have been discovered they have to be clustered. In the presented framework the requirements are categorized based on the executives in the product development process after Ehrlenspiel and Meerlenspiel [56]:

1. Marketing & Planning
2. Development
3. Materials Management & Logistics
4. Manufacturing & Assembly Planning
5. Component Manufacturing
6. Assembly
7. Test
8. Sales
9. Shipping
10. Commissioning

Further requirements are classified in the categories of the operative systems (product, technology, organization, employee, politics & society and procurement & sales). This enables an assignment to the responsible departments, which convert the requirements into measures. Additionally, the assignment to the operative systems remains intact and enables an efficient iteration, if required (Fig. 4).

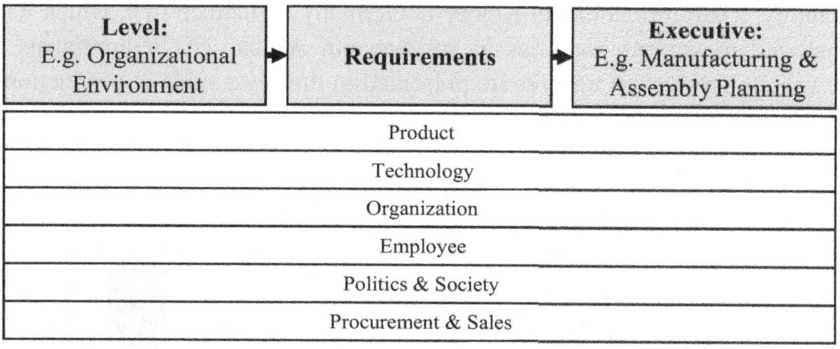

Fig. 4. Assignment and classification of requirements.

4.2.3 Requirements Prioritization and Specification

According to Ponn and Lindemann [57], requirements have to be prioritized to enable a focusing and allocation of resources. Hence, they propose a subdivision into demands and wishes [57]. The IEEE 830-1998 suggests a classification into essential, conditional, and optionally for the developed requirements [49]. This rating of requirements according to relevance is comparable to the classification according to the Kano model. Further possibilities are the House of Quality, evaluating according to effort or costs (requires an objectification of the requirements), as well as processing from outside to inside (environmental requirements to production requirements). The last procedure benefits from known environmental influences before detail levels are processed. Another part of the ranking of requirements should be based on production targets: Optimum quality at the lowest possible cost and time. These indicators extend the qualitative classification of the requirements, using quantitative indicators. The specification of the requirements must be consistent with the standards described in Sect. 3.3.

5 Application – The 'aCar Mobility' Project

The following case describes the 'aCar mobility' project where a light electric vehicle (LCEV) was developed especially for customers in rural areas of sub-Saharan Africa. Their requirements, e.g. the vehicle's robustness, range, ease of maintenance, payload of 1,000 kg, and multi-purpose use have been implemented in the vehicle. The aCar (Fig. 5) consists of a leader frame where the components module traverse, driver's cabin, and chassis are mounted. The batteries are attached in and under the cabin to ensure maximum ground clearance and safety. Power electronics, control units, high-voltage fuses and chargers are attached to the module traverse, which is located under the dashboard. With an interior and exterior made of bent sheet metal, the vehicle is easy to assemble and repair in the case of damage. The body allows the use of various loading platforms, which can be supplied with 230 V from the traction batteries. The aCar was presented at the international automotive show (IAA) in Frankfurt in 2017. The vehicle is planned be produced locally in developing countries in the terms of sustainability. Therefore, a model factory in Germany is planned first, which will then be transferred to various countries in sub-Saharan Africa. The requirements of the electric vehicle production transfer are presented in this case study at production level.

Fig. 5. The aCar – Side and top view.

5.1 Requirements for Assembly Planning at Production Level – Category Employee

The application of the framework is presented here exemplary for the aCar production transfer. We present the results of the requirements development in the area of employees at production level which must be executed by the assembly planning (Fig. 6).

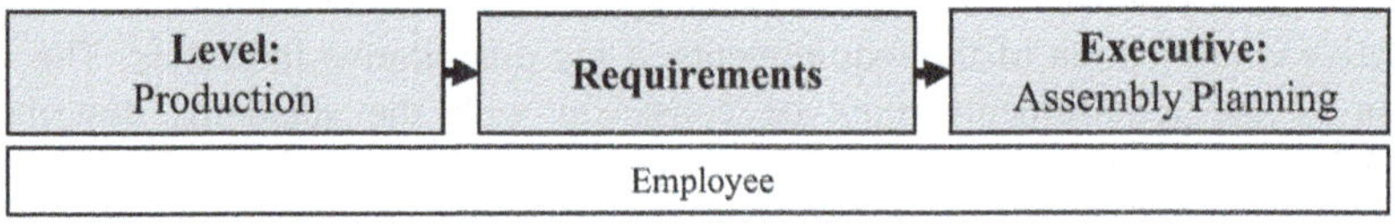

Fig. 6. Application – Level, category and executive.

In order to determine the requirements, we have included in the requirements discovery phase qualitative interviews according to a circular approach presented by Witt [58, p.6]. In total 12 interviews were conducted, with experts who work or have worked in developing countries. They included four experts working for German Original Equipment Manufacturers (OEMs), two working for African OEMs, two consultants, two experts working for suppliers, one expert in worker training, and one entrepreneur running a startup.

5.1.1 Requirements Discovery

According to the interviewees, the lack of qualified employees, the fluctuation and the culture in developing countries pose challenges to assembly planning (Fig. 7). This problem results from inadequate qualification of potential employees. Furthermore, there is a lack of employees with basic knowledge in joining technologies, electrical engineering skills as well as reading, and writing skills. Additionally, there are no employees with implicit knowledge in electric vehicle assembly, commissioning or work safety. These circumstances challenge the conduction of knowhow transfer. Further, our interviewees mention the fluctuation of skilled employees, the protection of intellectual property and training as difficulties in developing countries. Trained workers often leave the companies to take on better paid jobs. As examples of cultural differences, the interviewees point to the attitude towards time in the context of work culture. Especially mentioned is the position to work safety. To overcome these cultural differences, the employment of local managers is recommended, even if their availability is low.

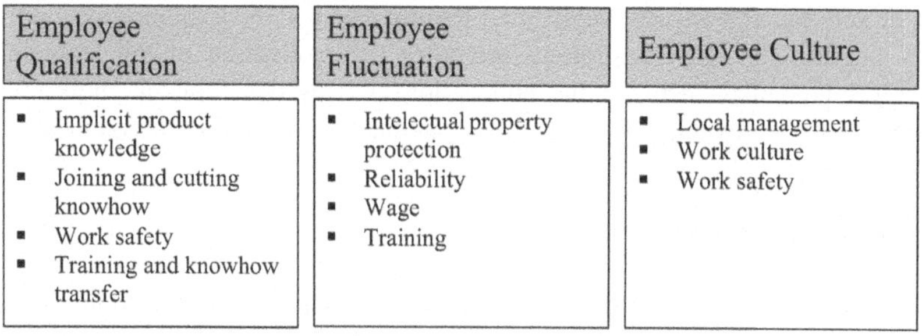

Fig. 7. Selected results of the interviews.

Based on these statements, the requirements for the qualification method for the assembly training are engineered. This training method is intended to impart professional competence near-the-job, whereby no level of employee qualification can be assumed so far. Temporal requirements are low training time because of the high fluctuation of trained employees. Secondary is the development time of the method. The quality must be independent of the trainer and the trainee. Furthermore, the motivation of the trainee is to be promoted and the quality of the training result is to be ensured. To keep costs low, the method should be scalable and the initial investment low. These requirements are summarized in (Table 3).

Table 3. Requirements for the qualification method.

Time	Low development time
	Low training time
Quality	High quality independent of the trainer and trainee
	Promotion of the employee motivation
	Ensure the training results
Costs	Low initial investment
	Scalability of the qualification method

5.1.2 Requirements Prioritization and Selection of a Qualification Method

We used a paired comparison to prioritize the requirements on the assembly qualification method. With this, most important is to secure the training result, followed by the level of initial investment, scalability and encouragement of motivation. The training period is in fourth place due to the low wages, followed by the constant quality of training. The development time is mentioned last. With these requirements, we have selected the method of employee qualification from 12 methods. These included the provision method, frontal teaching, quality circle, learning island, virtual learning and augmented reality. Because of these requirements the qualification of the employees using augmented reality is proposed for the qualification of employees to assemble aCars in Sub-Saharan Africa. Augmented Reality enables low training times with high quality output. The new technology promotes the motivation of the learners. The relatively high initial costs for the conception and the development time are compensated by these advantages and the low personnel input during the qualification.

6 Summary and Outlook

In this paper we present the basics of sustainable production in developing countries and approaches to plan production transfers to developing countries and their scientific deficiency are discussed in detail. This highlights the need for a methodology that develops the requirements for production in developing countries. The method must integrate the environmental factors in less developed countries in order to define the requirements of the production strategy and then plan the production location on this basis. Subsequently, the framework for a requirements engineering-based planning is presented. This enables the classification of the requirements into the level organizational environment, business operation and production. Further the levels are detailed in organization, employee, product, and technology. Dividing requirements planning into levels, the requirement discovery can be conducted more easily. Subsequent we suggest qualitative and quantitative methods for prioritization. Finally, we recommend within the specification of requirements to assign the requirements to departments. With the aCar project we show that the selection of a qualification method is supported by the presented framework. To discover the requirements, we use semi-structured qualitative interviews with experts in assembly training. Here the advantage of reducing complexity through dividing the transfer task into different levels and categories illustrates its benefits as the interview questions could be selected properly. The interviewees were able contribute their expertise in a systematic way through the presented division in three levels (organizational environment, business operation and production) and categories (product, technology, organization, procurement & sales and politics & society). With this framework requirements are developed based on verifiable facts. Remaining are tests on different levels of the framework. It must be established whether the framework enables the development of options for action in the production strategy and decision-making. Further expansion is planned concerning the linkages between the requirements. Additionally, implementation as a questionnaire should increase user-friendliness and efficiency.

Contributions and Acknowledgements. Matthias Brönner as the first author developed the presented framework, evaluated the interviews and leads the presented aCar project. Valerie Baumgartner conducted the interviews within her student thesis. Markus Lienkamp made an essential contribution to the conception of the research project. He revised the paper critically for important intellectual content. Markus Lienkamp gave final approval of the version to be published and agrees to all aspects of the work. As a guarantor, he accepts responsibility for the overall integrity of the paper. Many thanks to our interviewees and Svenja Kalt and Sebastian Wolff of the Institute for Automotive Technology for inspiring discussions and feedback. This paper is a result of the aCar mobility project. Further information can be found at http://www.acar.tum.de/. The research was conducted with basic research funds from the Institute of Automotive Technology, Technical University of Munich.

References

1. World Health Organization, Local production and technology transfer to increase access to medical devices: addressing the barriers and challenges in low-and middle-income countries, Geneva, Switzerland, WHO, vol. 17 (2012)
2. Leontiade, J.: Planning strategy for world markets. Long Range Plann. **3**(2), 40–45 (1970)
3. Xie, W.: Technological learning in China's colour TV (CTV) industry. Technovation **24**(6), 499–512 (2004)
4. Karabag, S.F., Tuncay-Celikel, A., Berggren, C.: The limits of R&D internationalization and the importance of local initiatives: Turkey as a critical case. World Dev. **39**(8), 1347–1357 (2011)
5. Filho, E.R.: Brazilian design for sustainability: in search of a local approach. J. Cleaner Prod. **107**, 467–474 (2015)
6. Bendul, J.C., Rosca, E., Pivovarova, D.: Sustainable supply chain models for base of the pyramid. J. Cleaner Prod. **162**, S107–S120 (2017)
7. Dane, A.: The potential of electric vehicles to contribute to South Africa's greenhouse gas emissions targets and other developmental objectives: How appropriate is the investment in electric vehicles as a NAMA? (2013). https://open.uct.ac.za/bitstream/11427/17148/1/Dane_potential_electric_2013.pdf. Accessed 05 Apr 2018
8. Dombrowski, U., Schmidt, S., Wrehde, J.: Challenges in the Automotive Production. (inGerman) ZWF, vol. 101, no. 5, pp. 254–259 (2006)
9. Lund-Thomsen, P., Nadvi, K., Chan, A., Khara, N., Xue, H.: Labour in global value chains: work conditions in football manufacturing in China, India and Pakistan. Dev. Change **43**(6), 1211–1237 (2012)
10. World Trade Organization, Trade, Technology and Jobs. WTO Publications, Geneva (2017). https://www.wto.org/english/res_e/booksp_e/world_trade_report17_e.pdf. Accessed 07 Jan 2019
11. Abele, E., Meyer, T., Näher, U., Strube, G., Sykes, R. (Eds.): Global Production: A Handbook for Strategy and Implementation. Springer, Heidelberg (2008)
12. Dombrowski, U., Vollrath, H.: Optimization of CKD-Processes, ZWF, vol. 103, no. 1–2, pp. 48–52 (2008). (in German)
13. Herrigel, G.: Globalization and the German industrial production model. J. Labour Mark. Res. **48**(2), 133–149 (2015). https://doi.org/10.1007/s12651-014-0170-5

14. Rauch, E., Dallasega, P., Matt, D.T.: Sustainable production in emerging markets through Distributed Manufacturing Systems (DMS), Bridges for a more sustainable future: Joining Environmental Management for Sustainable Universities (EMSU) and the European Roundtable for Sustainable Consumption and Production (ERSCP) conferences, vol. 135, pp. 127–138 (2016)
15. Radjou, N., Prabhu, J.C., Ahuja, S.: Jugaad Innovation: Think Frugal, be Flexible, Generate Breakthrough Growth, 1st edn. Jossey-Bass, San Francisco (2012)
16. Kampker, A., Förstmann, R., Ordung, M., Haunreiter, A.: Prototypes in agile development management, ATZ - Automobiltechnische Zeitschrift, vol. 118, no. 7, pp. 72–77 (2016). (in German)
17. Göpfert, I., Braun, D., Schulz, M.: Automotive Logistic; State and Future Trends. Springer Fachmedien Wiesbaden, Wiesbaden (2017). (in German)
18. Gastrow, M.: A review of trends in the global automotive manufacturing industry and implications for developing countries. Afr. J. Bus. Manage **6**(19) (2012)
19. Pawellek, G.: Holistic Factory Planning. Springer, Heidelberg (2014). (in German)
20. Stewart, F.: Technology and employment in LDCs. World Dev. **2**(3), 17–46 (1974)
21. Bångens, L., Araujo, L.: The structures and processes of learning. a case study. J. Bus. Res. **55**(7), 571–581 (2002)
22. Mattson, C.A., Wood, A.E.: Nine principles for design for the developing world as derived from the engineering literature. J. Mech. Des. **136**(12), 121403 (2014)
23. Bohn, R.: The economics of yield-driven processes. J. Oper. Manage. **18**(1), 41–59 (1999)
24. Alcorta, L.: The diffusion of advanced automation in developing countries: factors and adoption process. Technovation **19**(3), 163–175 (1999)
25. Uddin, M.K.: The role of diffusion of innovations for incremental development in small enterprises. Technovation **26**(2), 274–284 (2006)
26. King, K.J.: Kenya's informal machine-makers: a study of small-scale industry in Kenya's emergent artisan society. World Dev. **2**(4–5), 9–28 (1974)
27. Bair, J., Gereffi, G.: Local clusters in global chains: the causes and consequences of export dynamism in torreon's blue jeans industry. World Dev. **29**(11), 1885–1903 (2001)
28. Gereffi, G., Fernandez-Stark, K.: Global Value Chain Analysis: A Primer, 2nd edn. Duke Center on Globalization (2016)
29. Smith, A., Fressoli, M., Thomas, H.: Grassroots innovation movements: challenges and contributions. J. Cleaner Prod. **63**, 114–124 (2014)
30. Rosca, E., Arnold, M., Bendul, J.C.: Business models for sustainable innovation – an empirical analysis of frugal products and services. J. Cleaner Prod. **162**, S133–S145 (2017)
31. Athreye, S., Tuncay-Celikel, A., Ujjual, V.: Internationalisation of R&D into emerging markets: fiat's R&D in Brazil, Turkey and India. Long Range Plann. **47**(1–2), 100–114 (2014)
32. Teitel, S.: On the concept of appropriate technology for less industrialized countries. Technol. Forecast. Soc. Chang. **11**(4), 349–369 (1978)
33. Browa, H.: The Causes, Conditions and Patterns of Appropriate Technology: An Attempt of Explanation and Description by means of System-Analytical Macro-Approach, IFAC Proceedings Volumes, vol. 16, no. 2, pp. 11–20 (1983)
34. Ivarsson, I., Alvstam, C.G.: Technology transfer from TNCs to local suppliers in developing countries: a study of AB Volvo's truck and bus plants in Brazil, China, India, and Mexico. World Dev. **33**(8), 1325–1344 (2005)
35. Thomas, R.J., Wind, Y.: Symbiotic innovation: getting the most out of collaboration. In: Brem, A., Viardot, É. (Eds.) Evolution of Innovation Management: Trends in an International Context, pp. 1–31. Palgrave Macmillan UK, London (2013)

36. Simona, G.-L., Axèle, G.: Knowledge transfer from TNCs and upgrading of domestic firms: the polish automotive sector. World Dev. **40**(4), 796–807 (2012)
37. Altenburg, T., Meyer-Stamer, J.: How to promote clusters: policy experiences from Latin America. World Dev. **27**(9), 1693–1713 (1999)
38. Ponton, P., Jaehne, M., Mueller, E.: Relocation in production networks of a multi-national-enterprise. In: ElMaraghy, H.A. (Ed.) Enabling Manufacturing Competitiveness and Economic Sustainability, pp. 488–492. Springer, Heidelberg (2012)
39. Galbraith, C.S.: Transferring core manufacturing technologies in high-technology firms. Calif. Manage. Rev. **32**(4), 56–70 (1990)
40. Fredriksson, A.: Materials supply and production outsourcing, Göteborg, Chalmers University of Technology (2011)
41. Terwiesch, C., Bohn, R.E.: Learning and process improvement during production ramp-up. Int. J. Prod. Econ. **70**(1), 1–19 (2001)
42. Mefford, R.N., Bruun, P.: Transferring world class production to developing countries: a strategic model. Int. J. Prod. Econ. **56–57**, 433–450 (1998)
43. Madu, C.N.: Transferring technology to developing countries—critical factors for success. Long Range Plann. **22**(4), 115–124 (1989)
44. Helleiner, G.K.: The role of multinational corporations in the less developed countries' trade in technology. World Dev. **3**(4), 161–189 (1975)
45. Mogos, M.F., Alfnes, E., Swahn, N.: Prerequisites for Successful Production Transfers in the Electronics Industry (2018). http://ils2016conference.com/wp-content/uploads/2015/03/ILS2016_SC03_1.pdf
46. Cheng, Y., Skov, M.E., Liangsiri, J.: Transferring knowledge in the relocation of manufacturing units. Strateg. Outsourcing: Int. J. **3**(1), 5–19 (2010)
47. Rebentisch, E.S., Ferretti, M.: A knowledge asset-based view of technology transfer in international joint ventures. J. Eng. Technol. Manage. **12**(1), 1–25 (1995)
48. Bruun, P., Mefford, R.N.: A framework for selecting and introducing appropriate production technology in developing countries. Int. J. Prod. Econ. **46–47**, 197–209 (1996)
49. Mogos, M.F., Sjøbakk, B., Alfnes, E.: A structured outsourcing procedure. In: Advances in Production Management Systems. Initiatives for a Sustainable World, pp. 739–747 (2016)
50. Filip, F.-C.: Efficient management methods of production and product projects transfer. In: Recent Advances in Economics, Management and Development, pp. 138–145 (2014)
51. Riis, J.O., Waehrens, B., Madsen, E.: The knowledge dimension of production transfer. In: Advances in Production Management Systems, pp. 441–448 (2009)
52. Lampón, J.F., Cabanelas, P., Carballo-Cruz, F.: A model for international production relocation: multinationals' operational flexibility and requirements at production plant level. J. Bus. Res. **77**, 95–101 (2017)
53. IEEE Std 830-1998 (1998)
54. IEEE: ISO/IEC/ IEEE 29148 engineering (2011)
55. Sommerville, I.: Software Engineering, 9th edn. Pearson, Boston, (2011)
56. Ehrlenspiel, K., Meerkamm, H.: Integrated Product Development. Carl Hanser Verlag GmbH & Co. KG, München (2013). (in German)
57. Ponn, J., Lindemann, U.: Concept Development and Design of Technical Products. Springer, Heidelberg (2011). (in German)
58. Witt, H.: Strategies in qualitative and quantitative research. In: Forum: Qualitative Social Research, vol. 2, no. 1, art. 8 (2001)

Structural Indicators for Business Process Redesign Efficiency Assessment

Benjamin Urh(✉), Maja Zajec, Tomaž Kern, and Eva Krhač

Faculty of Organizational Sciences, UM, Maribor, Slovenia
benjamin.urh@um.si

Abstract. In this paper, we discuss some available methods for assessment of business process redesign efficiency. Business process efficiency can be assessed with operational (e.g. time, cost or quality) as well as structural indicators (on the basis of business process model's structure). The use of operational OR structural indicators depends on the context: operational indicators can be used only when the process has already been established or implemented in the company, where structural indicators can be used already in the phase of business process model design. The advantage of structural indicators is their possibility to assess business process efficiency before the process starts and costs increase (e.g. with implementation). In literature, many various structural indicators are presented, which give different efficiency assessment for the same business process model. Here, we propose a uniform assessment method for measuring business process model structural efficiency. The process redesign and the way of uniform structural efficiency assessment is presented using as a case study selected core processes in one of the most successful Slovenian health services organization.

Keywords: Business process redesign · Structural efficiency indicators · Operational efficiency indicators · Process redesign efficiency

1 Introduction

Business process is a combination of inter-connected events, activities and decision points that include many actors and information carriers. They all together contribute to a value-added result for at least one customer [1]. The development and rapid expansion of the information technology use (e.g. web services and mobile applications), leads to the increasing and more frequent adaptation and redesign of business processes. New technology may significantly improve effectiveness and efficiency, but it may also make existing process more complex, reduce usability, and cause more integration problems. As processes become more complex, the difficulty of locating and correcting problems rises dramatically [2]. Consequently, changes in organizing business processes are becoming every day's practice - we could also call it "continuous evolution of changes". These are not just occasional changes in product design or the implementation of a new information system. Practically everything is changing - e.g. developed products, organization models and acquired knowledge … [3].

A. Hamrol et al. (Eds.): *Advances in Manufacturing II* - Volume 3, LNME, pp. 16–32, 2019.
https://doi.org/10.1007/978-3-030-17269-5_2

Business processes adjustment to environment change is often carried out through business process redesign projects. Business process redesign projects in companies engage in the same goal: to achieve more efficient operations. The redesign of business processes can be made through different approaches. According to Vila [4] over 50 different approaches have been created over the past decades. The difference between them lies in the method they propose for achieving this goal. Some methods resulted in fast and radical (revolutionary) changes, where others resulted in slow and gradual (evolutionary) changes.

After business process redesign is over, top managers in business systems often ask next rhetorical questions: "Have we met our objectives?", "Is this what we needed?", "Where do we go from here and how?" [5]. At this point, top managers must answer to important questions for their organization system:

- "What is the level of business process performance efficiency?",
- "Is it necessary or reasonable to adapt or to change the process?",
- "What changes or adjustments must be made in the business process performance?",
- "How will the projected changes influence on the business process performance efficiency?".

In literature review we found a lot of recommendations for process performance efficiency assessment [2, 6, 7]. Process performance efficiency assessment can be made based on the operational and structural efficiency indicators. Operational efficiency indicators show spending time, expenditure and/or quality [8, 9], whereas structural efficiency indicators are connected with a structural complexity of business processes [10–13]. Business process structural complexity, according to Sarnikar and Deokar [14], is characterized based on the number of activities, number of employees included in the process and correlation between activities and employees. Also, it is important if process is dynamically changing or developing during the time. Structural complexity assessment allows to show the degree of business process model understandability. In the case of difficult understandability, the business process model needs to be redesigned – for example, with decomposition into smaller modules [15, 16].

According to operational indicators, the estimation of the current and future state of process performance efficiency is based on "on real-time" data collecting. That means that the process must be performed in practice. In the case of structural efficiency indicators, we can assess the potential of future process operational efficiency on the basis of process model structure [10]. Business process assessment, based on the structural efficiency indicators, can be very useful, especially when we have high implementation costs. The advantage of structural efficiency indicators is that executives can easily take a decision about new process implementation. The assessment of the process performance efficiency based on structural efficiency indicators is coarser than the assessment based on operational efficiency indicators. Therefore, the input to obtain assess is considerably lower [2]. In their research, authors [10, 17] also argue that the structural complexity of a process is one of the major sources of error as well as rapid increments in business process performance costs.

In this paper we will introduce successful business process redesign, which was carried out on the basis of process assessment with structural efficiency indicators (SEI). In Sect. 2, we will present the research problem (structural efficiency indicators and their advantages, criteria and selection of suitable indicators; methods used – business process modeling, business process structural efficiency assessment). The research results will be covered in Sect. 3 (current AS-IS process model, future TO-BE process model, AS-IS and TO-BE process model analysis, unrelated structural efficiency indicators, final assessment of process structural efficiency), followed by Sect. 4 with discussion and Sect. 5 with the research conclusions.

2 Research Methodology

According to Irani et al. [18], measuring SEI before process implementation is practical for the preparation of process changes proposals due to better:

- understanding of process path or flow. That can reflect in better communication among process owners and executors when preparing process changes,
- process proposals. They are prepared in a way that we can focus on parts of process structure that have greater impact on process results,
- comparison of prepared new process solutions with the old one. Therefore, we can anticipate if the new solution gives added value to the old process and where in the model it is located,
- recognition of the main problems in the process. We can evaluate the new process design with the best practice which means we can avoid repeating bad design approaches,
- understanding processes SEI in advance. This gives us a competitive advantage when planning the implementation of process changes.

We would also like to mention other beneficial uses of SEI according to Sánchez González et al. [19]:

- SEI help to prevent the occurrence of certain errors in the early stages of the business process lifecycle and are easier to maintain,
- in the field of business process measurement, the most measured part is the complexity (authors assume that the importance of complexity lies in its direct connection with the characteristics of comprehensibility and variability, which according to ISO 9126 are most used to define indicators - characteristics are the basic activities to ensure, that models are useful for communication between stakeholders and that business processes are in a state of continuous improvement,
- in recent years, the number of research (related to business process indicators) has been increasing because scientists have discovered the importance of measurement processes in order to improve the overall organization.

Based on the literature review, we have formed Table 1, which presents possible structural efficiency indicators and their short description.

Table 1. Structural efficiency indicators.

	Indicators	Description
Gruhn and Laue [16]	Cognitive weight of a BPM	The sum of the cognitive weights of BPM elements (e.g. sequence: W = 1; parallel split: W = 4…)
Rolón et al. [13]	Structural complexity of business process	They are based on the BPMN model elements: • base metrics (e.g. Number of Start Event, Number of Task, Number of Data Objects-In of the Process…), • derived metrics (e.g. Total Number of Start Events of the Model; Total Number of Task of the Model; Proportion of Data Objects, as Outgoing Product and the total of Data Objects…)
Mendling et al. [12]	Split-join-ratio	Calculation of misfit between split and join connectors in BPM
Cardoso et al. [10, 11]	Activity complexity	*NOA* – number of activities in a process *NOAC* – number of activities and control-flow elements in a process *NOAJS* – number of activities, joins and splits in a process
	Control-flow complexity	*CFC* – control-flow complexity (the number of mental states, considered when a designer is developing a process) *HPC* – Halstead-based process complexity (indicators for estimating process length, volume, and difficulty) *CNC* – the coefficient of network complexity (number of arcs divided by the number of nodes)
	Data-flow complexity	*IC* – interface complexity of an activity: length * (number of inputs * number of outputs)^2
	Resource complexity	Resource (e.g. Human resources, IT resources…) complexity analysis
Gruhn and Laue [20]	Nesting depth	Nesting depth of an element is a number of decisions, necessary to reach this element
	Process patterns	Examples, which show how to connect activities to solve common problem
	Anti-pattern	Uncovering anti-patterns is useful to define whether the model has a good modeling style
Ghani et al. [15]	Modularization	Structural complexity based on the information flow impact (measurement of input and output data)
Reijers and Mendling [21]	Model factors	The number of a particular type of elements, diameter, token splits, average & maximum connector degree, control flow complexity, mismatch, depth, connectivity, density, cross-connectivity, sequentiality, separability, structuredness, connector heterogeneity
Sun and Hou [22]	IF	Information flow complexity of business process

Authors classify structural efficiency indicators into several groups:

- process flow indicators; process connection indicators; process indicators of organizational structure elements - e.g. role, organization unit, etc.; process indicators of support objects - e.g. documents, software solutions, etc. [13];
- activity complexity, control-flow complexity, data-flow complexity, resource complexity [10, 11].

In the following years, several authors emphasize various aspects (properties) of business processes structural complexity and define few categories of indicators:

- size, coupling, complexity [23];
- size measures, connection, modularity, connector interplay, complex;
- behavior [24];
- size measures, connection, modularity/structuredness, gateway interplay/control structures, syntax rules [25].

A literature review showed the existence of a large number of structural efficiency indicators. The disadvantage of a large number of possible SEI is that the evaluation result of the business process by different criteria (in this case - by different SEI) can be a different assessment of the same business process. The question raised at this point is: "How we know which SEI or group of SEI are more appropriate for a more relevant business process assessment?"

Cardoso in paper [11] argues that, according to various modeling techniques, the indicator must be general (standard) and valid for all techniques. The indicator must be easy to learn, calculated, consistent and objective. Moreover, the characteristics of automation and the indicator as an additive are also desirable (if two parts are grouped into a sequence, the overall complexity must be at least the sum of the complexity of the two parts). Based on the recommendations made by Cardoso in paper [11], we decided to use Rolón's et al. [13] version of structural efficiency indicators in future research. Rolón's et al. [13] structural efficiency indicators are independent of the business process modeling technique and are easy to understand, calculate and apply.

On the basis of above findings, we decided to demonstrate how to combine various assessment of business process structural efficiency (the model of the current process and the model of the renewed process). They usually have very different values, so we want to combine them in a uniform assessment that must be comparable between process.

2.1 Methodology of Analysis

2.1.1 Business Process Modeling

The business process structural efficiency assessment is conditioned by previously performed process mapping and its record in the appropriate repository.

For modeling of the business process model, the ARIS methodology was used. More specifically, Event-driven Process Chain model (EPC model) type was used, because it presents a user perspective of the process [26]. A research by Sánchez González in paper [19] showed that EPC model is the second most often used business process modeling technique. This model is based on the logic that an event triggers an activity (task) or several activities. Consequently, the activity ends with a new event or with several events.

At mapping the current and future situation we used the symbols that are displayed in the Table 2. The rules for using logical operators are also shown in the table.

Table 2. Description of symbols used at business process mapping with EPC model type [27].

Symbol	Graphics	Description
Event	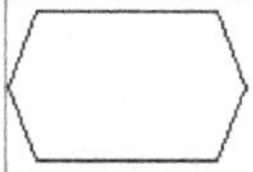	An event represents a state that influences or controls the further flow of one or more business processes Events trigger functions and are results of functions
Function	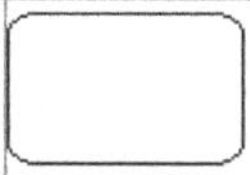	A function is the technical task or activity performed on an object in order to support one or several business objectives
Rules	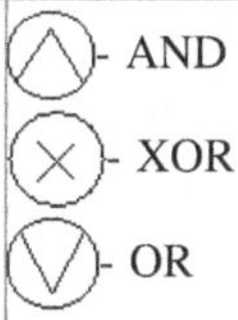	Rules represent logic operators which allow specifying the logical links in process chains AND – process continues by several ways XOR – process continues only by one of the possible ways OR – process continues by any combination of possible ways
Process interface	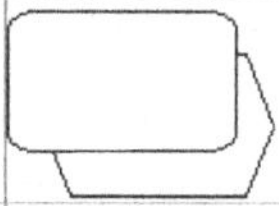	A symbol represents the process connection with another process or continuation of the process in the next process
Organizational unit	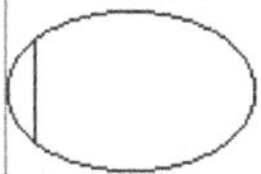	Organizational units are the performers of the tasks required to attain the business objectives
Position	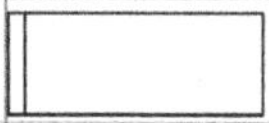	The smallest organizational unit in a company assigned to employees – persons
Application system	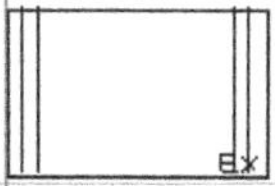	A symbol represents the individual application system which has exactly defined technological properties
Document	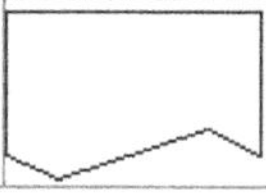	A symbol represents the information carrier such as pre-printed form, fax… which is used in the process or generated

2.1.2 Business Process Structural Efficiency Assessment

The process performance efficiency assessment can be done in several steps. In the first step (by analyzing the process model), we collect the basic data to assess the business process performance efficiency by individual structural indicators. The mentioned basic data are:

- n_E; *The number of events in a process,*
- n_{SE}; *The number of initial events in a process,*
- n_{FE}; *The number of completed and/or terminated events in a process,*
- n_{PA} *The number of activities in a process (functions and process interfaces),*
- n_{PI}; *The number of activities with connections to other processes (pro cess interfaces),*
- n_{PD}; *The number of decisions during process performance,*
- n_{AT}; *The number of possible transitions between activities in a process,*
- n_{LB}; *The number of loopbacks in a process,*
- n_{VAA}; *The number of activities in a process where added value is created,*
- n_{CPA}; *The number of connections between work positions and process activities,*
- n_{PP}; *The number of performers (work positions) participating in a process,*
- n_{HLP}; *The number of hierarchical levels of performers participating in the process,*
- n_{PAP}; *The number of work positions participating in performing all the business processes in a given business system,*
- n_{PBS}; *The number of performers (work positions) in a business system,*
- n_{DP}; *The number of documents used in a process,*
- n_{POD};*The number of documents that are to be created within a process,*
- n_{PID}; *The number of documents entering the process,*
- n_{SWP}; *The number of software solutions used in the process,*
- n_{SWA}; *The number of process activities whose performance is supported by software solutions.*

On the basis of collected data, individual structural efficiency indicators (SEI), are calculated [28, 29], so we get 15 indicators (different assessment) of the process structural efficiency [30].

All types were classified of selected SEI in Rolón's et al. [13] four categories of indicators:

1. Process flow indicators:

   - Initial process event indicator (SEI01)

$$K_{SE} = \frac{n_{SE}}{n_E} \cdot 100 \tag{1}$$

- Process activity indicator (SEI02)

$$K_A = \frac{1}{(n_{PA} - n_{PI})} \cdot 100 \tag{2}$$

- Process decisions indicator (SEI03)

$$K_D = \frac{n_{PD}}{(n_{PA} - n_{PI})} \cdot 100 \tag{3}$$

- Process added value indicator (SEI04)

$$K_{VAP} = \frac{n_{VAA}}{(n_{PA} - n_{PI})} \cdot 100 \tag{4}$$

2. Process connection indicators:
   - Process connection indicator (SEI05)

$$K_{PI} = \frac{n_{PI}}{n_{PA}} \cdot 100 \tag{5}$$

- Number of transitions between activities indicator (SEI06)

$$K_{PAT} = \frac{n_{AT}}{(n_{PA} - n_{PI})} \cdot 100 \tag{6}$$

- Loopback indicator (SEI07)

$$K_{LB} = \frac{n_{LB}}{(n_{PA} - n_{PI})} \cdot 100 \tag{7}$$

3. Process indicators of organizational structure elements
   - Level of inclusion of the performers indicator (SEI08)

$$K_{CLP} = \frac{n_{CPA}}{(n_{PA} - n_{PI}) \cdot n_{PP}} \cdot 100 \tag{8}$$

- Process performers indicator (SEI09)

$$K_{PP} = \frac{1}{n_{PP}} \cdot 100 \tag{9}$$

- Included performers indicator (SEI10)

$$K_{PPA} = \frac{n_{PP}}{n_{PAP}} \cdot 100 \tag{10}$$

- Hierarchy of process performers indicator (SEI11)

$$K_{HP} = \frac{1}{n_{HLP}} \cdot 100 \tag{11}$$

4. Process indicators of support objects

- Relationships of outgoing documents indicator (SEI12)

$$K_{POD} = \frac{n_{POD}}{n_{DP}} \cdot 100 \tag{12}$$

- Relationships of outgoing documents and process activities indicator (SEI13)

$$K_{PODA} = \frac{n_{POD}}{(n_{PA} - n_{PI})} \cdot 100 \tag{13}$$

- Software solutions for the process indicator (SEI14)

$$K_{SWP} = \frac{1}{n_{SWP}} \cdot 100 \tag{14}$$

- Information technology support for process activities indicator (SEI15)

$$K_{PSWA} = \frac{n_{SWA}}{(n_{PA} - n_{PI})} \cdot 100 \tag{15}$$

According to research of Urh et al. [30], it was found that a large number of the structural efficiency indicators (SEI) can be replaced by seven unrelated structural efficiency indicators (USEI). Authors found that they can use USEI and retain more than 77% of the variability of the basic variables structural efficiency indicators (SEI). The final estimate of the business processes structural efficiency can be calculated based on the value of an individual USEI and its share of the explained variance.

The indicator of business system organization (USEI01)

$$USEI01 = -6,947 + 0,016 * SEI09 + 0,013 * SEI11 + 0,005 * SEI08 + 0,060 * SEI10 \tag{16}$$

The indicator of business processes complexity (USEI02)

$$USEI02 = -10,602 + 0,048 * SEI03 + 0,064 * SEI07 \tag{17}$$

The indicator of performed work documentation (USEI03)

$$USEI03 = -1,913 + 0,021 * SEI13 + 0,017 * SEI12 \tag{18}$$

The indicator of scope of the business processes (USEI04)

$$USEI04 = -2,222 + 0,071 * SEI02 + 0,022 * SEI06 \tag{19}$$

The indicator of business processes interconnectedness (USEI05)

$$USEI05 = -5,381 + 0,044 * SEI01 + 0,030 * SEI05 \tag{20}$$

The indicator of information technology support (USEI06)

$$USEI06 = -1,545 + 0,018 * SEI14 + 0,021 SEI15 \tag{21}$$

The indicator of added value creation (USEI07)

$$USEI07 = -0,335 + 0,095 * SEI04 \tag{22}$$

3 Results

Below are presented the results of the structural efficiency assessment in the case of core process in the selected company. In the past, this company performed BPR project on all business processes except production. The project proved to be successful, so they decided to continue with the selected core process optimization. The main reason for this step was numerous patient's complaints about long waiting times.

They performed analyses of all current production processes with operational efficiency indicators and they found that:

- processing times were longer than expected,
- waiting times were too long in comparison with production times,
- there was a lot of administrative activities (tasks),
- a lot of physical documents were prepared and dispatched.

The following Fig. 1 shows the model of the current state of the process (As-Is process model).

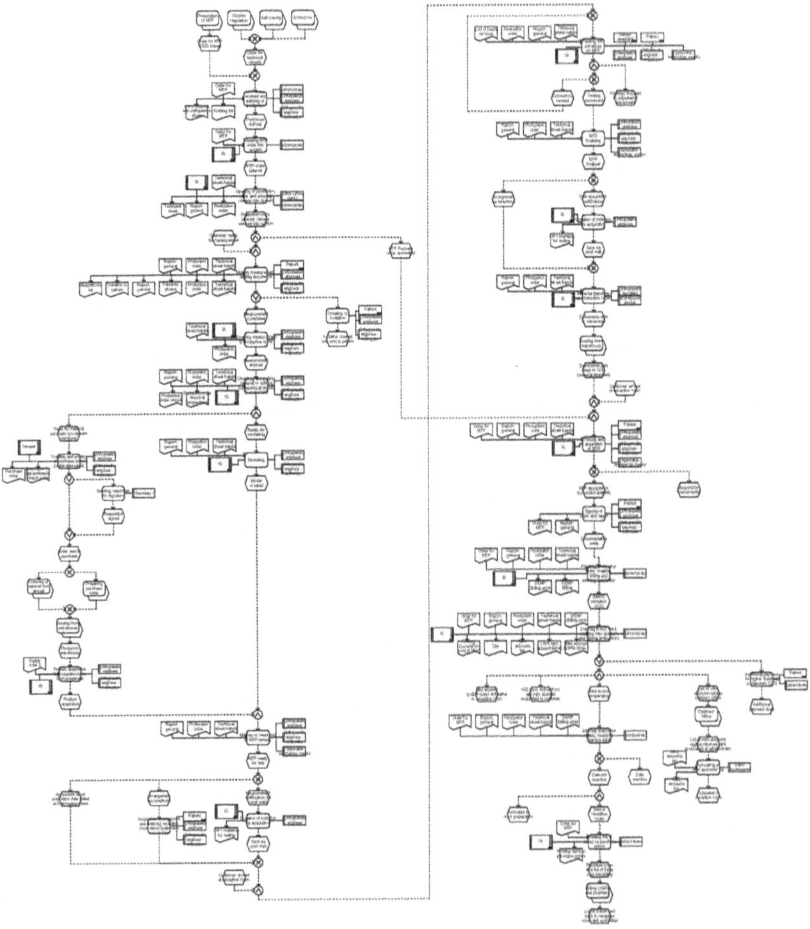

Fig. 1. The As-Is process model of the selected process.

(Figure 2) shows the model of the proposed state of the same process after the redesign (To-Be process model).

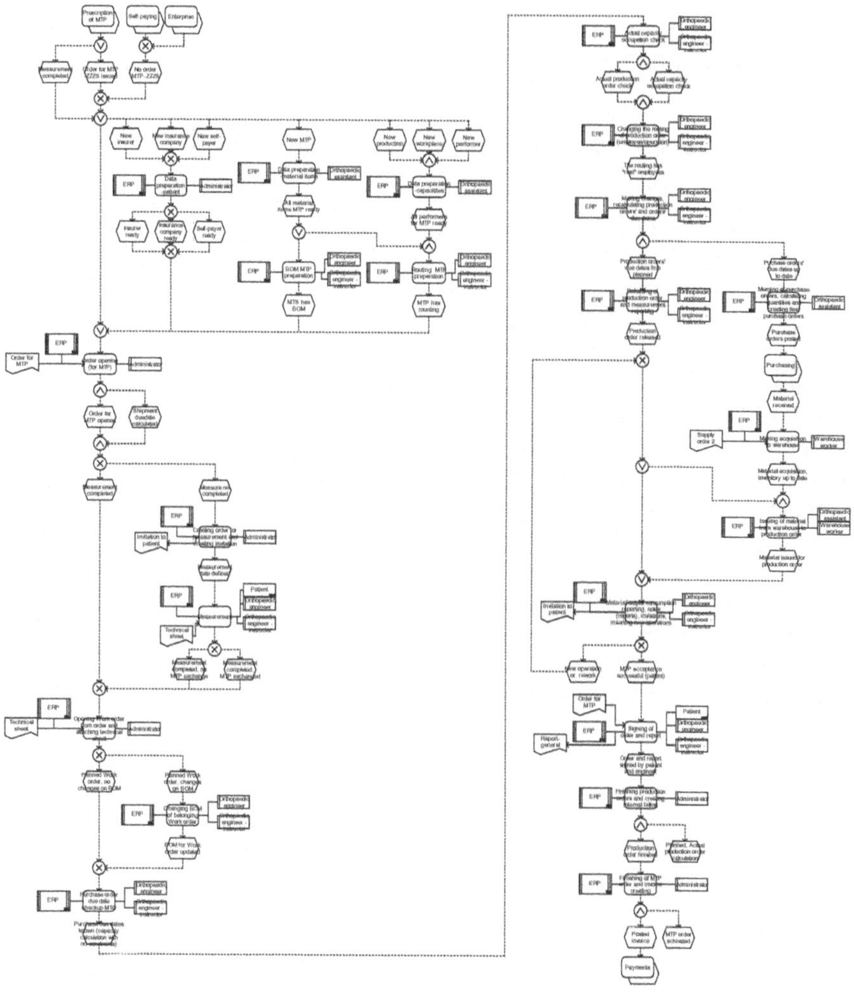

Fig. 2. The To-Be process model of the selected process.

3.1 Current As-Is and Future To-Be Process Model Analysis

We did two analyses: first for As-Is process (Fig. 1) and another for To-Be process (Fig. 2). Because of space limitation the description of symbols on models cannot be seen, but distinguish between structural complexities of models can still be seen. When we were modeling To-Be process, our goal was to eliminate as many disadvantages (identified in As-IS process) as possible. Therefore, in the new process we:

- eliminated those administrative activities that do not added value,
- reduced number of used and exchanged physical documents,
- eliminated work duplication,
- took into account the IT and ERP possibilities.

For calculating fundamental structural efficiency indicators (Table 3), we collected data from model of the current process and from model of the future process.

Table 3. Basic data about As-Is and To-Be process model.

Basic data	Label	AS-IS model	TO-BE model
The number of events in a process	n_E	50	45
The number of initial events in a process	n_{SE}	8	3
The number of completed and/or terminated events in a process	n_{FE}	8	3
The number of activities in a process (functions and process interfaces)	n_{PA}	36	27
The number of activities with connections to other processes (process interfaces)	n_{PI}	10	5
The number of decisions during process performance	n_{PD}	1	0
The number of possible transitions between activities in a process	n_{AT}	31	27
The number of loopbacks in a process	n_{LB}	0	0
The number of activities in a process where added value is created	n_{VAA}	6	8
The number of connections between work positions and process activities	n_{CPA}	47	34
The number of performers (work positions) participating in a process	n_{PP}	11	6
The number of hierarchical levels of performers participating in the process	n_{HLP}	3	2
The number of work positions participating in performing all the business processes in a given business system	n_{PAP}	88	88
The number of performers (work positions) in a business system	n_{PBS}	141	141
The number of documents used in a process	n_{DP}	25	6
The number of documents that are to be created within a process	n_{POD}	23	2
The number of documents entering a process	n_{PID}	8	3
The number of software solutions used in a process	n_{SWP}	2	1
The number of process activities whose performance is supported by software solutions	n_{SWA}	14	22

3.2 Unrelated Structural Efficiency Indicators

Based on the collected data from the selected core process structure and based on the equations for the USEI calculation [30], estimates for USEI were calculated for the As-Is and To-Be process models (Table 4). Because of differences in the possible values interval for the individual USEI, the obtained values were converted to the same evaluation interval. The presented results are calculated to an interval values (from 0 to 5 - value 0 is a poor structural efficiency, value 5 is very good structural efficiency) of the selected business process, regarding to an individual USEI.

Table 4. Structural efficiency assessment of As-Is and To-Be model.

Unrelated structural efficiency indicators	Label	AS-IS model	TO-BE model
Business system organization	USEI01	3,12	3,53
Business processes complexity	USEI02	2,94	4,51
Performed work documentation	USEI03	0,51	3,86
Scope of the business processes	USEI04	3,44	3,35
Business processes interconnectedness	USEI05	3,96	4,41
Information technology support	USEI06	2,61	5
Added value creation	USEI07	1,15	1,8

3.3 Final Assessment of Process Structural Efficiency

The final structural process efficiency assessment (Table 5) was calculated based on each unrelated structural efficiency indicator and their proportion of the explained variance.

Table 5. Final structural efficiency assessment of As-Is and To-Be model.

Unrelated structural efficiency indicators	Label	% variance	AS-IS model	TO-BE model
Business system organization	USEI01	24,237	3,12	3,53
Business processes complexity	USEI02	12,565	2,94	4,51
Performed work documentation	USEI03	10,328	0,51	3,86
Scope of the business processes	USEI04	9,303	3,44	3,35
Business processes interconnectedness	USEI05	8,643	3,96	4,41
Information technology support	USEI06	6,286	2,61	5
Added value creation	USEI07	5,706	1,15	1,8
Final estimation of process structural efficiency			2,69	3,8

4 Discussion

In the selected core process redesign, we showed practical implication of SEI. The results (Tables 4 and 5) show that proposed changes will have a positive impact on process efficiency. With those changes at the end of the redesign process, we reach higher final structural efficiency assessment (from 2.69 to 3.80), which is the result of:

- Medium improvement in the *business system organization* (from 3.12 to 3.53). The improvement is the result of a small number of different performers in the process and a small number of different hierarchy levels where those performers are employed.
- High improvement in *business processes complexity* (from 2.94 to 4.51). The improvement is the result of decisions number decrease that are taken by performers in a process.
- High improvement in *performed work documentation* (from 0.51 to 3.86). The improvement is the result of paper documents decrease (created during process performance).
- Poor *scope of the business processes* (from 3.44 to 3.35). A little bit worse scope is the result of more complex connections between many possibilities in process performance.
- Medium improvement in *business processes interconnectedness* (from 3.96 to 4.41). The improvement is the result of clearer connections and transitions into» connected «processes (preceding and consecutive).
- High improvement in *information technology support* (from 2.61 to 5.00). The improvement is the result of enterprise resource planning (ERP) system implementation.
- Medium improvement in *added value creation* (from 1.15 to 1.80). The improvement is the result of process activities number decrease (that do not add value for a customer or business system).

With presented example of the process assessment, we not only assessed the business process structural efficiency according to a particular indicator (as suggested by [2, 10, 11, 13] etc.). We also combined obtained assessment in a uniform assessment of business process structural efficiency. With a uniform assessment, we managed to avoid a great divergence between the values of structural efficiency assessment by individual structural indicators. This method of assessment can be highlighted as the most important added value of the presented research. We also believe that the presented method enables business processes efficiency assessment with a more appropriate set of indicators, from the point of view of: usability according to the modeling technique and manageability according to the number of indicators and the possibility of obtaining the necessary data.

5 Conclusion

In this article we introduced how the business process performance efficiency can be checked within process model. SEI and process model help us to assess the process performance efficiency. The assessment can be done for process model that is already executing or preparing for new process development or renewed for implementation. With calculation of unrelated structural efficiency indicators, we obtained structural performance efficiency assessment of selected core process in selected Slovenian company. Finally, we calculated the assessment of process structural efficiency according to the percentage of known variance for each unrelated structural efficiency indicator.

The introduced procedure of structural efficiency analysis can be very useful for executives when they must take a decision about business process performance efficiency improvement. It should be considered in case of: organizational changes in the business system, delegating, business process interconnectedness, improvement of information system, number of non-added value activities decrease, implementing e-commerce. Any of them can increase business process performance efficiency but with the introduced procedure, they can check in advance which one is more appropriate for each case. Of course, the actual effect of the implemented changes needs to be verified after a specified period (e.g. after six months).

However, it should be noted that in the described case is not clearly shown the interdependence (or impact) of the individual indicator change on the change of other indicators (e.g. the impact of improving the *Information technology support indicator* on the *Business processes complexity indicator*). On this question, we will try to answer in further research.

References

1. Dumas, M., La Rosa, M., Mendling, J., Reijers, H.A.: Fundamentals of Business Process Management. Springer, Heidelberg (2013)
2. Cheng, C.Y.: Complexity and usability models for business process analysis (2008)
3. Davenport, T.H.: Process Innovation: Reengineering Work Through Information Technology. Harvard Business Press, Boston (1993)
4. Vila, A.: Synthesized organization. Change Management: Proceedings of the 25th International Conference on the Development of Organizational Sciences, pp. 1–12 (2006)
5. Urh, B., Kern, T., Roblek, M.: Business process modification management. In: Putnik, G. (ed.) Encyclopedia of Networked and Virtual Organizations, Hershey, Information Science Reference, pp. 112–120 (2008)
6. Frederiksen, H.D., Mathiassen, L.: A contextual approach to improving software metrics practices. IEEE Trans. Eng. Manage. **55**(4), 602–616 (2008)
7. Sharma, A.: Implementing balance scorecard for performance measurement. ICFAI J. Bus. Strategy **6**(1), 7–16 (2009)
8. Valiris, G., Glykas, M.: Business analysis metrics for business process redesign. Bus. Process Manage. J. **10**(4), 445–480 (2004)
9. Van der Aalst, W.M.P.: Business process management: a comprehensive survey. ISRN Softw. Eng. **2013**, 1–37 (2013)

10. Cardoso, J., Mendling, J., Neumann, G., Reijers, H.A.: A discourse on complexity of process models. In: International Conference on Business Process Management, pp. 117–128. Springer, Heidelberg (2006)
11. Cardoso, J.: Business process control-flow complexity: metric, evaluation and validation. Int. J. Web Serv. Res. **5**(2), 49–76 (2008)
12. Mendling, J., Moser, M., Neumann, G., Verbeek, H.M.W., Van Dongen, B.F., van der Aalst, W.M.P.: A quantitative analysis of faulty EPCs in the SAP reference model. BPM Center Report BPM-06-08, BPMcenter.org. (2006)
13. Rolón, E., Ruiz, F., Garcia, F., Piatiini, M.: Applying software metrics to evaluate business process models. CLEI Electron. J. **9**(1), 1–15 (2006)
14. Sarnikar, S., Deokar, A.V.: A design approach for process-based knowledge management systems. J. Knowl. Manage. **21**(4), 693–717 (2017)
15. Ghani, A.A.A., Wei, K.T., Muketha, G.M., Wen, W.P.: Complexity metrics for measuring the understandability and maintainability of business process models using goal-question-metric. Int. J. Comput. Sci. Network Secur. **8**(5), 219–225 (2008)
16. Gruhn, V., Laue, R.: Adopting the cognitive complexity measure for business process models. In: 5th IEEE International Conference on Cognitive Informatics, ICCI 2006, vol. 1, pp. 236–241 (2006)
17. Mendling, J.: Detection and Prediction of Errors in EPC Business Process Models. Doctoral dissertation, Vienna University of Economics and Business Administration (WU Wien), Austria (2007)
18. Irani, Z., Hlupic, V., Giaglis, G.M.: Business process reengineering: an analysis perspective. Int. J. Flex. Manuf. Syst. **14**(1), 1–5 (2002)
19. Sánchez González, L., García Rubio, F., Ruiz González, F., Piattini Velthuis, M.: Measurement in business processes: a systematic review. Bus. Process Manage. J. **16**(1), 114–134 (2010)
20. Gruhn, V., Laue, R.: Approaches for business process model complexity metrics. In: Technologies for Business Information Systems, pp. 13–24. Springer, Dordrecht (2007)
21. Reijers, H.A., Mendling, J.: A study into the factors that influence the understandability of business process models. IEEE Trans. Syst. Man Cybern. Part A Syst. Hum. **41**(3), 449–462 (2011)
22. Sun, H., Hou, H.: Study on complexity metrics of business process. In: International Conference on Computer Science and Service System (CSSS 2014) (2014)
23. Aguilar, E.R., Garcia, F., Ruiz, F., Piattini, M., Visaggio, C.A., Canfora, G.: Evaluation of BPMN models quality-a family of experiments. In: 3rd International Conference on Evaluation of Novel Approaches to Software Engineering, Funchal, pp. 56–63 (2008)
24. Mendling, J., Sánchez-González, L., García, F., La Rosa, M.: Thresholds for error probability measures of business process models. J. Syst. Softw. **85**(5), 1188–1197 (2012)
25. Figl, K.: Comprehension of procedural visual business process models. Bus. Inf. Syst. Eng. **59**(1), 41–67 (2017)
26. Pavlović, I., Kern, T., Miklavčič, D.: Comparison of paper-based and electronic data collection process in clinical trials: costs simulation study. Contemp. Clin. Trials **30**(4), 300–316 (2009)
27. Davis, R.: ARIS design platform: advanced process modelling and administration. Springer Science & Business Media (2008)
28. Fitz-enz, J.: Predicting people: from metrics to analytics. Employ. Relat. Today **36**(3), 1–11 (2009)
29. Poniatowski, S., Wichser, J.D.: A better metric for IT efficiency. Optimize **5**(5), 43–46 (2006)
30. Urh, B., Kokalj, Š., Zajec, M.: The importance of structural indicators in assessing the efficiency of business process performance. V: KERN, Tomaž (ur.), RAJKOVIČ, Vladislav (ur.). People and sustainable organization. Frankfurt am Main [etc.]: Peter Lang, pp. 248–270 (2011)

Examination of the Mediating Effects of Physical Asset Management on the Relationship Between Sustainability and Operational Performance

Damjan Maletič[1], Matjaž Maletič[1], Basim Al-Najjar[2], and Boštjan Gomišček[3(✉)]

[1] Faculty of Organizational Sciences, Enterprise Engineering Laboratory, University of Maribor, Maribor, Slovenia
[2] School of Engineering, Linnaeus University, Växjö, Sweden
[3] Faculty of Business, University of Wollongong in Dubai, Dubai, United Arab Emirates
BostjanGomiscek@uowdubai.ac.ae

Abstract. This study examines the mediating effects of physical asset management on the relationship between sustainability and operational performance. Using empirical data based on survey data from six European countries (i.e. Greece, Poland, Slovakia, Slovenia, Sweden and Turkey), this study utilized mediation analysis in order to address the research problem. A macro for SPSS was used to estimate the size of an indirect effect of sustainability on operational performance through a mediator (physical asset management). Results of this study show mediator effect of physical asset management on the relationship between sustainability and operational performance. The paper provides valuable insights into mechanism that have a potential to enhance operational performance. The results contribute to a better understanding on how organizations could achieve higher operational performance outcomes by implementing sustainability and physical asset management practices.

Keywords: Sustainability · Physical asset management · ISO 55001 · Operational performance · Mediation

1 Introduction

Recently, physical assets management is gaining the attention of both researchers and practitioners. Rapidly changing business environment, strong competition, requirements for minimizing the losses are some of the conditions in which organizations operate today [1]. This had triggered organizations to continuously search for new ways to improve their processes and gain a competitive advantage [2], taking into account the rising relevance of Industry 4.0 [3, 4]. Over the last two decades, there has been a steady increase in the demand for an effective physical asset management [5]. As such, physical asset management became an important area, especially in asset intensive industry [6]. It is widely recognized that physical asset management has

A. Hamrol et al. (Eds.): *Advances in Manufacturing II* - Volume 3, LNME, pp. 33–43, 2019.
https://doi.org/10.1007/978-3-030-17269-5_3

impact on various operational measures considering different stages of an asset life cycle. For example, Haider, Koronios, and Quirchmayr [7] argued that design of an asset has a direct impact on its productivity, which is concerned with minimizing the disturbances such as unplanned failure. Performance benefits can therefore be reflected in terms of creating and sustaining value during each life cycle stage, and throughout the asset's life [8], as well as in achieving factors such as quality, cycle time, employee skills, and productivity [9]. While many studies have highlighted the importance of physical asset management, to the best of our knowledge, only a few of them have examined the link between physical asset management practices and different aspects of organizational performance (e.g. [10–12]). As such, studies on physical asset management in relation to organizational performance are rather limited. Most studies are devoted on defining the field of physical or engineering asset management (e.g. [13]), exploring the asset management practices in industry (e.g. [14]), proposing models for managing life cycle of physical assets (e.g. [15]), asset risk management (e.g. [16]) or physical asset management implementation [17, 18]. As regards the sustainability in relation to physical assets, previous studies (e.g. [15]) posited that asset management practices are increasingly required to demonstrate potential benefits other than costs from a sustainable development perspective. Accordingly, Liyanage et al. [19] stated that asset and maintenance management has a crucial part in achieving the status of a sustainable company. Similarly, Garetti and Taisch [20] reported that the role of maintenance should be revised by taking into account a life cycle management oriented approach, which will allow maintenance to provide effective contributions towards sustainable performances. However, none of the prior studies has taken under consideration sustainability and physical asset management together as a way to improve performance outcome of an organization. As a way to better understand these differing views, this paper intends to fill this gap by exploring the mediating effect of physical asset management on the relationship between sustainability and operational performance.

2 Research Model Development

Much has been written in recent years about sustainability activities that concern industry. It is widely recognized that sustainability is and will be a crucial issue in asset centric industry [21]. It is further argued that the management of a physical asset as an entity is capable of creating, enhancing and/or sustaining the TBL [8]. Sustainability is also an important part of ISO 55001:2014 standard [21]. According to ISO 55001:2014 strong link between effective deployment of asset management and long-term sustainability - from the economic, environmental and social point of view exists [21]. In support of this statement, it has been acknowledged in the literature that manufacturing has an enormous impact on all the key challenges of sustainability. For example, previous research (e.g. [22]) indicates that energy efficiency is influenced by maintenance and usage, and running costs are influenced by energy consumption efficiency, energy costs, maintenance costs and downtime costs. Moreover, an effective asset maintenance may decrease environmental risks and reduce greenhouse gas emissions, water pollution and soil contamination [23]. In fact, higher performance in respect to resource and energy consumption could be achieved through effective and efficient maintenance, making this enterprise function an important issue for sustainability [20]. The latter supports the discussion that effective asset care is vital for achieving sustainability goals. Furthermore, it has been evident in the literature that physical asset management affects all three main pillars of corporate social sustainability, namely economic, environmental, and social. Accordingly, Maletič et al. [11] presented empirical evidence linking physical asset management and sustainability performance (considering environmental, social and economic aspects). This means that physical asset management has an important role in achieving sustainability goals. On these grounds, it could be interesting to explore the role of AM in affecting the link between sustainability practices and organizational performance dimensions. As such, our research suggests that that asset management mediates the relationship between sustainability practices and organizational performance dimensions. In this regard, we propose research framework shown in (Fig. 1).

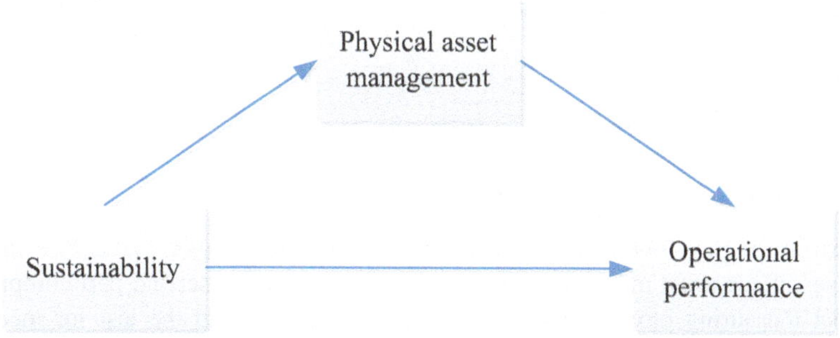

Fig. 1. The mediating effect of physical asset management.

3 Methods

3.1 Sample and Data Collection

The data used in this study are obtained from a research project conducted by a team of international researchers in the field of maintenance and asset management. The target survey population consisted of international e-mail lists of managers across a wide range of functions. In total, 138 usable responses were collected during the given time window. The questionnaire was responded by organizations that are located in located in Poland, Slovenia, Greece, Sweden, Turkey and Slovakia, in portion of 34.1%, 31.9%, 16.7%, 6.5%, 5.8% and 5.1%, respectively. In terms of organizational size (following the guidelines of the Statistical Office of the Republic of Slovenia), 12.2% of the sample was composed of micro-enterprises having five or fewer employees, 17.4% were small-sized organizations employing 50 or less employees, 31.3% were medium sized organizations, employing 51–250 employees, 21.7% organizations were with 251–500 employees and 12.2% organizations were with more than 500 employees. Based upon Slovenian Standard Industrial Classification Codes (SIC), (Table 1) shows the industry structure of the organizations.

Table 1. Sample distribution by industry type.

Industry (standard industrial classification)	Share (%)
Agriculture, Forestry and Fishing	1.7
Mining and Quarrying	6
Manufacturing	39.3
Electricity, Gas, Steam and Air Conditioning Supply	2.6
Water Supply, Sewerage, Waste Management and Remediation Activities	0.9
Construction	6.8
Wholesale and Retail Trade, Repair of Motor Vehicles and Motorcycles	16.2
Transportation and Storage	5.1
Accommodation and Food Service Activities	0.9
Information and Communication	3.4
Financial and Insurance Activities	0.9
Other	16.2
Total	100

3.2 Measures

The instrument developed in this study consists of three major parts. The first part comprises of construct measuring sustainability practices, the second part comprises of construct measuring physical asset management practices and the last for measuring operational performance. The comprehensive literature was conducted (e.g. [24, 25]) in order to develop measurement items relating to sustainability. Items for measuring these construct were derived from past studies on PAM (e.g. [10, 14, 26]). The items relating to operational performance were based on a review of the literature (such as [27, 28]). The list of all items can be seen in the Appendix A.

3.3 Mediation Analysis

In order to test the mediation effects of physical asset management on the relationship between sustainability and operational performance, we used SPSS procedure (SPSS macro) for estimating indirect effects in simple mediation models proposed by Preacher and Hayes [29].

The macros provide unstandardized coefficients as required to test mediation [30]. Path a represents the effect of X on the proposed mediator, whereas path b is the effect of M on Y partialling out the effect of X (Fig. 2B). All of these paths would typically be quantified with unstandardized regression coefficients. The indirect effect of X on Y through M can then be quantified as the product of a and b (i.e., ab). The total effect of X on Y is quantified with the unstandardized regression weight c (Fig. 2A). The total effect of X on Y can be expressed as the sum of the direct and indirect effects: $c = c' + ab$.

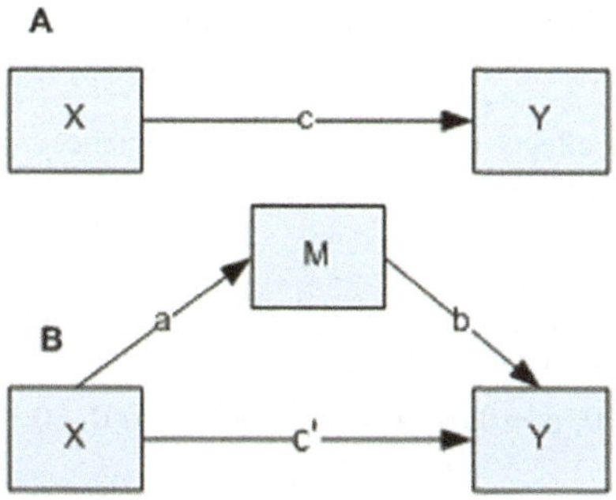

Fig. 2. (A) Illustration of a direct effect. X affects Y. (B) Illustration of a mediation design. X is hypothesized to exert an indirect effect on Y through M.

4 Analysis and Results

4.1 Measurement and Validation of Constructs

The scales for measuring study variables were subjected into validity and reliability tests. The construct validity was assessed merely using exploratory factor analysis (EFA) based on Varimax rotation. The scale reliability was tested by calculating its Cronbach's alpha. In summary, the results of the validity tests provide sufficient evidence regarding the validity of the measurement model and, therefore, supported empirical justification for combining the asset management constructs into composite variable. Accordingly, mean scores were calculated from the scale's items to generate the composite scores for the asset management construct as well as for other studied variables. The results of the validity assessment are presented in Appendix A.

4.2 Results of Mediation Analysis

By following Baron and Kenny [31] who recommend that a mediator, rather than a moderator function, is better when there is a strong relationship between the predictor and the criterion variable. As such, we consider that the predictor variable 'physical asset management' is related with the criterion variable 'operational performance' and we take the position that physical asset management has mediator function on the relationship between sustainability and operational performance.

As described above mediation analysis was tested within the following model (Table 2).

Table 2. Description of the analyzed mediation model.

Model	Independent variable	Proposed mediator	Dependent variable
Model 1	Sustainability practices	Physical asset management	Operational performance

The results of the mediation analysis are summarized in Table 3.

Table 3. Mediation of the effects of the sustainability practices on operational performance through physical asset management

Coefficients			
(a paths)	(b paths)	Total effect (c path)	Direct Effect (c-prime path)
Model 1 (F = 15.1833, p < 0.01)			
0.6161, p = 0.0000	0.4110, p = 0.0009	0.4124, p = 0.0001	0.1592, p = 0.1901

It appears that (Model 1) direct effect becomes insignificant after controlling for mediator ($p = 0.1592$, $p > 0.05$). The latter is consistent with the Baron and Kenny [31] who suggest that perfect mediation occur if c' becomes insignificant after controlling for M, which is so in our case.

As can be seen from results (Table 4), the true indirect effect of physical asset management is estimated to lie between 0.1129 and 0.4388 considering Model 1.

Table 4. Bootstrap estimates of the mediated effect and its standard error.

Point estimate	Product of coefficients		Bootstrapping BCa 95% CI	
	SE	Z	Lower	Upper
Model 1				
0.2532	0.0798	3.1719	0.1129	0.4388

Bca-Bias Corrected and Accelerated Confidence Intervals, 1000 bootstrap samples

Overall, the directions of the a and b paths are consistent with the interpretation that greater engagement in sustainability leads to greater mastery of the asset management, which in turn leads to greater operational performance.

5 Discussion and Conclusion

Although there is a growing literature focused on physical asset management, little empirical evidence regarding the impact of physical asset management practices on different aspects of organizational performance has been presented in the literature. The subject is of great importance, especially if we consider that physical asset management can be seen as a source of substantial strategic and competitive advantages [6].

The main goal of this research was to examine the role of sustainability, physical asset management and maintenance performance, in which physical asset management plays a role as a mediator. Our research produced several interesting findings, which add to our knowledge in this area. First and foremost, our findings suggest that physical asset management completely mediates the effect of sustainability practices on operational performance. Although researchers have discussed the relevant issues of maintenance and physical asset management form as explained above, so far none of the prior studies examined the possible role of physical asset management as a mediator. Therefore, this study contributes to the growing research stream by examining the effects of physical asset management on the relationship between sustainability practices and operational performance. In this regard, above arguments support the notation that sustainability can deliver value to assets' owner through improving performance of physical assets in terms of operational performance. As such, in some way our findings are in line with research stream supporting the notation that physical asset management contribute to company's performance [6, 8, 11–15]. Maletič et al. [11] showed that there is a strong link between physical asset management and sustainability performance. Prior studies have also dealt with sustainability practices and operational performance. For example, Green et al. [23] showed that green supply chain management practices implemented by manufacturing organizations lead to improved environmental performance and economic performance, which, in turn, positively affect operational performance. Therefore, our study contributes to the understanding of the benefits of implementing sustainability practices as well. In particular, our findings suggest that investment in sustainability leads to better exploited physical asset management and consequently better operational performance. This could be also explained through the broadness of physical asset management. It has important role in the management of life cycle of an asset as a whole, paying attention to economic as well as physical performance and risk measures, appreciating the broader strategic and human dimensions of the asset management environment, with the objective of improving both efficiency and effectiveness of resources [13]. In addition, these are all important elements to be considered when incorporating sustainability into business. We believe that physical asset management as a mediator could be explained through the aforementioned argument.

As such, by testing the proposed relationships, this study advances the previous work on the physical asset management area. The empirical results show the benefits of different physical asset management and sustainability practices on operational performance as well as suggest a framework to guide decision makers in customizing physical asset management and sustainability practices.

Acknowledgment. We would like to express our deepest gratitude to many people, without whom we could not have realized the international survey, Assoc. Prof. Dr. Katerina Gotzamani (University of Macedonia, Greece), Maria Gianni (University of Macedonia, Greece), Assist. Prof. Dr. T. Bartosz Kalinowski (University of Lodz, Poland), Prof. Dr. Hana Pačaiová (Technical University of Kosice, Slovakia), Dr. Anna Nagyová (Technical University of Kosice, Slovakia) and Onur Altekin (Turkey).

Appendix A: Measurement Scales

The items marked with the symbol (*) were excluded from further analysis. The value in parenthesis for each retained item indicates the standardized factor loadings.

Sustainability Practices

Respondents were asked to indicate how much emphasis is placed on each of the following activities where 1 means totally disagree and 5 means totally agree.

SU1: We incorporate the interests of key stakeholders (customers and suppliers) in our business decisions (0. 581)
SU2: We incorporate the interests of employees in our business decisions (0. 641)
SU3: We incorporate environmental concerns in our business decisions (0. 801)
SU4: We minimize the environmental impact of all our organization's activities (0. 785)
SU5: We provide a safe and healthy working environment to all employees (0. 822)
SU6: We emphasise the importance of our social responsibilities to the society (0. 797)

Physical asset management (consisted of 4 constructs, namely Risk Management, Performance Assessment, Life Cycle Management, Policy & Strategy).

Risk Management

Respondents were asked to indicate how much emphasis is placed on each of the following activities where 1 means totally disagree and 5 means totally agree.

RM1: We embed risk into all activities which could affect assets performance (0.947)
RM2: We analyse IT-system, business system, human resources, competence, etc. and address risk (0.799)
RM3: We analyse operation, production, quality and logistic process and address risk (0.792)
RM4: We perform risk assessment in order to minimize business losses (0.767)
RM5: Risk management is an integrated part of asset management strategy (0.756)
RM6: We analyse equipment failure causes and effects to address risk (0.657)

Performance Assessment

Respondents were asked to indicate how much emphasis is placed on each of the following activities where 1 means totally disagree and 5 means totally agree.

PA1: We exploit asset history to enhance asset knowledge (0.848)
PA2: We regularly review overall effectiveness of asset management activities (0.830)
PA3: We undertake benchmarking to support asset management activities (0.813)
PA4: We monitor key performance indicators (KPIs) to verify the achievement of organization's asset management goals (0.812)
PA5: We proactively pursue continuous improvement of asset management activities (0.721)
PA6: Company collects and analyses data related to asset management activities (0.681)
PA7: We regularly review overall efficiency of asset management activities (0.673)
PA8: We exploit information systems to support asset management activities (ERP, CMMS, AMS, or similar ones) (0.584)
PA9: We monitor condition of critical assets (0.567)

Life cycle Management

Respondents were asked to indicate how much emphasis is placed on each of the following activities where 1 means totally disagree and 5 means totally agree.

LM1: We continuously modernise our assets in accordance with our renewing/revision plans (0.874)
LM2: We continuously rationalise our assets to reduce production cost (0.866)
LM3: We assure quality of our assets during the whole life cycle phases (0.582)
LM4: We assure execution of maintenance processes within all assets' life cycle phases (0.581)
LM5: We execute disposal of assets in accordance with the asset management plan (0.573)

Policy & Strategy

Respondents were asked to indicate how much emphasis is placed on each of the following activities where 1 means totally disagree and 5 means totally agree.

PS1: We execute asset management strategy (0.624)
PS2: We undertake analyses of asset management policy to determine future production capacity (0.468)
PS3: We apply asset management policy (0.822)
PS4: We develop asset management objectives (0.463)

Operational Performance

Respondents were asked to select the number (on a 5 point Likert scale) that accurately reflects the extent of their organization's overall performance over the last three years on each of the following.

OP1: Flexibility to change product mix has improved during the last 3 years (0.826)
OP2: Percentage of internal scrap and rework has decreased during the last 3 years (0.813)
OP3: On-time delivery performance has improved during the last 3 years (0.806)
OP4: Cost of poor quality has decreased during the last 3 years (0.780)
OP5: Average lead time (from order to delivery) has improved during the last 3 years (0.765)
OP6: The production volume has increased during the last 3 years (0.550)
OP7: Unit cost of manufacturing has decreased during the last 3 years (0.541)

References

1. Pačaiová, H., Sinay, J., Nagyová, A.: Development of GRAM–a risk measurement tool using risk based thinking principles. Measurement **100**, 288–296 (2017)
2. Hamrol, A.: A new look at some aspects of maintenance and improvement of production processes. Manag. Prod. Eng. Rev. **9**(1), 34–43 (2018)

3. Algabroun, H., Iftikhar, M.U., Al-Najjar, B., Weyns, D.: Maintenance 4.0 Framework using self-adaptive software architecture. In: Proceedings of 2nd International Conference on Maintenance Engineering, IncoME-II 2017, The University of Manchester, UK (2017)
4. Rodič, B.: Industry 4.0 and the new simulation modelling paradigm. Organization **50**(3), 193–207 (2017)
5. Wijnia, Y., de Croon, J.: The asset management process reference model for infrastructures. In: 9th WCEAM Research Papers, pp. 447–457. Springer International Publishing (2015)
6. Komonen, K., Kortelainen, H., Räikkönen, M.: Corporate asset management for industrial companies: an integrated business-driven approach. In: Asset Management, pp. 47–63. Springer, Dordrecht (2012)
7. Haider, A., Koronios, A., Quirchmayr, G.: You cannot manage what you cannot measure: an information systems based asset management perspective. In: Mathew, J., Ma, L., Tan, A., Anderson, D. (eds.) Proceedings of Inaugural World Congress on Engineering Asset Management, Gold Coast, Australia, 11–14 July 2006
8. Amadi-Echendu, J.E.: Managing physical assets is a paradigm shift from maintenance. Paper presented at the IEEE International Engineering Management Conference (2004). https://doi.org/10.1109/iemc
9. Bond, T.C.: The role of performance measurement in continuous improvement. Int. J. Oper. Prod. Manag. **19**(12), 1318–1334 (1999)
10. Maletič, D.: Interaction between quality management, production and maintenance performance. Doctoral dissertation, Faculty of Organizational Sciences, University of Maribor, Kranj (2015)
11. Maletič, D., Maletič, M., Al-Najjar, B., Gomišček, B.: Development of a model linking physical asset management to sustainability performance: an empirical research. Sustainability **10**(12), 4759 (2018)
12. Alsyouf, I., Alsuwaidi, M., Hamdan, S., Shamsuzzaman, M.: Impact of ISO 55000 on organisational performance: evidence from certified UAE firms. Total Qual. Manag. Bus., 1–9 (2018)
13. Amadi-Echendu, J.E., Willett, R., Brown, K., Hope, T., Lee, J., Mathew, J., Vyas, N., Yang, B.S.: What is engineering asset management? In: Proceedings of 2nd World Congress on Engineering Asset Management and the 4th International Conference on Condition Monitoring, pp. 116–129 (2007)
14. Emmanouilidis, C., Komonen, K.: Physical asset management practices in industry: comparisons between Greece and other EU countries. In: Prabhu, V., Taisch, M., Kiritsis, D. (eds.) Advances in Production Management Systems. Sustainable Production and Service Supply Chains. IFIP Advances in Information and Communication Technology, pp. 509–516. Springer, New York (2013)
15. Schuman, C.A., Brent, A.C.: Asset life cycle management: towards improving physical asset performance in the process industry. Int. J. Oper. Prod. Manag. **25**(6), 566–579 (2005)
16. Komljenovic, D., Gaha, M., Abdul-Nour, G., Langheit, C., Bourgeois, M.: Risks of extreme and rare events in Asset Management. Saf. Sci. **88**, 129–145 (2016)
17. Roda, I., Macchi, M.: Studying the funding principles for integrating Asset Management in Operations: an empirical research in production companies. In: 3rd IFAC Workshop on Advanced Maintenance Engineering, Services and Technology, AMEST 2016, Biarritz, France, 19–21 October 2016. IFAC-PapersOnLine 49(28), 1–6
18. Maletič, D., Maletič, M., Al-Najjar, B., Gotzamani, K., Gianni, M., Kalinowski, T.B., Gomišček, B.: Contingency factors influencing implementation of physical asset management practices. Organizacija **50**(1), 3–16 (2017)

19. Liyanage, J.P., Badurdeen, F., Chandima Ratnayake, R.M.: Industrial asset maintenance and sustainability performance: economical, environmental, and societal implications. In: Ben-Daya, M., Duffuaa, S.O., Raouf, A., Knezevic, J., Ait-Kadi, D. (eds.) Handbook of Maintenance Management and Engineering, pp. 665–693. Springer, Heidelberg (2009)
20. Garetti, M., Taisch, M.: Sustainable manufacturing: trends and research challenges. Prod. Plan. Control **23**(2–3) (2012)
21. ISO 55001:2014, Asset Management—Management Systems—Requirements. ISO, Geneva (2014)
22. Baglee, D., Knowles, M., Yau, C.Y.: Development of techniques to manage asset condition using new tools. In: Asset Management, pp. 143–154. Springer, Dordrecht (2012)
23. Yan, J., Hua, D.: Energy consumption modeling for machine tools after preventive maintenance. In: 2010 IEEE International Conference on Industrial Engineering and Engineering Management (IEEM), 7 December 2010, pp. 2201–2205. IEEE (2010)
24. Fairfield, K.D., Harmon, J., Behson, S.J.: Influences on the organizational implementation of sustainability: an integrative model. J. Organ. Manag. **8**(1), 4–20 (2011)
25. Maletič, M., Maletič, D., Dahlgaard, J.J., Dahlgaard-Park, S.-M., Gomišček, B.: The relationship between sustainability–oriented innovation practices and organizational performance: empirical evidence from Slovenian Organization. Organizacija **47**(1), 3–13 (2014)
26. EFNMS-EAMC: How organizations manage their physical assets in practice, EFNMS Asset Management Survey (EFNMS) (2012). http://www.hms-gr.eu/t/files/EAMCSurvey2011ReportFinal02122012.pdf
27. McKone, K.E., Schroeder, R.G., Cua, K.O.: The impact of total productive maintenance practices on manufacturing performance. J. Oper. Manag. **19**(1), 39–58 (2001)
28. Kaynak, H.: The relationship between total quality management practices and their effects on firm performance. J. Oper. Manag. **21**(4), 405–435 (2003)
29. Preacher, K.J., Hayes, A.F.: SPSS and SAS procedures for estimating indirect effects in simple mediation models. Behav. Res. Methods Instrum. Comput. **36**(4), 717–731 (2004)
30. Preacher, K.J., Hayes, A.F.: Asymptotic and resampling strategies for assessing and comparing indirect effects in multiple mediator models. Behav. Res. Methods **40**(3), 879–891 (2008)
31. Baron, R.M., Kenny, D.A.: The moderator-mediator variable distinction in social psychological research: conceptual, strategic, and statistical considerations. J. Pers. Soc. Psychol. **51** (6), 1173–1182 (1996)
32. Green Jr., K.W., Zelbst, P.J., Meacham, J., Bhadauria, V.S.: Green supply chain management practices: impact on performance. Supply Chain. Manag. **17**(3), 290–305 (2012)

Assessment of the Small Enterprise's Maturity to Improvement Projects Based on the Lean Six Sigma Concept

Ewa Marjanska(✉), Piotr Grudowski, and Anna Wendt

Faculty of Management and Economics,
Department of Quality Management and Commodity Science,
Gdansk University of Technology, Narutowicza 11/12, 80-233 Gdansk, Poland
ewa.marjanska@zie.pg.gda.pl

Abstract. This paper explores the methodology to assess the maturity of small enterprise of building masonry sector to implement integrated approach of Lean Six Sigma. For this purpose survey data from senior managers were used. Quantitative method of the case organization's needs and capabilities in the scope of Lean Six Sigma project implementation was applied. The value of maturity indicator for Lean Six Sigma projects dedicated to the sector of small and medium size enterprises was calculated and the specific guidance for the management was proposed.

Keywords: Lean Six Sigma · Maturity index · Small and medium enterprises

1 Introduction

Small and medium enterprises (SMEs) are the main driving force in the field of innovation, development and employment in all economies [1–3]. According to European Commission there are two main criteria for an enterprise to be an SME. These are number of employees and annual turnover. The maximum number of employees should not exceed 250 in case of medium and 50 in case of small enterprises. The annual turnover should not be greater than 50 million EUR and 10 million EUR respectively [4]. However, these criteria and their values vary depending on the country [5].

Maintaining competitiveness and a stable position on the global market requires small and medium size enterprises to constantly improve their products and services [6–10]. In their operation SMEs are exposed to risk, especially as subcontractors of larger companies. Poor quality or quality deficiencies may have particularly dangerous consequences involving raised costs, loss of customers and poor reputation [11–13]. Therefore, implementation and maintenance of quality management systems together with regard to customer orientation and process approach is highly recommended in case of SMEs. Business practitioners and researchers emphasize the effectiveness of using in such cases the integrated Lean and Six Sigma approach, called Lean Six Sigma (LSS) [14–19].

There is quite large body of literature about implementation, functioning and maintenance of Lean Six Sigma in large organizations and recently also in SMEs [20–27]. It is known that implementation of LSS in SMEs requires application of additional,

A. Hamrol et al. (Eds.): *Advances in Manufacturing II* - Volume 3, LNME, pp. 44–55, 2019.
https://doi.org/10.1007/978-3-030-17269-5_4

dedicated support [28]. In order for the change process to be effective, it is necessary for the organization to represent an adequate level of maturity for its implementation. The concept of organizational maturity in the implementation of quality-oriented strategy results from the principles of TQM [28]. The first model, so-called Quality Maturity Grid (QMG) was developed by P. Crosby [29]. In 1989 a book "Managing the Software Process" by W. Humphrey was published. This work was based on previous studies of W. Shewhart and W.E. Deming and created the foundations for development of Capability Maturity Model (CMM) [30]. The model of excellence created by European Foundation for Quality Management (EFQM) and Common Assessment Framework (CAF) may also be applied to diagnose maturity of the organization. These models enable assessment of effectiveness and efficiency of an organization with respect to its goals [31].

However, there is a lack of scientific studies on the readiness and maturity of SMEs to implement an integrated approach to quality management. There is particularly little literature on SMEs dedicated maturity assessment. The results of Polish research show that although almost all tools associated with the Lean Management concept can be easily applied in SMEs, it is very difficult to implement Six Sigma tools there. The reason for this is that the latter seem to be too complicated for SMEs and advanced knowledge is required to use them [28]. The decision which and how many LSS tools should be chosen by an organization is of individual character. The demands of an organization depend mainly on its maturity [28].

The main goal of the authors adopted in this article is to assess the organizational maturity of the small enterprise of building masonry sector to implement an improvement project based on the Lean Six Sigma concept using the original LSS Plutus methodology. First we present literature review regarding integrated approach of Lean and Six Sigma. This leads us to explain the need for special support in LSS implementation for SMEs. The LSS Plutus methodology, research design and the enterprise case are discussed next. Then the empirical findings connected with the maturity assessment, their implications and discussion are presented. Conclusions, limitations and future research indications are finally described.

This paper is focused on one small-sized enterprise which has not implemented any of formalized quality management approaches. We consider the maturity of this particular organization to implement LSS. For this purpose, information obtained from representatives of the management of this organization was used. The application of the diagnostic tool as part of the LSS Plutus methodology allowed to set recommendations for improvement for the management of the surveyed organization.

2 Literature Review

2.1 Lean Management Approach

LSS concept integrates two concepts of operational improvement. The idea of Lean Production, also called Lean Thinking, is to constantly improve the main process efficiency. It is achieved by identification of all non-value adding activities (called wastes, jap. *muda*) and their sources and eliminating them [32, 33]. The examples of wastage cause may be: overproduction, unnecessary transport, downtime, unnecessary inventory, wasted manpower, space or efforts. The concept of waste *(muda)* was originated by T. Ohno who created the production philosophy of Toyota in 1950s [34]. This philosophy, also called Toyota Production System (TPS), together with *muda* reduction became one of the most important quality improvement ideas [35, 36]. Lean production was given its name by J. Womack in 1986 [36]. Within Lean production concept there are several systems for *muda* reduction, e.g. Kanban, Just-in-Time (JIT), Jidoka etc. Lean emphasizes the key role of the human factor involvement in activities aimed at improving the management system, emphasizing the importance of appropriate leadership, knowledge management in organization and communication [37].

2.2 Six Sigma

Six Sigma is considered to be "the foundation to drive breakthrough improvements" [38]. The idea was developed in Motorola company in 1980s and was called "the six steps to Six Sigma" [39, 40]. Six Sigma defines the global standard of variation characterizing processes, expressed in the standard deviation (sigma), meaning that no more than 3.4 defects/errors per million possibilities of their occurrence can be expected in the process. Six Sigma is also a framework for the implementation of improvement strategies, providing methods, techniques and tools to support the change process in the organization. This approach focuses on the characteristics of processes that are most important from the point of view of customers' needs. Six Sigma, on the one hand, is a synonym for the highest global quality standard, referring to the characteristics of products or services and parameters of activities as a result of which these products or services are obtained. On the other hand, it is a multi-stage, cyclical process focused on improvements enabling achievement of the aforementioned, close to perfection standard [28].

As a key mechanism of the management system based on the Six Sigma concept, process improvement included in the DMAIC cycle (Define, Measure, Analyze, Improve, Control) should be considered as a variation of the classic PDCA (Plan, Do, Check, Act) cycle of continuous improvement - the foundation of TQM (Total Quality Management) [28].

2.3 Lean Six Sigma Approach

There is the clear evidence in the literature for successful implementation of LSS as a business improvement methodology not only in manufacturing but also in many other sectors, e.g. banking, pharmaceutical [28]. Hence, it seems that LSS has displaced Lean and Six Sigma as the separate methodologies [41]. Integration of those two methodologies leads to focusing on costs and wastes reduction by variation reduction and elimination of non-value adding activities in a company's key processes.

In case of SMEs, Lean Six Sigma is considered to be the leading methodology for effectiveness improvement [15–18, 42, 43]. The results of various studies carried out in the SMEs all over the world show that Lean and Six Sigma methodologies are not distinguished, but implemented as one integrated methodology [15, 18, 44]. Many researchers agree that the evolution of LSS implementation knowledge requires further exploration of SMEs sector [45]. In Poland there is still a large number of SMEs which have not implemented any business improvement methodology. The maturity of SMEs to implement this methodology remains unexplored. Due to individual character of small and medium-sized organization's needs and possibilities in the field of LSS tools application choice, the need for special support in LSS implementation for SMEs exists. In this paper we have focused particularly on Polish small company and on its maturity to implement LSS as the leading methodology.

3 The Research Methodology

The single case study was conducted by authors of this paper. This research method allows detailed analysis of a case organization and deep insight into its structure and key processes [46–49]. LSS Plutus original methodology was used to assess the maturity of the case organization to implement LSS [50].

LSS Plutus methodology originally involves four stages based on DMAIC analysis. These are: (1) definition of the project's aim, (2) improvement activities, (3) evaluation of project results, (4) ensuring durability of introduced improvements. The original element of LSS Plutus methodology in stage 1 is calculation of two indicators. These are: customer's emotional value indicator and maturity indicator for LSS projects dedicated to the SMEs sector. In this study we focus on the latter one. This indicator allows to adjust the scope of the project to the actual needs and capabilities of the company [28]. It is based on the SMEs evaluation criteria in terms of the organization's maturity to conduct the LSS project. These criteria are divided into 8 groups. Three of them define the organization's needs, and the rest its capabilities (Table 1).

Table 1. SMEs evaluation criteria in terms of the organization's maturity to conduct the LSS project.

Criteria	Description
A. *The number of processes*	Medium-sized business and/or Numerous and complicated business processes and/or Complex relationships between processes
B. *Need for change*	High costs of the production or service process and/or Money frozen in surplus stocks and/or High price of products or services compared to the competition and/or Strong competition and/or Low customer satisfaction, loss of clients or numerous complaints and/or Intention to acquire new customers and/or Low quality of products or services and/or Need for the development of new products or services and/or Need for streamlining the current manufacturing process or the introduction of a new production line and/or

(*continued*)

Table 1. (*continued*)

Criteria	Description
	Long order execution time and/or Long reaction time to fluctuations in demand and/or Need for introducing production to order and/or LSS tools unsuitable or implemented selectively and improperly and/or Demand for introducing a management system from a corporation and/or Inability to permanently maintain improvement and/or Low use of the competence and experience of personnel and/or Need for improving working conditions and/or Strong motivation to develop the company and/or Intention to implement a quality system and industry standard and/or Motivation to win awards for quality and/or Intention to cooperate with partners within a logistics chain and/or Intention to cooperate within a cluster or with research institutions and universities
C. *Difficulties in obtaining customer satisfaction*	Lack of knowledge on the level of customer satisfaction or considerable difficulties in identifying it and/or Products or services not fulfilling the expectations of clients or lack of knowledge on customer needs and/or Loss of clients or lack of regular clients and/or Numerous complaints and/or Long-time order execution or untimely deliveries and/or High price in comparison with competition
D. *Acquired competence regarding quality oriented production management*	Implemented methods of organization and production management (including Lean Six Sigma, TQM) and/or Experience in the independent running of an improvement project and/or Appropriate self-evaluation of company results (e.g. self-evaluation according to the criteria of the EFQM or CAF models, corporation criteria, agreements with business partners, own criteria) and/or Performed audits of operation (e.g. internal audits in compliance with ISO type standards, industry standards) and/or Completed training on production management
E. *Knowledge of Lean Six Sigma*	Knowledge of the Lean Thinking concept and the practical application of tools and/ or Six Sigma Capability of the appropriate selection and effective implementation of LSS tools and/or

(*continued*)

Table 1. (*continued*)

Criteria	Description
	Capability of the autonomous running of an improvement project based on Lean Six Sigma and/or Capability of independently maintaining the introduced LSS solutions and/or Presence of a specialist responsible for continuous improvement
F. *Organizational culture conducive* to the implementation of permanent development	Organizational culture based on trust, respect, recognition, motivation and cooperation and/or Project awareness in the organization and/or Permanent involvement of all personnel in continuous improvement, including the most senior management and/or Interest in and recognition for the efforts of the staff, shown by the senior management and/or Permanent support for the involvement and/or the functioning of a quality circle in a company and/or awarding achievements and/or Autonomy of work connected with delegating competences and increasing the sense of responsibility for the completed task and/or Versatility and/or team work and/or Sharing skills within one team (cross training) and organization (internal benchmarking) and/or Lack of communication barriers between departments and/or Announcing results to staff and/or
G. *Time availability for conducting an LSS project*	Time reserves for conducting an improvement project and/or Time availability for training
H. *Availability of financial resources for activities connected with improvement*	Possibility of devoting considerable financial resources to improvement and/or Low cost of improving activities and/or Availability of considerable financial resources from the mother company and/or obtained the EU or state funding and/or Ease of obtaining a loan and/or financial resources for training

Source: [28].

The quantitative method as the supporting tool was used – each criterion from Table 1 was evaluated by the senior manager of the case organization in the five-point Likert scale. In the next step, the criteria presented in Table 1 were used to identify the needs and capabilities of the enterprise in terms of implementing the concept elements according to Table 2.

Table 2. Evaluation of maturity indicator of SMEs for the LSS project scale.

SME evaluation categories	SME maturity evaluation criteria for running the LSS project	Company maturity evaluation for LSS (1 = definitely no, 2 = no, 3 = partly/maybe/hard to say, 4 = yes, 5 = definitely yes)				
		1	2	3	4	5
Needs	A. Large number of processes					
	B. Need for change					
	C. Difficulties in obtaining customer satisfaction					
Capabilities	D. Competence in production management					
	E. Knowledge of LSS					
	F. Organization culture conducive to the implementation of continuous improvement					
	G. Time availability for the implementation of the LSS project					
	H. Availability of financial resources for improvement activities					
Assessment weight		0.2	0.4	0.6	0.8	1.0
Needs or capabilities		Small		Medium-sized	Big	

Source: [28].

Information from Table 2 was used for determination of a maturity indicator for LSS projects dedicated to the SMEs sector according to Eqs. (1–2).

$$\text{Enterprise needs indicator} = \frac{\sum_{i=1}^{criteria=A,B,C} \text{weight of the criterion}}{3} \times 100\% \tag{1}$$

$$\text{Enterprise capabilities indicator} = \frac{\sum_{i=1}^{criteria=D,E,F,G,H} \text{weight of the criterion}}{5} \times 100\% \tag{2}$$

$$\text{Enterprise maturity indicator} = \frac{\text{needs indicator value} + \text{capabilities indicator value}}{2} \tag{3}$$

The maturity indicator for LSS projects dedicated to the SMEs is the arithmetic mean of the sum of enterprise needs and capabilities level (3). In (Table 3) the interpretation of the maturity indicator value is presented.

Table 3. The interpretation of the enterprise needs and capabilities indicators' values making up the value of maturity indicator value for LSS projects dedicated to SMEs.

Enterprise needs (capabilities) indicator value	Interpretation
0–50%	Low needs *(capabilities)* level
51–69%	Average needs *(capabilities)* level
70–100%	High needs *(capabilities)* level

Source: [28].

4 Case Organization

The case organization is the small Polish company (40 employees) representing building masonry sector located in the northern part of the country. The company has been successfully operating on the Polish, Swedish, Danish, Norwegian and German markets for 25 years. The company, due to its modern technology, is capable of providing non-standard and unique services and products. Therefore, it is very competitive on the local market. The company's professional staff consists of experienced engineers and production employees, mostly employed in this lineup for 15 years. However, the company has not implemented any of quality management approaches so far. Current problems of the company comprise too high costs of raw and production materials. The company also has problems regarding production time planning. Despite being very competitive, as a small company, the organization experiences many threats as a subcontractor of larger companies. The existing intuitive management method has provided the company with perfect development. Currently, however, it would be advisable to look closely at the processes in the company in order to increase focus on acceleration of the production process, elimination of waste, and simplification of the way of doing business. In case of such small organizations as studied one it is especially recommended to precede the implementation of the system with analysis of the maturity of organization. Assessing the company's needs and capabilities can prevent unnecessary costs.

5 Empirical Findings and Discussion

The results of the case study are presented in (Table 4).

Basing on data given in Table 4 enterprise needs index and enterprise capabilities values were calculated. They are respectively 67% (average needs level) and 72% (high capabilities level).

The maturity indicator of case organization for the LSS project is presented graphically in (Fig. 1).

Table 4. The results of evaluation of maturity indicator of SMEs for the LSS project.

SME evaluation categories	SME maturity evaluation criteria for running the LSS project	Company maturity evaluation for LSS (1 = definitely no, 2 = no, 3 = partly/maybe/hard to say, 4 = yes, 5 = definitely yes)				
		1	2	3	4	5
Needs	A. Large number of processes					X
	B. Need for change			X		
	C. Difficulties in obtaining customer satisfaction		X			
Capabilities	D. Competence in production management			X		
	E. Knowledge of LSS			X		
	F. Organization culture conducive to the implementation of continuous improvement				X	
	G. Time availability for the implementation of the LSS project					X
	H. Availability of financial resources for improvement activities			X		
Assessment weight		0.2	0.4	0.6	0.8	1.0
Needs or capabilities		Small		Medium-sized	Big	

Source: self elaboration based on [28].

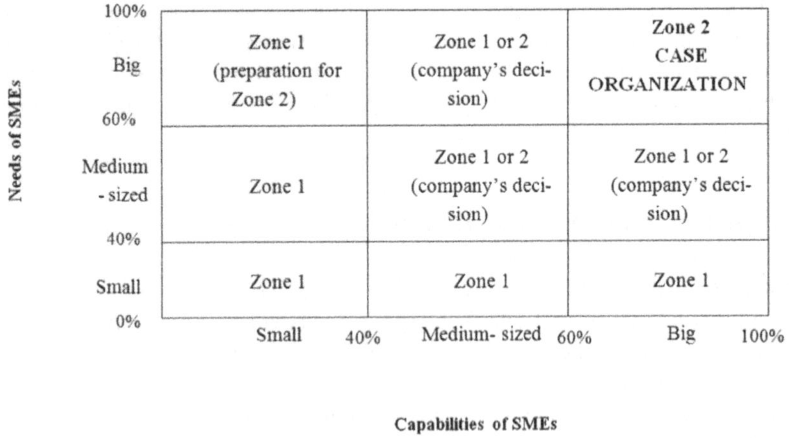

Fig. 1. Interpretation of maturity indicator of case organization for the LSS project.

The case organization was assigned to Zone 2 according to maturity index value. This is the result of high levels of needs and capabilities of this enterprise. Location in this zone indicates the company's readiness to apply LSS methodology at an advanced level. The company is mature enough to take advantage of the entire LSS methodology repository available, including *Poka-Yoke*, Kanban. However, it should be noticed, that enterprise's needs indicator is equal to 67%, which is only 7% points above the border separating big and medium needs. It is worth emphasizing that if big capabilities of the enterprise are accompanied by medium or low needs, implementation of LSS on basic level is recommended. In case of studied enterprise, despite its maturity, a gradual implementation would be desirable. Due to this excessive use of procedures and tools in relation to the uncomplicated organization processes and transparent relationships between them will be avoided.

6 Conclusions

Small enterprises are characterized very often by simple processes, better ability to react quickly and effectively to market changes and low costs of development activities comparing to big enterprises. Therefore, small organizations require relatively simple improvement initiatives. It is important to define needs and capabilities of the organization before incurring costs for implementation. It can be done quantitatively using the maturity indicator for the LSS project. However, due to their complexity LSS projects are not very often used in SMEs. In this paper the possibility of maturity evaluation of the case organization belonging to SMEs was shown. The results obtained might be used by the leaders of the enterprise to take the decision on the implementation of LSS toolbox. These results might also provide important information on future needs of internal customers of the studied company with respect to quality related trainings and possible savings. The tool proposed might offer such possibility also to smaller organizations.

7 Limitations and Areas for Future Study

This paper is based on a single case study which is a limitation in the sense that general conclusions cannot be drawn. However, further research into organizational culture and its impact on LSS implementation strategy in the case organization may result in establishing new, extended measures for determining the organization's potential for LSS implementation.

References

1. Neneh, N.B., Vanzyl, J.: Towards establishing long –term surviving small and medium enterprises (SMEs) in South Africa. An entrepreneurial approach. Afr. J. Bus. Manag. **6**(28), 8327–8343 (2012)
2. Storey, D.: optimism and care: the elephants in the entrepreneurship room. Int. Small Bus. J. **29**(4), 303–321 (2011)
3. Uchikawa, S.: Small and medium enterprises in Japan: surviving the long-term recession. In: ADBI Working Paper 169. Asian Development Bank Institute, Tokyo (2009)
4. Commission Regulation (EC) No 800/2008 of 6 August 2008 declaring certain categories of aid compatible with the common market in application of Articles 87 and 88 of the Treaty (General block exemption Regulation) (Text with EEA relevance)

5. Tyagi, D., Soni, V.K., Khare, V.K.: A review on issues for the implementation of lean six sigma by small and medium enterprises. Int. J. Latest Trends Eng. Technol. **3**(4), 94–98 (2014)
6. Lyu Jr., J., Liang, C.-C.: Effective approach to quality control for small-medium software companies. Total Qual. Manag. Bus. Excellence, Spec. Issue: TQM: Qual. Costs Qual. Improv. Roadmaps Qual. Manag. Syst. **25**(3–4), 296–308 (2014)
7. Teixeira, H.N., Lopes, I.S., Sousa, S.D.: Diagnosis of quality problems in a footwear SME. In: Proceedings of the World Congress on Engineering, III, pp. 1438–1443. WCE, London (2012)
8. Mendes, L.: Employees Involvement and quality improvement in a manufacturing small and medium enterprise (SME): a comparative analysis. Afr. J. Bus. Manag. **6**(23), 6980–6996 (2012)
9. Blanco-Callejo, M., Gutierrez-Broncano, S.: Application of the total quality management approach in a spanish retailer: the case of Mercadona. Total Qual. Manag. Bus. Excelence **21** (12), 1365–1381 (2010)
10. Kureshi, N.L., Mann, R.: Quality management practices of SMEs in developing countries: a survey of manufacturing SMEs in Pakistan. J. Qual. Technol. Manag. **5**(2), 63–89 (2009)
11. Kumar, M,. Khurshid, K.K., Dora, M.K., Timas, W., Antony, J.: Lean/Six Sigma implementation in SMEs: key findings from international research, pp. 1–10 (2012)
12. Antony, J.: Can Six Sigma be effectively implemented in SMEs? Int. J. Prod. Perform. Manag. **57**, 420–423 (2008)
13. Wattanapruttipaisan, T.: SME subcontracting as a bridgehead to competitiveness: an assessment of supply-side capabilities and demand-side requirements. Asia Pac. Dev. J. **9**(1), 65–87 (2002)
14. Byrne, G., Lubowe, D., Blitz, A.: Using a lean six sigma approach to drive innovation. Strategy Leadersh. **35**(2), 5–10 (2007)
15. Thomas, A., Barton, R., Chuke-Okafor, C.: Applying lean six sigma in a small engineering company – a model for change. J. Manuf. Technol. Manag. **20**(1), 113–129 (2009)
16. Laureani, A., Antony, J.: Lean six sigma in a call centre: a case study. Int. J. Prod. Perform. Manag. **59**(8), 757–768 (2010)
17. Assarlind, M., Gremyr, I., Backman, K.: Multi-faceted views on lean six sigma application. Int. J. Qual. Reliab. Manag. **29**(1), 21–30 (2012)
18. Enoch, O.K.: Lean six sigma methodologies and organizational profitability: a review of manufacturing SMEs in Nigeria. Am. J. Ind. Bus. Manag. **3**(6), 573–582 (2013)
19. Jie, J.C.R., Kamaruddin, S., Abd Azid, I.: Implementing the lean six sigma framework in a small medium enterprise (SME) – a case study in printing company. In: Proceedings of the 2014 International Conference on Industrial Engineering and Operations Management, Bali, Indonesia, pp. 396–398, 7–9 January 2014
20. Vinodh, S., Vaudham, S.G., Anesh Ramiya, R.: Implementing lean six sigma framework in an indian automotiver valves manufacturing organisation: a case study. Prod. Plan. Control **22**(7), 708–722 (2011)
21. Wang, F.-K., Chen, K.-S.: Application of lean six sigma to a panel equipment manufacturer. Total Qual. Manag. Bus. Excellence **23**(3), 417–429 (2012)
22. Monteiro de Carvahlo, M., Lee, H.L., Boarin, P.S.H.: The Six Sigma program: an empirical study of Brazilian companies. J. Manuf. Technol. Manag. **25**(5), 602–630 (2014)
23. Zaman, D.M., Zerin, N.H.: Applying DMAIC methodology to reduce defects of sewing section in RMG: a case study. Am. J. Ind. Bus. Manag. **7**(12), 1320–1329 (2017)
24. Pearce, A., Pons, D., Neitzert, T.: Implementing Lean – outcomes from SME case studies. Oper. Res. Perspect. **5**, 94–104 (2018)

25. Syathputri, K., Sari, M.R., Anizar, I., Tarigan, R., Siregar, I.: Application of lean six sigma to waste minimization in cigarette paper industry. In: IOP Conference Series, Materials Science and Engineering, pp. 309–313 (2018)
26. McLean, R.S., Antony, J.: A conceptual continuous improvement implementation framework for UK manufacturing companies. Int. J. Qual. Reliab. Manag. **34**(7), 1015–1033 (2017)
27. Hamrol, A.: Strategies and practices of efficient operation. In: Lean Six Sigma and Other, Warsaw (PWN) (2015)
28. Grudowski, P.: How to assess the maturity of small and medium-sized enterprises to lean six sigma projects. Argumenta Oeconomica **2**(37), 312–327 (2016)
29. Maria, L., Calingo, R.: The evolution of strategic quality management. Int. J. Qual. Reliab. Manag. **13**(9), 19–37 (1996)
30. Herbsleb, J., Zubrow, D., Goldenson, D., Hayes, W., Paulk, M.: Software quality and the capability maturity model. Commun. ACM **40**(6), 30–40 (1997)
31. Bou-Llusar, C.J., Escrig-Tena, A.B., Roca-Puig, V., Beltran-Martin, I.: To what extent do enablers explain results in the EFQM excellence model? An empirical study. J. Qual. Reliab. Manag. **22**(4), 337–353 (2005)
32. Bicheno, J.: The Quality 7S: Towards Six Sigma Performance in Service and Manufacturing. PICSIE Books, Johannesburg (2002)
33. Womack, J.P., Jones, D.T.: Lean Thinking – Banish Waste and Create Wealth in Your Corporation. Simon & Schuster, London (1996)
34. Dahlgaard-Park, S.M.: From ancient philosophies to TQM and modern management theories. Thesis No. FiF-a 31, Linkoping University, Sweden (2000)
35. Udagawa, M., Satoo, H., Nakamura, K., Nonaka, I, Nihon, K.H.K.: Quality Management in the Japanese Enterprises, Yuhikaku, Tokyo (1995)
36. Womack, J.P., Jones, D.T., Roos, D.: The Machine that Changed the World. Maxwell Macmillan International, New York (1990)
37. Dahlgaard, J.J., Dahlgaard-Park, S.M.: Lean production, six sigma quality, TQM and company culture. TQM Mag. **18**(3), 263–281 (2006)
38. Dedhia, N.S.: Six sigma basics. Total Qual. Manag. Bus. Excellence **16**(5), 567–574 (2005)
39. George, S.: The Baldrige Quality System. Wiley, New York (1992)
40. Fukuda, R.: Managerial Engineering. Productivity Inc., Stanford (1983)
41. Marsh, J., Perera, T.: Lean Six Sigma: Exploring Future Potential and Challenges, Lean Six Sigma Research and Practice, pp. 35–45. Ventus Publishing APS, Frederiksberg (2011)
42. Wessel, G., Burcher, P.: Six Sigma for Small and medium-sized enterprises. TQM Mag. **16** (4), 264–272 (2004)
43. Wiśniewska, M.: Quality in crisis times. Works Mater. Fac. Manag. Univ. Gdańsk **3**(2), 265–275 (2009)
44. Timans, W., Antony, J., Ahaus, K., van Solingen, R.: Implementation of lean six sigma in small and medium enterprises in the Netherlands. J. Oper. Res. Soc. **63**(3), 339–353 (2012)
45. Zhong, O., Ifran, M., Khattak, M.A.O., Zhu, X., Hassan, M.: Lean six sigma. A literature review. Interdisc. J. Contemp. Res. Bus. **3**(10), 599–605 (2012)
46. Yin, R.K.: Case Study Research: Design and Methods, 2nd edn. Sage, Thousand Oaks (1994)
47. Yin, R.K.: Application of Case Study Research, 2nd edn. Sage, Thousand Oaks (2003)
48. Meredith, J.: Building operations management theory through case and field research. J. Oper. Manag. **16**, 441–454 (1998)
49. Saunders, M., Lewis, P., Thornhill, A.: Research Methods For Business Students, 4th edn. Pearson Education, Harlow (2007)
50. Grudowski, P., Leseure, E.: LSS Plutus - Lean Six Sigma for Small and Medium Enterprises. WNT, Warsaw (2013)

Technical Culture Maturity as a Manifestation of Implementation of Lean Management Principles – Situation in Agricultural Machinery Sector

Przemysław Niewiadomski[1], Agnieszka Stachowiak[2(✉)], and Natalia Pawlak[2]

[1] University of Zielona Góra, Zielona Góra, Poland
[2] Poznan University of Technology, Poznań, Poland
agnieszka.stachowiak@put.poznan.pl

Abstract. The paper is qualitative-quantitative research designed and conducted to analyze technical culture maturity in companies representing agricultural machinery sector. Authors believe that technical culture maturity proves that company implemented and is following lean management principles. The research conducted showed high level of technical maturity, proving that companies within the sector are mature in technical culture context, nevertheless there are some aspects that could be improved. The high level of maturity is on one hand consequence of lean management principles implementation, on the other it creates environment that supports and strengthens leaning the companies. The developed method can be used by companies for self-assessment and reference when improving their maturity and implementing lean.

Keywords: Technical culture · Lean management · Agricultural machinery sector

1 Introduction

Efficient management of organizations depends on many factors characterizing contemporary organization, as well as on the increasingly rapidly changing environment [1]. The issue of efficiency is particularly important for enterprises, as it allows to assess the level of their development and the impact of individual departments, positions, etc. on overall results [2]. As a result, in management more and more attention is paid to operational activities concerning core processes [3]. The management processes are one of the areas to be improved in the production enterprise. Their level of their development can be referred to as maturity of processes or of the entire management system [4]. It is suggested, therefore, that on the basis of existing experiences, new concepts and management methods striving for counteracting the waste of the organization's activities, contributing to the reduction of its operating costs should be sought.

A. Hamrol et al. (Eds.): *Advances in Manufacturing II* - Volume 3, LNME, pp. 56–74, 2019.
https://doi.org/10.1007/978-3-030-17269-5_5

Such a modern trend - according to the authors of the paper - is lean management that sets ever more stringent criteria for companies' assessment together with specific factors determining their activity, requiring changes in all processes implemented in the enterprise, reaching for comprehensive concepts that primarily involve cultural management criteria. Lean management is a holistic approach to business [5]. It allows to approach problem-solving in a systemic manner, included in the organization's strategy. It takes into account many aspects of the organizational operation and at the same time is not limited to establishing a rigid framework of areas that can be subjected to a process analysis. Hamrol postulates that lean is associated primarily with improving the efficiency of actions by eliminating waste, minimizing waste, and controlling the flow of added value [6]. Elimination of waste leads to the increase of work efficiency and effectiveness with the maximum focus on the value added for the client [7–14]. The term that is naturally associated with lean management is a organizational culture defined as a set of values, traditions, beliefs, attitudes and behaviors that are the essence of everything that one does and thinks in an organization. It is powered by a system of routines, rituals, communication patterns, and informal structures [15]. In contemporary companies, dealing with continuous technical advance, scientific and technical aspects of culture are more and more important. The technical culture can be defined as the set of technical solutions and their usage patterns [16] and in authors opinion it contributes significantly to companies well-being. However, the issue is not widely discussed as when searching WoS database, the authors have found only about 30 publications the term "technical culture" in the title, and most of them referred to education issues.

As technical culture is associated with technical solutions, it deals with technical resources, nevertheless, from the resource-based point of view, more important are the resources that cannot be imitated nor renewed (VRIN approach [17]), for example professionalism, competences and mentality of leaders. They are the basic element of the organization's culture and through it they influence the way the company operates [18]. In the context of the above, the maturity of the organization's technical culture based on values, concepts or management methods is crucial to meeting the challenges of the future in business.

Hence, the research question was formulated: is it possible to implement methods or lean management concepts without absorbing a sufficiently high level of technical culture maturity? In reference to the above, the goal of the paper was defined as an attempt to assess the level of maturity of technical culture as symptom of the implementation of lean management principles in manufacturing enterprises operating in the agricultural machinery sector.

In reference to such a goal, the following research procedure was recommended: on the theoretical level - using the method of reconstruction and literature of the subject - selection of questions expressing the level of technical culture maturity; on the design level - compiling the research tool in the form of an evaluation sheet being the result of exploration of the literature on the subject and discussion among experts (including answer to questions: what aspects of lean management reveal the technical culture of the enterprise); on the empirical level - practical use of the method; surveying agricultural machinery manufacturers.

In the context of such a goal, it seems accurate to plan the research based on the literature review method, expert knowledge, creative discussion and direct interviews with selected representatives of the sector under investigation. The theoretical research, for management practitioners, can become the basis for evaluation and inspiration to build their own strategies in the field of implementation of management concepts.

2 Technical Culture and Lean Management – Research Background

2.1 Technical Culture

The term technical culture is defined as the conformity of development and, above all, implementation and use of certain techniques [16]. It is connected with the skillful use of technical means according to their purpose. The term also refers to a certain level of experience with technology. The most characteristic approaches to technical culture include those in which it is combined with the appropriate technical facilities. It manifests then in the ways of using these objects because the need to use the technique causes changes in the operation [19].

Technical culture is also defined as entire technical knowledge, assessments, opinions and views on the technical system, its individual elements and their effects, as well as the skills of the society (group) in the use of technology. Technical culture is a very important part of the culture of society, and it acquires more value, the more value a technician has in a given society [20].

There are also approaches in which attention is paid to connections of technical culture with such technical objects that are characterized by modernity. Technical culture also refers to professional environments involved in development and implementation of technology due to their education and profession [21]. Its connection with other types of culture, for example, political, aesthetic are considered, because it is not being developed in isolation, but through the habits of everyday life, as well as through aesthetic tastes and standards accepted in the society.

Technical culture can also be perceived as part of the culture of the society. Firstly, due to the role that it plays in technical activities and secondly because of the role of human activity, which often has a technical aspect [22].

Raising the level of technical culture is not only important for itself, but it is above all a social meaning because it is the general social balance that determines the level of society's technical culture.

2.2 Lean Management and Lean Culture

Currently, many Polish and foreign representatives of agricultural machinery manufacturers[1] [23–30] and foreign [31–35] show interest in solutions from the broadly understood scope of the so-called "lean management". Lean production refers to the

[1] SaMASZ company may be an example, it has implemented SPS – SaMASZ Production System, i.e. a production management system based on the philosophy of Lean manufacturing. This company implements all the best models and tools of lean manufacturing, such as: Poka-yoke, Kaizen, 5S, Kanban, SMED. After: [44].

continuous elimination of waste and continuous improvement [36]. This concept strongly emphasizes the need for elimination of waste and use of such forms of production units that the effectiveness and efficiency of the enterprise would not be reduced [37]. In addition, it solves the elimination of waste and makes the flow process more streamlined and efficient [38]. Therefore, the elimination of waste has become one of the decisive elements helping to build an effective and lasting competitive advantage of a production company [39]. Thus, a lot is said about the system that allows increased productivity and efficiency, reducing production costs and the broadly understood Lean Manufacturing philosophy is being considered [40].

Hence, Lean Management is a management methodology that creates a kind of work culture in an organization. It comes down to the fact that all people associated with a given organization are interested in a constant reduction of costs, shortening the delivery cycle and improving quality [41]. Lean management is achieving such efficiency that makes the company flexible, lean, and trained. A lean company builds its organization and manages the process so that the client actually pays for its production, not for the functioning of, for example, huge organizational structure, warehouses, means of transport or overly extensive administrative work, etc. Therefore Lean Management is flexible production that allows o achieve above-average profits. This concept, which is referred to as "lean" production that consumes less resources - less human effort, less equipment, less time and space, while striving to supply the client exactly what he expects [42]. Lipecki [43] believes that the main goal of Lean Management is to achieve a high level of economic efficiency, quality and flexibility at the same time. The complexity of activities related to this makes the chain of various undertakings aimed at "lean organization" should never end or close [44]. In addition, the Lean Manufacturing concept emphasizes the simplification of business management by building simple, relatively flat and transparent structures, and eliminating processes that do not create added value [45].

3 Research Methodology and Data Sources

3.1 Preliminary Research

Using the method of reconstruction and interpretation of the Polish and foreign literature on the subject, a set of questions evaluating the level of maturity of a technical culture oriented towards lean management was selected. Such action - at the project level - made it possible to compile a research tool in the form of an assessment sheet being the result of literature exploration and discussion among experts in the field.

Qualitative research was focused on identifying factors reflecting the maturity of the technical culture of production companies operating in the agricultural machinery sector. From authors' point of view maturity is a symptom of the implementation of the principles of lean management. The goal of the research was to understand the causes of a particular perception by understanding and interpreting the action and thinking pattern correctly. Information collected in this way has been presented in a descriptive manner. Having a phenomenological character (based on experience) - qualitative research - made it possible to recognize opinions, feelings and associations that were triggered in the analyzed case by a number of factors referring to the technical culture problem. Based on the results of qualitative research, guidelines for quantitative research have been depicted. The conducted research helped to identify the areas of the next stage of research, formulate problems and specific issues. They provided interesting information about the language, which is used by "industry experts" to describe phenomena of interest. In the opinion of the authors, this allowed to avoid errors at the level of constructing questions and adapting the language to potential respondents. These studies have, to a great extent, facilitated the researchers' closeness to the natural environment being under research, understanding their attitudes and language, which greatly assisted quantitative research, while ensuring full understanding of the phenomena among potential respondents.

The research technique that was selected to conduct research and collect primary data was an in-depth individual interview conducted among experts in the field. During the qualitative survey, the respondent was in direct contact with the researcher, called an indagator in this type of research. The individual interview was of a relaxed nature, although it followed the previously developed scenario. In the first place, general questions were presented, which gradually turned into a more detailed issues. The selected method of recording data from the conducted qualitative study consisted in observing the course of individual thematic sessions included in the study immediately after its completion. 28 representatives related to the agricultural machinery sector too part in the discussion, including:

- 16 owners and co-owners - 57.14% of the population surveyed - enterprises operating in the agricultural machinery sector;
- 8 managers - 28.57% of all respondents - employed under a contract of employment;
- a representative of the Industrial Institute of Agricultural Machines (3.57%);
- a representative of a university specializing in modern management concepts (3.57%);
- an expert for over 12 years associated with production companies operating in the automotive sector (3.57%); directly coordinates lean projects; is qualified as a Certified Risk Manager ISO 31000 and is a plenipotentiary for the Integrated Management System (ISO 9001, ISO 14001, PN-EN 18001) in one of the enterprises of the machine sector (production of trailers);
- an adviser with many years of professional experience, full-time employee in a unit offering comprehensive, long-term support for consulting projects in the field of Lean Management (3.57%).

The selected characteristics of the surveyed population are presented below.

The group of people with secondary and higher education prevailed among those surveyed (over 92%); 60.71% of experts represented higher education, level, 32.15% - secondary school level, 7.14% - vocational (Table 1).

Table 1. The selected characteristics of the surveyed population.

	Education	
	Number of participants	%
Primary school	N = 0	0
Vocational school	N = 2	7,14
Secondary school	N = 9	32,15
University	N = 17	60,71
Total	N = 28	100,00

Source: Own work

The age of the respondents was between 26–69 years (including 7.14% of respondents in the age group up to 30 years, 25% in the range from 31 to 40 years, 39.29% in the age group 41–50 years, 17, 86% in the range of 51–60 years, 10.71% of experts were over 60 years old) (Table 2).

Table 2. Age of the surveyed population.

	Age	
	Number of participants	%
Up to 30 years	N = 2	7,14
From 31 to 40 years	N = 7	25,00
From 41 to 50 years	N = 11	39,29
From 51 to 60 years	N = 5	17,86
Over 60 years	N = 3	10,71
Total	N = 28	100,00

Source: Own work

When deciding on the selection of respondents, their direct acquaintance with researchers was an important criterion. It allowed to determine whether the respondents are independent in the presented views and whether they have sufficient knowledge and experience in the field of the subject matter. In addition, taking into account communication barriers, people with whom the authors have good relations have been invited to the study.

Taking into account the suggestions of experts, a research tool was developed in the form of a questionnaire. The questionnaire was structured into two layers: the general one with aspects of corporate culture, and the detailed with specified elements of the aspects identified. The questions were brainstormed by the authors. The developed questionnaire includes forty-two closed questions[2] referring to five spaces of technical culture geared to lean management (Table 3). The need to limit the number of questions that were included in the research emerged from the difficulty of carrying out the research for too many of them; it resulted, among others from the limited time of the meeting planned as part of the research. In addition, according to the expert assessment,

[2] Each of the questions is one-choice.

Table 3. Lean management areas – structure of the questionnaire.

NR	Area	Accronim
A. [1]	Culture oriented to innovation	[S1] K_{INNOV}
A. [2]	Culture oriented to technology	[S2] K_{TECH}
A. [3]	Culture oriented to staff	[S3] K_S
A. [4]	Culture oriented to client	[S4] K_C
A. [5]	Culture oriented to development	[S5] K_{DEV}

Source: Own work

the indicated areas were considered sufficient to make a comprehensive assessment of the maturity.

In order to carry out the assessment, a five-point scale was adopted describing the level of technical culture maturity of individual descriptions corresponding to the areas of lean management (Table 4).

Table 4. Level of maturity of culture of selected lean management area.

	Intensity	Description
Maturity Level	5	Culture functions perfectly, f.ex. in enterprises losses emerging from undesired actions and behavior
	4	Culture functions well, however there can be found some areas that can be improved
	3	Culture functions in practice, however there are many areas that could be improved
	2	Culture functions only theoretically, there are no signs of them in business practice
	1	Culture does not function at all, there are actions against lean management principles

Source: Own work

The recognition of maturity of technical culture and development of its final model allows for the analysis of the gap between the desired and current level of implementation of its individual descripts. The developed method of assessment, in principle, is to be a universal and useful tool for assessing the level of technical culture maturity, and also an important instrument in the process of its improvement. Thanks to this method, enterprises can make self-assessments and determine which cultural attributes can be associated with great opportunities, and which of them should be subject to improvement. To make this possible, it is necessary to know the current status of culture, as well as the ability to state and distinguish priority issues from what is irrelevant in an enterprise.

3.2 General Research

The main stage of the research was carried out from 30 June to 22 October 2018. Initially, the research was planned to be carried out in two stages, as direct meetings, planned for 30 June–1 July 2018 during the "Agro Tech" Agro and Industrial Fair, and between 20–23 September 2018 during the International Agricultural Exhibition "Agro Show". However, in order to obtain more representative scope of the target group (enterprises using Lean management tools) and to obtain as many answers as possible, from September 23 to October 20, 2018, an additional survey was conducted among selected companies cooperating with the Spare Parts and Agricultural Machinery Manufacturing Plant "Fortschritt" as a research partner.

The research was conducted on a sample of 63 enterprises representing the agricultural machinery sector, selected basing on the maturity criterion perceived as the ability of the company to effectively select lean management tools supporting the organization's goals and strategy in order to achieve high quality processes, products, repeatability of successes and avoiding mistakes. From among the surveyed companies, 25 enterprises indicated a very high level of maturity in the implementation of the principles of lean management, 27 enterprises high, and 11 companies moderate. Respondents were owners and managers representing: micro - 5 people (7.94%), small - 20 people (31.75%), medium - 31 people (49.21%) and large - 7 people (11.11%) manufacturing enterprises operating in the agricultural machinery sector. Small and medium-sized enterprises play an important role in the agricultural machinery sector, hence such entities constituted the majority (over 80%). In the case of large companies, the share of foreign capital was declared.

There were 36 owners (57.14% of all respondents) and 27 managers (42.86% of all respondents) that took part in the survey. The age of the respondents was between 25–69 years (of which 31.74% were respondents in the age range of 31–40 years, 26.98% in the age group 41–50 years and 3% in the interval 51–60 years). The youngest participant was 25 years old and the oldest was 69 years old; 47.22% of owners were over 50 years old, the age of 27.78% of owners was 40–50 years, while 25% of owners were under 40 years of age.

In the case of managers, 14.82% of the respondents were over 50, the age of 33.33% was in the range of 40–50 years, 44.4% of managers were between 30 and 40, while 7.41% had less than 30 years. Detailed characteristics are depicted in (Table 5).

Table 5. Age characteristics of surveyed population (N = 63).

	Age					
	Owners		Managers		Total	
	57,14%		42,86%		100%	
	Number of participants	%	Number of participants	%	Number of participants	%
Under 30 years	N = 1	2,78	N = 2	7,41	N = 3	4,76
From 31 to 40 years	N = 8	22,22	N = 12	44,44	N = 20	31,74
From 41 to 50 years	N = 10	27,78	N = 9	33,33	N = 19	30,16
From 51 to 60 years	N = 13	36,11	N = 4	14,82	N = 17	26,98
Over 60	N = 4	11,11	N = 0	0	N = 4	6,36
Total	N = 36	100,00	N = 27	100,00	N = 63	100,00

Source: Own work

The majority of respondents were people with secondary and higher education (over 90%), of which 50% of the owners represented higher education level, 36.11% - secondary school, 13.89 – vocational. In the case of managers, 70.73% had higher education, 25.93% secondary, and 3.70% vocational. Detailed characteristics are illustrated in (Table 6).

Table 6. Education characteristics of surveyed publication (N = 63).

Level of education	Education					
	Owners		Managers		Total	
	57,14%		42,86%		100%	
	Number of participants	%	Number of participants	%	Number of participants	%
Primary	N = 0	0	N = 0	0	0	0
Vocational	N = 5	13,89	N = 1	3,70	N = 6	9,52
Secondary	N = 13	36,11	N = 7	25,93	N = 20	31,75
Higher (University)	N = 18	50,00	N = 19	70,37	N = 37	58,73
Total	N = 36	100,00	N = 27	100,00	N = 63	100,00

Source: Own work

Analysis and interpretation of research results is one of the most important stages of the research process. The paper attempts to interpret the results and present deeper analysis based on respondents' declarations. The analysis proceeded according to previously assumed stages. The first of them was the proper development of the original data obtained and their proper organization. Such a method of data processing allowed for the segregation of material, compilation of designates into appropriate groups, and rejection of these data, which from the point of view of the problem under research turned out to be insignificant. The next step, presented in the following chapter was to describe the data obtained and to make their interpretation.

4 Results of the Research

While assessing the level of technical culture maturity of manufacturing enterprises operating in the agricultural machinery sector, which is a symptom of the implementation of the principles of lean management, the commonly used practice in the analysis of survey results was used. The diagnosis was made on the basis of the mean value calculated on the basis of responses of the respondents taking part in the research. Generally, for ordinal scales, the average value of a given characteristic should not be taken into account. Nevertheless, it is used in survey questionnaires and it enables getting an answer regarding the degree of acceptance of some phenomenon or view.

An effective strategy is to guarantee the uniqueness of the company, so that it stands out from among competitors. Innovation plays a very important role in the long-term development of the enterprise. The research results show clearly that innovation strategies are being developed in the enterprises covered, including objectives, ways and scope in which innovations (product, process or organizational) will be used to build a strategic advantage permanently embedded in the company's technical culture (average rating 4.27, 42.9% of indications for the assessment of 5 points)[3] (Table 7).

Table 7. Culture oriented to innovation – maturity assessment.

Nr	Culture oriented to innovation	1	2	3	4	5	Pts.
1.	The strategy of development implemented considers including innovation in the set of corporate goals?	1	1	5	29	27	4,27
		1,6	1,6	7,9	46,0	42,9	
2.	Does the company have ability to adapt to influence of changing environment?	-	-	4	28	31	4,43
		-	-	6,3	44,4	49,2	
3.	Interest in development, change and longtime horizon plans	-	1	2	27	33	4,46
		-	1,6	3,2	42,9	52,4	
4.	Is there a multi-direction flow of information in areas connected with innovativeness of company?	1	2	6	25	29	4,25
		1,6	3,2	9,5	39,7	46,0	
5.	Orientation towards future	-	1	4	23	35	4,46
		-	1,6	6,3	36,5	55,6	
6.	Extraversion of risk	1	2	7	26	27	4,21
		1,6	3,2	11,1	41,3	42,9	
7.	High tolerance of (affirmation of) uncertainty	1	3	8	27	24	4,11
		1,6	4,8	12,7	42,9	38,1	
8.	Ability to acquire and implement new ideas	-	1	4	26	32	4,41
		-	1,6	6,3	41,3	50,8	

Source: Own work

Variable and uncertain market environment mobilizes the surveyed enterprises to search for safe areas of operation and develop skills in quick capture of threats[4]. Companies are trying to adapt to these difficult conditions, creating open and flexible strategies that allow them to quickly face the challenges and market opportunities (average rating 4.43, 49.2% indications for the assessment of 5 points) (Table 7). They show great interest in development, change and long-term plans (average rating 4.46, 52.4% indications for the assessment of 5 points). A well-functioning technical culture is a matter of engaging all employees and values that are close to everyone and form the basis for action. It is technical culture where the information systems are built in the

[3] The authors believe that maintaining status of declared, high level of technical culture maturity within innovation area requires continuous work on improvement.

[4] It is characterized by relatively high level of extraversion to risk (average rating 4,21; 42,9% indications for the assessment of 5 points) and high affirmation of uncertainty (average rating 4,11; 38,1% indications for the assessment of 5 points).

enterprises surveyed in areas related to the company's innovativeness (average rating 4.25, 46.0% indications for the assessment of 5 points) (Table 7). It is becoming more and more important, especially in the context of the continuous increase in the value of intangible assets in relation to tangible assets.

Considerations towards the fourth generation era are conducted by contemporary entrepreneurs. They result in a comparative analysis of old and new desiderata and the emergence of a new overall paradigm, taking into account the building of electronic channels of business activity, i.e. the virtual dimension of doing business associated with networking of key (often strategic) business areas. Attention is paid to the automation and the ability to process and exchange data. It is assumed that companies will implement new technologies allowing the creation of so-called cyber-physical systems and change of production methods in the nearest future [20, 25, 26]. In the context of the above, it should be stated that modern enterprises are characterized by a high level of maturity in terms of new horizons (average rating 4.46, 55.6% indications for the assessment of 5 points), accented by the ability to acquire and implement new ideas (average mark 4.41, 50.8% indications for the assessment of 5 points) (Table 7).

The production activity requires equipping the production system with resources that will create the opportunity of manufacturing products in accordance with the needs of the market. Hence, an important criterion verifying the level of technical engineering maturity - completed at a very high level - access to technologies determining the effectiveness of the production process was nominated (average score 4.46, 58.7% indications for the assessment of 5 points) (Table 8). The need to ensure a proper level of quality of products and services, while maintaining the condition of profitability of production, puts Polish enterprises in need of modernization of the machine park. An important element of this type of activities is the investment process, which requires a solid preparation, both from the managers and engineers, whose task is, among others, to control the needs for the expansion or reconstruction of the machine park. There is often a conflict between the technical aspect of the issue and economic conditions. Sometimes it is forgotten that the profitability of production is determined not so much by the cost (depreciation) of the machinery, but to a large extent the way of organizing the production process. In this respect, according to the surveyed enterprises, the number of implemented solutions is adequate to the current demand (average rating of 4.33, 52.4% of indications for the assessment of 5 points) (Table 8). Realization of this state of affairs is the tendency to invest in modern machines, devices and technologies (average rating 4.41, 55.6% indications for the assessment of 5 points). New technologies bring new chances and new opportunities for the development of the enterprise. Their implementation is a great challenge as stated by the surveyed enterprises; the executive staff is characterized by the desire to learn about and implement new solutions (average mark 4.11, 41.3% of indications for the assessment of 5 points) (Table 8). In this respect, the great responsibility lies with managers who must interest and properly prepare serial employees to use such opportunities. The problem is that it requires the transformation of technical culture. Those enterprises that effectively change the way of perceiving will be beneficiaries of the new generation era.

Table 8. Culture oriented to technology – maturity assessment.

Nr	Culture oriented to technology	1	2	3	4	5	Pts.
1.	Number of implemented technological solutions	1	3	3	23	3	4,33
		1,6	4,8	4,8	36,5	52,4	
2.	Access to technologies influencing efficiency	1	1	3	21	37	4,46
		1,6	1,6	4,8	33,3	58,7	
3.	Orientation towards purchasing new machines and devices	1	1	4	22	35	4,41
		1,6	1,6	6,3	34,9	55,6	
4.	Operational Staff is characterised by the will to learn about and implement new solutions	1	4	8	24	26	4,11
		1,6	6,3	12,7	38,1	41,3	
5.	The company has cooperative skills	1	1	5	24	32	4,35
		1,6	1,6	7,9	38,1	50,8	
6.	The company performs maintenance and supervising actions	2	4	9	24	24	4,02
		3,2	6,3	14,3	38,1	38,1	

Source: Own work

The research results predispose to the statement that introducing changes in enterprises is a conscious, organized and controlled process. According to the surveyed enterprises, changes in the production method correspond to the market needs and the volume of implemented technological solutions is sufficient. As part of the assessment, designations have been distinguished that allow to state a high level of technology-oriented culture maturity.

Many Polish companies in the agricultural machinery sector report problems in obtaining appropriate personnel. Often these problems can be minimized by using the appropriate employee acquisition tactics, because in the employee's market, non-standard activities bring the best results more and more often. The companies surveyed indicate the conscious building of business networks that help exchange information about potential candidates (average rating 3.81, 30.2% indications for the assessment of 5 points) (Table 9). Databases are containing current information about who is looking for a job are being developed, providing information on who is just free on the market, etc. Such activities are very reliable and very quickly result in recruitment.

Table 9. Culture oriented to staff – maturity assessment.

Nr	Culture oriented to Staff	1	2	3	4	5	Pts
1.	The company has an active database of potential employees	3	4	14	23	19	3,81
		4,8	6,3	22,2	36,5	30,2	
2.	The company assesses employees satisfaction on cooperation	3	5	13	19	23	3,86
		4,8	7,9	20,6	30,2	36,5	
3.	The company implements actions striving for increasing employees satisfaction on work performed	1	2	9	23	28	4,19
		1,6	3,2	14,3	36,5	44,4	

(*continued*)

Table 9. (*continued*)

Nr	Culture oriented to Staff	1	2	3	4	5	Pts
4.	The company implements actions striving for deepening or building relations with employees	-	1	6	23	33	4,40
		-	1,6	9,5	36,5	52,4	
5.	The company carries research on reasons for losing its employees	1	3	11	21	27	4,11
		1,6	4,8	17,5	33,3	42,9	
6.	Employees are aware of the role they play in the company	1	2	7	21	32	4,29
		1,6	3,2	11,1	33,3	50,8	
7.	Motivation system for employees is related to results of the company	1	1	5	22	34	4,38
		1,6	1,6	7,9	34,9	54,0	
8.	In the company there is a system of employees participation in actions striving for development	1	2	5	22	33	4,33
		1,6	3,2	7,9	34,9	52,4	

Source: Own work

In business it is very important to communicate the values on which the organization is based. Employees feel better in places where values are implemented into the company's technical culture and where they are approached in a serious way. In these work places they feel safer, they easier identify themselves with such work places they are more willing to participate in selected projects, they are the best ambassadors of the company and can recommend it to jobseekers. Therefore, it is important from the point of view of the surveyed enterprises to recognize the level of employee satisfaction implicating parameterized feedback on the processes occurring at the lower organizational levels, the level of satisfaction, identification with the company, level of motivation and professional development. To this end, these tools are used to assess employees' level of satisfaction and enable recognition of opinions in the area of selected areas of the company's activity (average rating of 3.86, 36.5% of indications for the assessment of 5 points) (Table 9). The research gives the possibility to parameterize selected areas, and thus to detect strengths and weaknesses. Based on the verified information, the companies implement activities aimed at increasing the satisfaction of employees with their work (average mark 4.19, 44.4% indications for the assessment of 5 points) (Table 9). From the point of view of maturity, it is important to create a culture based on feedback, build a well-coordinated team and jointly achieve results. A technical culture, in which a particular emphasis is placed on the management of relationships, within the organization, are implemented by the surveyed enterprises (average rating 4.40, 52.4% indications for the assessment of 5 points) (Table 9). Respondents declare a successive reconnaissance of the reasons for the loss of employees, who are aware of the role they play in the company, actively participate in activities aimed at development, the more so as the employee motivating system is related to the company's results.

Sustainable trade relations are not possible without favorable environment of technical culture and a similar value system. The cooperation of enterprises established exclusively for the commercial transaction does not bring the desired effects in relationships in which one of the cooperators wants to build lasting relationships. In the conditions of constant and growing competition, building long-term relationships with clients becomes possible only through the full satisfaction of both partners. The long-term relationship between the client and the supplier emerges from and results in trust and loyalty. It is an exchange of values not only material or immaterial, but also emotions. Satisfaction is closely related to the client's expectations, the satisfaction of which affects satisfaction with cooperation; producers are aware of this fact (average grade 4.46, 58.7% of indications for the assessment of 5 points) (Table 10). In the face of increasing client requirements, employment in companies, salespeople, consultants and sellers constitute a knowledge base including data on clients' preferences, by examining needs and changing trends. They are aware of the role they play in creating clients satisfaction (average rating 4.21, 38.1% indications for the assessment of 5 points) (Table 10).

Table 10. Culture oriented to clients – maturity assessment.

Nr	Culture oriented to clients	1	2	3	4	5	Pts
1.	The company is benefiting from active database of clients	-	-	5	19	39	4,54
		-	-	7,9	30,2	61,9	
2.	The company is assessing clients satisfaction	-	1	5	23	34	4,43
		-	1,6	7,9	36,5	54,0	
3.	The company is oriented towards actions striving for increased satisfaction from cooperation	-	-	4	21	37	4,46
		-	-	6,3	33,3	58,7	
4.	The company implements actions striving for deepening and building relations with clients	-		3	15	45	4,67
		-	-	4,8	23,8	71,4	
5.	The company carries research on the reasons for losing clients	1	3	6	22	31	4,25
		1,6	4,8	9,5	34,9	49,2	
6.	Employees are aware of the role they play in creating clients' satisfaction	1	2	4	32	24	4,21
		1,6	3,2	6,3	50,8	38,1	
7.	Motivation system is combined with sale results and clients satisfaction	-	2	4	24	33	4,40
		-	3,2	6,3	38,1	52,4	
8.	In company there is a system of participation of clients and employees in processes execution	-	2	4	33	24	4,25
		-	3,2	6,3	52,4	38,1	

Source: Own work

An important designation of the maturity of a client-oriented technical culture is building an active database of clients. Without it, reaching a larger number of clients would be impossible. It is thanks to the network of recommendations and references that provides satisfactory results. An appropriately constructed client base and dedicated offers affect the effectiveness of enterprises; reflects the high level of technical client maturity focused on the client (average rating 4.54, 61.9% indications for the assessment of 5 points) (Table 10).

The problem of client loyalty and retention is a significant issue in the process of assessing the maturity of technical culture. Due to the high market saturation, easy change of supplier and very high costs of gaining new clients, loyalty and client retention have a special dimension in the context of producers operating in the agricultural machinery sector. Hence, according to respondents, it is always worth analyzing what was the reason for the clients breaking the contract, what caused them to resign from cooperation (average mark 4.25, 49.2% indications for the assessment of 5 points).

The presented results predispose the authors to the statement that a culture that is client-oriented is a conscious, organized and controlled process. This is evidenced by, among others implemented motivating system, which is assumed to be related to sales results and client satisfaction with cooperation and system of employee participation in the client service process. The designata recognized in the study allow to determine the high level of maturity of the client-oriented culture.

People are a unique capital of the organization, worth special efforts, while an investment in human resources, although usually expensive, is highly profitable. Employee development should therefore be a basic value for both the organization and the employee. Among the surveyed companies, attention is paid to the relatively high activity of employees in the area of self-empowerment, their ability to seek self-development opportunities (average rating 4.06, 34.9% of indications for the assessment of 5 points) (Table 11). Despite the fact that enterprises allocate specific financial resources for this purpose (average rating 3.97, 30.2% of indications for the assessment of 5 points) (Table 11) and systematically perform monitoring in the scope of training needs (average mark 3.83, 27.0% of indications for the assessment of 5 points) (Table 11), it is difficult to determine whether their amount is adequate to the demand.

The attitude of creative activity and cooperation is highly valued in enterprises. The importance of creative activity should be recognized here in two dimensions, paying attention to the benefits of the individual as well as the enterprise. Employees activity is an expression of their developmental achievements, especially of qualitative changes. The technical culture, which is important from the point of view of the surveyed enterprises, approves the ability to constructively criticize the condition of developing a specific solution. At the same time it is pointed out that the disapproval of certain actions must result from good intentions. Structural criticism arouses respect and trust. Constructiveness, consisting in presenting your own ideas regarding the solution of specific problems, makes it possible to achieve the intended goal. Especially that the vast majority of the surveyed companies creates the possibility of acting independently, giving employees the opportunity to submit new ideas without fear of neglecting them.

The presented results predispose the authors to the statement that culture oriented towards development is a conscious and mature process. It is convinced by the actions of enterprises that are focused on acquiring, linking, shaping and using available knowledge and skills in the field of technology and other knowledge and skills to plan production and create and design new, changed or improved products, processes and services. The designata recognized in the study allow to determine the high level of maturity of a development-oriented culture.

Table 11. Culture oriented to development – ocena maturity assessment

Nr	Culture oriented to development	1	2	3	4	5	Pts
1.	Activity of employees in the area of self-improvement	-	3	6	28	26	4,22
		-	4,8	9,5	44,4	41,3	
2.	In the company there are tools for informing on opportunities of development implemented	1	4	7	29	22	4,06
		1,6	6,3	11,1	46,0	34,9	
3.	In the company constructive criticism is an attitude appreciated	1	3	5	31	23	4,14
		1,6	4,8	7,9	49,2	36,5	
4.	The company gives active employees opportunity to undertake free actions	-	2	5	29	27	4,29
		-	3,2	7,9	46,0	42,9	
5.	Active approach and cooperation between employees is appreciated	-	1	3	21	38	4,52
		-	1,6	4,8	33,3	60,3	
6.	The company finances training for its employees	1	4	10	29	19	3,97
		1,6	6,3	15,9	46,0	30,2	
7.	The company provides opportunity for benefiting from co-financed trainings	1	4	8	27	23	4,06
		1,6	6,3	12,7	42,9	36,5	
8.	The company systematically monitors need for training	2	4	14	26	17	3,83
		3,2	6,3	22,2	41,3	27,0	
9.	There is active system of searching for opportunities	1	3	6	28	25	4,16
		1,6	4,8	9,5	44,4	39,7	
10.	There is opportunity for presenting new ideas without the fear of neglecting	-	2	4	20	37	4,46
		-	3,2	6,3	31,7	58,7	
11.	Orientation towards research and development	1	2	4	21	35	4,38
		1,6	3,2	6,3	33,3	55,6	

Source: Own study

5 Summary

As new directions of research in management science are necessary for the creation of more durable and more effective development strategies, the development of the assessment method for the maturity of the organization's technical culture based on values, concepts or management methods was considered justified in this publication. Currently, many producers of the agricultural machinery sector as well as representatives of science show interest in lean management solutions. Despite the unquestionable popularity of this concept in the business environment and the constantly growing research interest in this issue, it turns out that there are many areas to be solved in the near future.

The method of assessing the maturity of technical culture presented in the article is part of a comprehensive approach to self-assessment of enterprise development and explanation of management mechanisms. It is used to indicate strengths and weaknesses and to identify areas for improvement. The research described in the publication was aimed at recognizing the actual level of technical culture. The research methodology adopted allowed the authors to identify the quantity and quality of its status among selected enterprises operating on the Polish agricultural machinery market. The companies surveyed declare a relatively high level of technical culture maturity, which, according to the authors, indicates the direction of their transformation in line with the concept of lean management. The collected research material made it possible to formulate general and cognitive conclusions. The paper presents a procedure and a tool enabling the identification of key designates of technical culture of lean companies, which, as the authors think, will contribute to the fragmented fulfillment of lack of knowledge in this area. We recognize the need for further, even more in-depth research on the tools, methods and conditions of good governance. There is a need to propose a tool to support the assessment of organizational structures that imply sound management, which will be the subject of further studies by the author team, and in the future will enable more and more appropriate designing, to the conditions of a specific company, interrelationships and cooperation between all its elements.

Hence, the authors recognize the need for further, even more in-depth research on the tools, methods and conditions of good governance. There is a need to develop a tool to support the assessment of organizational structures that imply sound management, which will be the subject of further studies by the authors, and in the future will enable designing, more and more appropriate to the conditions of a specific company, considering interrelationships and cooperation between all its elements.

References

1. Dostatni, E., Niewiadomski, P.: Operational competences in manufacturing process – case study. Bus. Manag. (1) (2013). (in Polish)
2. Rostek, M., Knosala, R.: Research on effectiveness of logistics processes in manufacturing company. Bus. Manag. (1) (2014). (in Polish)
3. Hamrol, A.: Quality Management and Engineering. PWN, Warsaw (2018). (in Polish)
4. Grabowska, M., Hamrol, A.: Research on maturity of quality management processes in manufacturing companies. Q. J. **1**(16) 2016. (in Polish)
5. Ball, D.R., Maleyeff, J.: Lean management of environmental consulting. J. Manag. Eng. (1) (2003)
6. Hamrol, A.: Strategy and Practice of Efficient Performance. Lean, Six Sigma and Others. PWN, Warsaw, Lean (2015). (in Polish)
7. Benjamin, C.T.: A study of behaviours that retard the implementation of lean operations. J. Assoc. Prof. Eng. Trinidad and Tobago **41**(1) (2013)
8. Bicheno, J.: The Lean Toolbox for Service Systems. PICSIE Books, Buckingham (2008)
9. Damrath, F.: Increasing Competitiveness of Service Companies: Developing Conceptual Models for Implementing Lean Management in Service Companies. Politecnico di Milano, Milan (2012)
10. Emiliani, M.L.: Lean behaviours. Manag. Decis. **30**(9) (1998)
11. Keyte, B., Locher, D.: The Complete Lean Enterprise. Value Stream Mapping for Administrative and Office Processes. Productivity Press, New York (2004)
12. Rich, N., Bateman, N., Esain, A., Massey, L., Samuel, D.: Lean Evolution. Lessons from Workplace. Cambridge University Press, Cambridge (2006)

13. Ravasi, D., Schultz, M.: Responding to organizational identity threats: exploring the role of organizational culture. Acad. Manag. J. **49**(3), 433–458 (2006)
14. Godin, B., Gingras, Y.: What is scientific and technological culture and how is it measured? A multidimensional model. Public Understand. Sci. **9**, 43–58 (2000)
15. Peteraf, M.A.: The cornerstones of competitive advantage: a resource-based view. Strateg. Manag. J. **14** (1993)
16. Lachowski, S.: Road More Important Than the Goal, p. 250. Studio Emka, Warszawa (2012)
17. Chałas, K.: Technical culture and human axjosphere. Adv. Sci. Technol. Res. J. **8**(24), 107–110 (2014)
18. Lib, W.: Technical language as an indicator of technical culture. Informatol. **43**(1), 54–57 (2010)
19. Sweeney, G.P.: Technical culture and the local dimension of entrepreneurial vitality. Entrep. Reg. Dev. **3**(4), 363–378 (1991)
20. Tierney, T.: The Value of Convenience: A Genealogy of Technical Culture. Suny Press, Albany (1993)
21. Fertsch, M.: Organization of production and logistics in automotive industry. Logistics (2) (2007). (in Polish)
22. Hadaś, Ł., Stachowiak, A., Cyplik, P.: Production-logistics system in the context of production planning at strategic and operational level and customer service level. LogForum High. School Logist. **3**(9) (2014)
23. Golińska, P.: Lean as a way to improvement of remanufacturing process. LogForum High. School Logist. **3**(5) (2014). (in Polish)
24. Domański, R., Fertsch, M.: Integration of production and supplies in Lean environment according to lot for lot principle – research results. LogForum High. School Logist. **4**(4) (2015). (in Polish)
25. Pawłyszyn, I.: Time-driven activity based costing as a basis for taking lean actions. LogForum High. School Logist. **2**(2) (2017). (in Polish)
26. Hajdul, M., Mindur, L.: Reliable lean e-supply chains – case study. LogForum High. School Logist. **1**(2) (2015). (in Polish)
27. Konecka, S.: Lean and flexible supply chain in the risk management aspect. LogForum High. School Logist. **4**(3) (2010). (in Polish)
28. Domański, R., Hadaś, Ł.: Technological and organizational similarity coefficient s a basis for shaping value streams in lean manufacturing. LogForum High. School Logist. **2** (2008). (in Polish)
29. Maleyeff, J.: Exploration of internal service systems using lean principles. Manag. Decis. **44** (5) (2006)
30. Hashmi, H., Khan Naveed, R., Haq, M.A.: Influence of lean management on operations. LogForum High. School Logist. **4**(6) (2015). (in Polish)
31. Mann, D.: Creating a Lean Culture. Tools to Sustain Lean Conversion. Productivity Press, New York (2005)
32. Duggan, K.J.: Facilities design for Lean manufacturing. IIE Solut. **30**(12), 30–34 (1998)
33. Lin, Z., Hui, C.: Should lean replace mass organization systems? A comparative examination from a management coordination perspective. J. Int. Bus. Stud. **30**(1), 45–80 (1996)
34. Sheridan, J.H.: Lean sigma synergy. Industry Week, pp. 81–82, 16 October 2000
35. Romm, J.J.: Keep your facility fit with Lean and cLean engineering. IIE Solut. **27**(6), 17–22 (1995)
36. Womack, J.P., Jones, D.T.: Leaning Companies: Waste Elimination as Key to Success. Manager Information Center, Warsaw (2001)
37. Nogalski, B.: Lean Management. In: Czerska, M., Szpitter, A. (eds.) Management Concepts, Warszawa (2010). (C. H. Beck)

38. Liker, J.K.: Becoming Lean. Free Press, New York (1996)
39. Niewiadomski, P., Sterna, K.: New approach to environment protection and waste elimination. In: Wyrwicka, M. (ed.) Waste, Symptoms and Minimization Methods, p. 92. Publisher of Poznan University of Technology, Poznan (2009). (in Polish)
40. Niewiadomski, P., Pawlak, N.: Analysis of raw material participation in the production process, Part II - practical aspects, Research and logistics & production, NR 1/2016, p. 72. Poznan University of Technology, 29 January 2016
41. Pawłowski, E., Pawłowski, K., Trzcieliński, S.: Lean Manufacturing: Methods and Tools. Publisher of Poznan University of Technology, Poznan (2010). (in Polish)
42. Koch, T., Kornicki, L., Sobczyk, T., Oleksy, S.: Lean management implementation in Poland. In: III Conference on Lean Manufacturing, Conference Proceedings, Wroclaw (2003). (in Polish)
43. Lipecki, J.: Lean Management as a Method of Management Restructuring, nr 8, pp. 12–15. Economics and Organization of the Company (1998). (in Polish)
44. Pawlak, N., Niewiadomski, P.: Lean product concept and its implications in terms of cost and quality. In: Hadaś, Ł. (ed.) Production Management – Contemporary Approaches, Selected Aspects, pp. 67–68. Publishing House of Poznan University of Technology, Poznan (2012)
45. Hamrol, A.: Quality Management and Engineering. PWN, Warsaw (2017). (in Polish)

The Meaning of Technological Culture in Manufacturing

Magdalena K. Wyrwicka(✉)

Faculty of Engineering Management, Poznan University of Technology, Poznań, Poland
magdalena.wyrwicka@put.poznan.pl

Abstract. The understanding of technological culture has been shown in this text based on the author's research. Manufacturing which is very important in contemporary economy needs some new solutions to give correct answer to customers' requirements. The reaction of people in manufacturing system should be very fast and typical in human behavior. Culture as a result of the human race adapting to its natural environment is connected with collective programming of the human mind. Technological culture is a system of permanent attitudes and efficiency of the human and it enables the proper use of existing products or technology, in order to change the pattern of quality of cooperation and supporting making changes. A comparative analysis of interpretation of the term 'technological culture' based on sex and activity of a person is the main topic of this paper. The results of the author's research present the most quoted descriptions for technological culture.

Keywords: Culture in manufacturing · Concept of technological culture · Research results of technological culture understanding

1 Introduction

When speaking of growth factors, economics points to technological progress as an important determinant (80% for Robert Slow MIT, 1957). A human being who has been "forever" a technical creature, has lived in the age of technology for a relatively short time [1]. The term "forever" applies to how technology is embedded in the very nature of the man-inventor and the creator of new, increasingly better solutions. It is not the humanity or society that are developing, but rather culture, being a lasting set of real and ideal systems [2]. K. Kelly claims that contemporary network economy is based on technology which always creates new possibilities of demand. Designing and shaping business relations are based initially on technology and conclude with trust [3].

The common technological changes and the introduction of new technical concepts leads to the emergence, in goods producing societies, of entirely new needs with regard to behaviors and competences of those who realize manufacturing processes. The contemporary requirements of new concept of manufacturing - Industry 4.0 - connected with the dynamics of market environment, time pressure, resource limitation and the rapid growth of complexity in new systems have caused that such a technological sort of culture should be searched.

A. Hamrol et al. (Eds.): *Advances in Manufacturing II* - Volume 3, LNME, pp. 75–82, 2019.
https://doi.org/10.1007/978-3-030-17269-5_6

The culture is ‘the customs and beliefs, art, way of life and social organization of a particular country or group’ [4]. According to F. Znaniecki the culture is the result of humankind adapting to natural environment, consisting in, on one hand, certain organic transformation within humans, creating new and increasingly more complex mechanisms of reaction to external influences, and on the other hand, on the transformation of natural environment items, resulting from these reactions [2].

Invoking the term ‘technological culture’, popular in 1980s in Poland, and researching this phenomenon since 2005, the author claims that one can have a major contribution into a correct operation of a production company through preventative research of the existing culture, treated as a premise for the growth of an organization that applies to the employees’ attitudes and expectations [5–7]. Contemporary manufacturing needs new solutions to give correct answer to customers’ requirements. The reaction of people in manufacturing system should be very fast and agile. Recently a new paradigm called Industry 4.0 (the fourth industrial revolution) has aroused in manufacturing. It refers to the process optimization which is driven by cloud computing, cloud-based services, big data analytics, robotics, Internet of Things, real-time sense-and-response technologies, artificial intelligence and 3D printing. It allows to create a smart network of machines, products, components, properties, individuals and ICT systems in the entire value chain to have the intelligent factory [8]. A manager should be a system-designer, catalyst of change, trainer, leader of a team and a winner, but first of all he should create a mastery and provide a pattern of attitudes and behavior.

A question about typical human behavior in technical and technological environment is still open. Authentically, rational managers make attempts to optimize the activity in long-term pointing efficiency. Still, management decisions are often distorted by non-rational factors such as overly optimistic forecasts, overly timid choices derived from excessive loss aversion, neglect of differences in the ways of processing of information or preferred values, and also by erasing memory about crises or troubles. Good policies and plans, together with careful monitoring of implementation will not prevent all mishaps.

The main goal of this paper was to establish whether the term of technological culture is being currently used and how it is interpreted.

2 The Concept of Technological Culture

The culture [9] is the result of complex learning process within a group. A set of basic beliefs developed, invented or adopted to solve difficulties that are faced by an organization when adapting to external conditions and internal integration is considered by E.H. Schein to be the essence of organizational culture. The culture is learned (not inherited) aspects of human societies, a common operating context [10], a collective program of a human mind that differentiates between one group of people and another [4, 11].

According to E. H. Schein, the culture is the result of behaviors and interactions which are based, on fundamental assumptions, standards, values and artefacts.

The assumptions are hidden and unconscious, difficult to identify. They determine the relation to the environment, the way of understanding the truth, beliefs about human nature, the ideas about the purpose or attractiveness of work and the regularities in putting social, formal and informal relations in order.

The standards and values form a set of principles of daily operation which are shaped under the influence of dominating values. The latter may be declared (formally accepted) or observed (visible during long, careful observation of a group).

The artefacts are the manifestation of culture in a language, activities (ceremonies, rituals, typical reactions), accepted concepts (organizational structure, spatial and visual solutions).

The processes connecting the constituents of culture listed above are reactive, two-sided and take place as:

- manifestation of assumptions of culture in the form of values,
- realization of values in the form of artefacts,
- symbolization of artefacts, or assigning special status to some of them,
- interpretation of symbols that supports assumptions.

The culture forms a basis for perception, adaptation and integration. It also determines the impact of internal arrangements on external contacts and the extent to which environmental parameters are taken into account when making decisions (strategic, tactical and operational).

Being a sort of creativity, the man's technical activity, especially when related to manufacturing products, is also the man's culture-building activity [12]. Knowledge resources are used (but also created) when designing, creating and using technology's products. According to F. Znaniecki, technical knowledge is the knowledge required to manage work processes leading to the achievement of objectives [13]. These considerations allow to return to terms, used at the end of 20th C., characterizing the desired, positive phenomenon referred to as the "technological culture".

J. Bańka claims [14] that the technological culture covers the behaviors of technical environments - of the people who have daily and direct contact with technology - the vision of future shaped within the field of activity of technology and the synthesis which unites humanistic and technical elements from the perspective of the needs of engineers and technicians. This term covers mental patterns of operations and technologies possible to achieve as human knowledge and practice develops.

H. Pochanke determined the technological culture as the entire achievements in the field of technical sciences and their application, and at the same time as all the knowledge and abilities determining the understanding of these achievements by means of using them, their transfer to the younger generation and the creation of new, related values [15]. According to Z. Wołk, the technological culture is a rational, aesthetic and socially useful relation of the man towards technology and the application of technology to increase the level of economic, social, spiritual and daily life of the society, according to the state of technological development [16].

For W. Furmanek, the technological culture is a system of the man's continued tendencies and will that allow using the products and creations of technology present in the surrounding reality in a worthy manner with the aim to alter the man's and others' quality of life [3]. It is also manifested in relatively permanent and positive attitudes towards technical phenomena, creators and producers of technology and the acquired technical knowledge, but mostly in the man's ethical behaviors and activities in various technical situations. The technological culture's subject - the man - is both its creator and user. One of the attributes of technical creativity is its dependence on human invention and work [17]. New technologies or technical solutions are considered from the point of view of improved efficiency or opening to new markets [18]. Thus special attention should be paid to the transfer of new technical knowledge and its usefulness in the context of improving the functioning of production systems. Such possibilities arise if a technological culture exists locally.

The systemic management of technological culture within an enterprise, aiming to shaping and improving it, requires, among else that a method of interpreting this concept in order to allow the reaction or shaping of this kind of culture, is identified.

Technological culture is a term eagerly invoked in the past by teachers of technical schools and universities and by employees of production companies. Development in contemporary meaning is the function of HRM concerned with preparing employees to work effectively and efficiently in the organization, and the creation of new patterns of the quality of organizational system. Understanding of technological culture in contemporary life could be a basis to identify the typical human imaginations, behaviors and attitudes about technique or products of technology. Without people implementing new technology and correctly using technique the progress would be impossible.

Technological culture is a term with a positive meaning, often used to describe the mastering of technical and technological systems, emphasizing the creativity and precision of designers, accuracy in performance and control, good maintenance of machines, devices, equipment and tools, as well as tidiness in the workspace. This is not an unambiguous notion. The author has been conducting surveys since 2005 to identify the meaning of the term 'technological culture'.

3 The Research of Technological Culture

The survey has been prepared to identify the perception of technological culture and the interpretation of this statement by workers and students.

Recent surveys on the perception of technological culture by graduates and students of technical universities were conducted in the autumn 2017 and the winter 2018. A form created in 2005 was used in the surveys (among other places, the form was described in: [7]). The last survey was conducted with the participation of the students of the Faculty of Engineering Management (FEM) at Poznan University of Technology, employees of production enterprises and working students (of the FEM). 540 surveys were distributed and 236 completed ones which were collected. The survey form was distributed directly or through managers of the corporation.

The completed surveys were mainly from young people (88% of the responses were from people in the age of 35 years and younger). 13% of the respondents represented workers, 53% working students, and 34% students only. 114 respondents were female (48.3%) and 122 male (51.7%).

Hitherto prevailing research of the author concludes that the term of technological culture in professional life is being used. The vast majority of the respondents stated that technological culture exists - 94% (222 responses). In the group that confirmed the existence of technological culture, the following suggested meanings of verbal associations with technological culture were chosen according to the sequence on the survey form:

- obeying norms - 76.4% of responses
- establishing procedures - 63% of responses
- creation of standards - 47% of responses
- unification of approaches to technological problems - 36% of responses
- anticipation of future requirements - 22% of responses
- thriftiness - 29% of responses
- caring for efficiency- 52% of responses
- solidarity in workplace and cooperation - 40% of responses
- reliability - 45% of responses
- discipline - 34% of responses
- involvement in one's job - 41% of responses.

(Figure 1) presents results by gender. In principle, the results show that women and men share opinions with regard to the existence and essence of a technological culture.

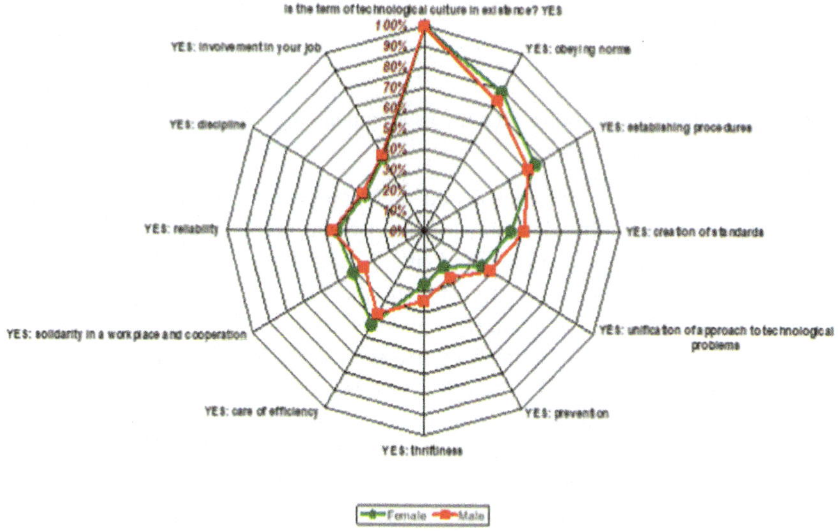

Fig. 1. Comparative analysis of research results according to gender groups (own research).

In-depth examination of results presented in (Fig. 1) indicate that women are more willing to obey norms, follow procedures, care for efficiency, maintain professional solidarity and cooperate. Men, on the other hand, are more focused on standardization, unification of approaches to technological problems, anticipation of future requirements, thriftiness, reliability, discipline and involvement in one's job.

Compiling the results according to groups (students, working students, employees) allowed to obtain information about differences of opinions on the phenomenon of technological culture.

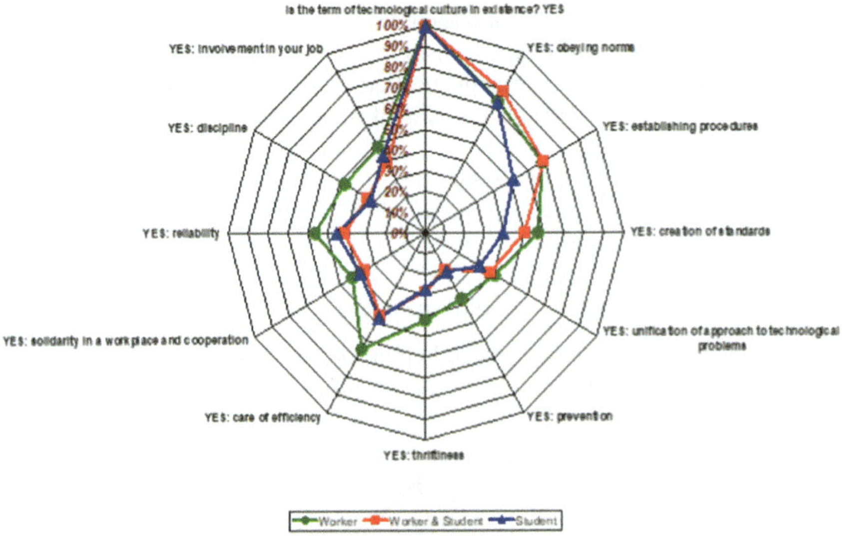

Fig. 2. Comparative analysis of research results according to groups of activity (own research).

According to (Fig. 2), workers stressed such associations as prevention, thriftiness, care of efficiency, reliability, discipline, solidarity and involvement in one's job more than other groups (students, working students). Establishing procedures and creation of standards are less important for students. In the case of associating technological culture with obeying norms, working students expressed stronger opinions than employees (78.7% of working students compared with 77.2% of employees, and 72.5% of students, for an average of 76.4% of all respondents).

Analysis of (Fig. 2) shows the importance of job experience which gives other attitudes towards technological and technical problems in company.

As many as 93.5% of those who expressed favourable opinions on the existence of technological culture are convinced that it can be learned. Consequently, its manifestations are likely to depend on the system of education, broadcasting information in the media, the way of introducing to work and coaching, as well as on training and professional development. It may be also alleged that being around modern technological solutions in manufacturing activities and innovative technologies in private life is not without consequence for the creation of technological culture.

4 Summary

Shaping the technological culture manifested in readiness to accept new solutions, understanding of functionalities and observing the principles of operation, a positive attitude towards the products of technology or new technologies related to the safety of humans and the environment seem of key importance in times of such a dynamic growth. Since people are responsible for the level of culture, it is worth to analyze how they interpret this phenomenon and to learn their expectations.

The fact that the term 'technological culture' is accepted and the belief that it can be taught is encouraging. The slight differences in the evaluation of the importance of individual discriminants of this phenomenon by various groups of respondents are quite typical given the different experiences of students, working students and employees. The survey results allow to see the expectations expressed by employees (or potential employees) towards the employer. If observing the standards is a manifestation of technological culture, one must know how to determine these standards in order to announce them and use as a basis to evaluate work efficiency. Procedures which the surveyed associate with technological culture should not be created solely on the basis of evidence of the actual states, often a consequence of historical and sometimes chaotic changes. The development of procedures that are to be observed and the creation of good habits should be based on proactive shaping of flexible work systems and optimization of processes. As long as it is conducted in a methodical manner, standardization can help manage the variety of things, events and activities, whereas caring for efficiency, reliability and involvement in one's job should foster innovations and create a company's identity on the basis of competences and common values.

Compiling the results obtained over a dozen years ago with the use of the same survey form [5–7] against the present ones, one can spot only minor differences (<5%) and the respondents exhibit converging beliefs, despite the flow of time. We can deduct on this basis that the features characterizing the phenomenon of technological culture have remained relatively stable.

The further research work about technological culture is needed and required about contemporary understanding of technology and its influence on development of manufacturing companies.

References

1. Piecuch, Cz.: Condition of Contemporary Human Being. UNIVERSITAS Press, Kraków (2006). (in Polish)
2. Znaniecki, F.: Introduction to Sociologie. Polish Scientific Publishing PWN, Warszawa (1988). (in Polish)
3. Furmanek, W.: Technological culture and information culture. Implication of the term. Methodological consequences. In: Morbitzer, J. (ed.) Computing Technologies in Education Transfer. Scientific Publishing AP, Kraków (2002). (in Polish)
4. Hornby, A.S.: Oxford Advanced Learner's Dictionary of Current English. Oxford University Press, Oxford (2006)

5. Wyrwicka, M.K.: Analysis of perception of technological culture in Wielkopolska. In: Lewandowski, J. (ed.) Theories and Praxis of Management in Business Organizations, pp. 347–355. University of Technology Press, Łódź (2005). (in Polish)
6. Wyrwicka, M.K., Prussak, W.: Research of technological culture in aspect of change. In: Lisiecki, M. (ed.) Companies, Organizations and Capital Markets in the Global Environment. Materials and Studies of Management and Marketing Institute, pp. 253–262. Catholic University of Lublin Publishing, Lublin, Nr 7 (2006). (in Polish)
7. Wyrwicka, M.K.: Perception of technological culture in Poland. In: Thomas, R.A. (ed.) Quality, Reliability and Maintenance, pp. 92–97. Coxmore Publishing Co., Oxford (2007)
8. Mrugalska, B., Wyrwicka, M.K.: Towards lean production in industry 4.0. In: Procedia Engineering, vol. 182, pp. 466–473 (2017)
9. Schein, E.H.: Organizational Culture and Leadership. Jossay-Bass, San Francisco (2004)
10. Giddens, A.: Socjologie. Polish Scientific Publishing PWN, Warszawa (2004). (in Polish)
11. Hofstede, G.: Cultures and Organizations. PWE Publishing, Warszawa (2000). (in Polish)
12. Senior, B., Fleming, J.: Organizational Change, 3rd edn. Prentice Hall, London (2006)
13. Znaniecki, F.: Cultural Sciences: Their Origin and Development. Scientific Publishing PWN, Warszawa (1992). (in Polish)
14. Bańka, J.: Humanisation of Technology. Main Questions of Eutyfronics. Publishing Śląsk, Katowice (1976). (in Polish)
15. Pochanke, H.: Didactic of Technic. Polish Scientific Publishing PWN, Warsaw (1985). (in Polish)
16. Wołk, Z.: Design of Technological Culture for Students at Secondary School. Wyższa Szkoła Pedagogiczna Publishing, Zielona Góra (1990). (in Polish)
17. Kapitańczyk, K.: A Need of Dialogs about Technology. Methodological Material, 1. Poznan University of Technology Press (1966). (in Polish)
18. Bouncken, R.B.: Cultural diversity in entrepreneurial teams: findings of new ventures in Germany. In: Creativity and Innovation Management, vol. 13, no. 4. Blackwell Publishing Ltd. Hoboken (2004)

Analysis of Continuous Improvement Projects in the Production Company

Marta Grabowska[1(✉)], Mariusz Bożek[2], and Marta Królikowska[2]

[1] Poznan University of Technology, Poznań, Poland
marta.grabowska@put.poznan.pl
[2] Aesculap Chifa Sp. z o.o., Nowy Tomysl, Poland

Abstract. This article presents the conclusions from process improvements activities implemented for the past 15 years in a company producing surgical instruments. Authors seek to prove that the selection of a proper improvement strategy should be based on an analysis of the baseline condition in the business area subject to improvement. In order to test this thesis improvement projects were analyzed taking into account five criteria: the baseline condition, beneficiaries of the project, the dominating improvement strategy used in the project, the tools and methods used, and improvement measures taken. There is a variety of unique improvement scenarios, inspired by the diversity of processes and production improvement strategies, methods, and tools, as well as the uniqueness of each process. It is therefore essential to share experience of best practice in these matters.

Keywords: Six Sigma · Lean Manufacturing · Continuous improvement

1 Introduction

It is practically a cliché to say that continuous improvement (CI) is the precondition for obtaining and maintaining competitiveness in any business [1]. Improvement of processes and resources is important – and in many cases critical – for the assurance of product quality at reasonable – typically the lowest possible – operating costs [2–4]. However, it should not be limited – as is often the case – to the introduction of better materials, more productive technologies, more effective forms of work organization, or more recent IT systems. This also needs to cover employee interactions, for example empowering employees to be more responsible and accountable for the outcomes of their actions [5].

Improvement is usually referred both to processes which are less sufficient then they could be according to their build in potential and to processes that fulfill their potential but they can be even more effective because of new opportunities emerging (e.g. organizational, technical). However it should be emphasized that the first case is nothing more than just a simple process correction. Only the second case can be referred as to a pure process improvement (Fig. 1).

A. Hamrol et al. (Eds.): *Advances in Manufacturing II* - Volume 3, LNME, pp. 83–100, 2019.
https://doi.org/10.1007/978-3-030-17269-5_7

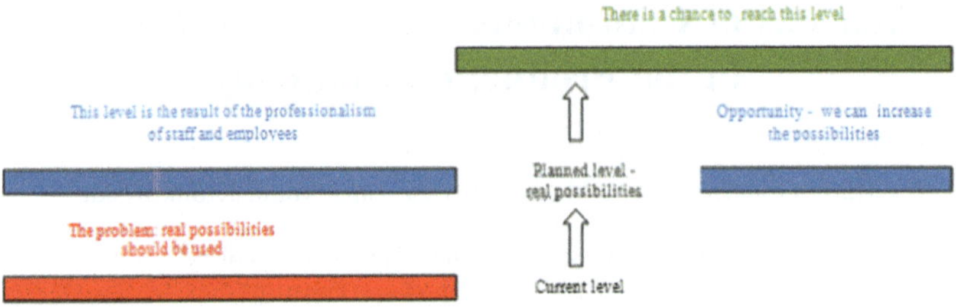

Fig. 1. Possible steps of continuous improvement. Source: own elaboration.

The idea of continuous improvement started to gradually disseminate throughout production companies in 1950s, while the modus operandi of Japanese businesses became its most recognizable symbol. Over time, continuous improvement concepts or continuous improvement strategies (CIS) have been developed, such as Kaizen, Lean Manufacturing (LM), Total Quality Management (TQM), Six Sigma (SS), and Theory of Constraints (ToC) [6, 7]. Also the arrival of ISO 9001 standards and the resulting development of standardized quality management systems (SSQM), with a focus on continuous improvement as one of the main tasks, have played a cross-cutting role in the promotion and dissemination of continuous improvement strategies in 1980s.

2 General Overview - Continuous Improvement Strategies

All continuous improvement strategies have specific rules, methods, and tools in common, which contribute to the increasingly better fulfilment of customer requirements, both internal and external. Each of the continuous improvement strategies has a unique character reflecting its basic assumptions [8–11]. It can easily be expressed in the following statements:

- TQM motivates all employees to engage in the continuous improvement process [2, 12],
- Kaizen highlights the need to take every opportunity to improve your work and the surrounding working environment [13],
- Lean Manufacturing calls to eliminate all sources of waste and to provide continuous flow of added value [14–16],
- Theory of Constraints helps identify 'bottleneck' processes, workstations, and resources which negatively affect the performance of the whole system in general [6],
- Six Sigma seeks to obtain process perfection according to the 'do it well for the first time' principle [17],
- SSQM supports the application of recognized good practices [12].

Reasonable use of each of these continuous improvement strategies reduces losses attributed to waste and improves the quality capability of individual processes and resources. LM and ToC are most useful in identifying and eliminating waste, defined as operations which contribute no added value to the product (NVA – Non Value Added and BVA – Business Value Added). Six Sigma and Kaizen support the capability of processes to add value to products (VA – Value Added). TQM and SSQM provide motivation and formal support for these measures (Fig. 2). It makes sense, therefore, to consider all of the discussed CIS as a set of complementary and mutually reinforcing strategies.

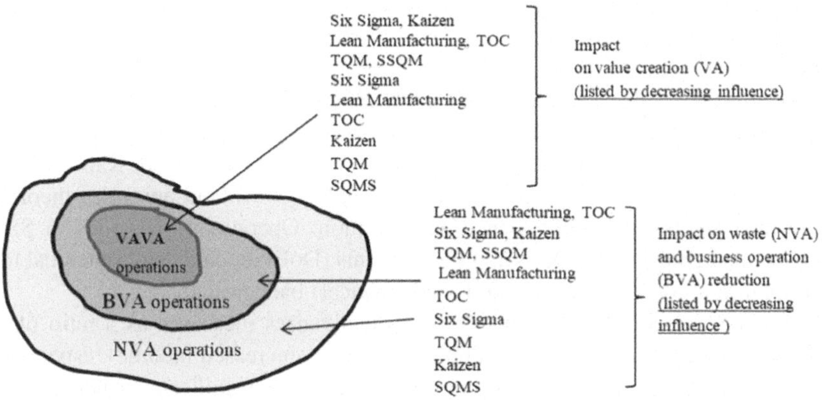

Fig. 2. The impact of improvement strategies on waste and added value. Source [18].

From all CIS used in practice as the most efficient has been considered from over 15 years a combination of Lean Manufacturing and Six Sigma. It is well known under the name Lean Six Sigma (LSS) [19–21].

There is a variety of definitions for Lean Manufacturing provided in the literature. What all of them have in common is the purpose of LM, namely systematic reduction of lead time, i.e. the period between the placing of an order by the customer, and product delivery to the final consignee [22, 23]. This goal can be achieved through identification and elimination of all signs and symptoms of waste from the value stream [24]. The extent of waste is reflected in the results of studies revealing that only 5–15% of operations directly contribute to the increase of product value. In other words, 85–95% of operations can be classified as waste, to a larger or lesser extent [25]. In this respect, it should be underlined that these findings refer to the total time a product remains in the manufacturing system [26, 27]. It includes the storage time of raw materials, semi-finished products, finished goods, as well as the waiting time for processing in individual production cells, etc. [28].

In its essence, Six Sigma places great emphasis on the need to systematically improve processes by optimizing the process conditions, and by limiting variability; it argues that the operating costs can be lowered by rigorously selecting the parameters for conducting processes, and by reducing process variability in time (the more stable the parameters, the better) [29, 30]. In SS strategy, organizations are recommended to accept a reference quality level defined as 6 sigma, meaning that the fraction of incompatibilities occurring throughout processes equals 3.4 per million of possible incompatibility cases. Incompatibilities cover not only finished (or semi-finished) products, but also operations taking place when processes are carried out (such as delay in the transfer of materials, parts, information, etc. from one place to another). It is relatively easy to come close to 6 sigma level if the workflow and the outcome of a process mainly depend on the condition and quality capabilities of machines and equipment (for example measured by Cp, Cpk, Pp, PPK, OEE indicators). Whenever human is the decisive factor affecting the outcome of a process, 6 sigma level is very difficult to attain (a maximum of 3 human errors per million of possible errors is almost impossible to accomplish) [31, 32].

Lean Manufacturing (LM) and Six Sigma (SS) strategies are both focused on creating an added value by improving processes and resources, and by engaging the company management in the improvement process. They differ in terms of methodologies, methods, and tools. LM mainly relies on methods such as Kanban, supermarket, SMED, and 5S. They can be used by employees without any prior theoretical training and are based on experience and intuition. Operational activities in SS are mainly conducted using the Design of Experiments (DoE) methods and statistical tools, which require a good theoretical (and mathematical) background.

LM and SS also differ in terms of cost effectiveness measured as a ratio of economic benefits (cost reduction, containment of losses, increased income), especially in the long-term perspective. It is often recognized that significant benefits can be obtained with LM within a relative short time [17, 33, 34]. However, the possibilities offered by LM are exhausted in the long run [35]. Due to a relatively formalized procedure for the conduct of improvement projects (the DMAIC cycle), it takes considerable time for the Six Sigma strategy to achieve the intended outcome, but the economic effects can be greater than with the LM strategy, while both strategies combined can generate synergies [36]. It should be also stressed that cost savings attributed to LM are difficult to demonstrate in view of cost accounting systems used by businesses, favouring savings made in value-adding processes.

3 Research Problem – Continuous Improvement Culture

Systematic application of CIS creates in a company a culture of continuous improvement (CIC). CIS means that improvement isn't conducted occasionally, or ad hoc; instead, it is implemented in the form of a well thought out, skilfully prepared projects, here referred to as continuous improvement projects, CIPs. In the beginning, all possible circumstances and determinants (problem, challenge) should be identified, and then progressively narrowed down and eliminated to arrive at a possible solution (decision) considered the best under the given circumstances. In order to avoid taking a wrong turn by implementing changes in a single place only (in a field, process, and workstation), the possible consequences of changes produced elsewhere should be envisaged. As a result, CIPs must fit with the company strategy and its strategic goals [12, 37]. They should contribute to the generation of increasingly favourable key performance indicators (KPI) as defined by the company management.

On a day-to-day basis, CIPs are mainly conducted at the operational level, in the most basic and auxiliary processes. However, if the company management is able to coordinate and take advantage of CI measures, the improvement of basic processes shall by itself contribute to a more effective and efficient achievement of the strategic business objectives (improvement of strategic KPIs). The engagement of the company management in CI and in that regard, undertaking joint actions with the company's employees is an issue of key strategic importance [38]. First and foremost, employees should be empowered to take appropriate measures.

Regardless of the place, field, or process to which CIP applies, CIP can be initiated by both managers at the operational level (bottom-up), and by top executives (top-down). It is difficult to definitely state which of these approaches is better. The advantage of the first one (bottom-up) is that projects are located in areas where problems are encountered on a daily basis. However, such projects may be inconsistent or only partially consistent with the company's strategic goals. This disadvantage can be overcome with the top-down approach, which nevertheless involves a different risk: the projects may not reflect the actual operational problems. Low effectiveness of the projects and poor engagement of medium-level management and employees in the implementation of these processes are the undesirable consequences.

The reflections on the assumptions and goals of improvement strategies and ways to implement improvement projects discussed in the previous sections, compiled on the basis of literature review and the authors' own experience, have been confronted with practice in Aesculap Chifa (ACP), one of production companies operating in Poland. ACP started to implement some quality techniques almost 15 years ago. It can be seen as a beginning of continuous improvement culture building. Then the company decided to use Lean Manufacturing methods and to extend it to Six Sigma. Till now about 90 Lean Six Sigma projects were implemented in ACP according to the DMAIC methodology. The scope and purpose of projects were both defined by employees and managers (according to the bottom-up principle) with only discreet support of the company management. The question occurs "*if ACP' atmosphere and practices can be called continuous improvement culture?*". If their actions are well organized and planned. *What should be done to develop continuous improvement culture in ACP*? Therefore the Lean Six Sigma leaders decided to conduct a survey to identify possible regularities in the selection and running of projects. This kind of information can be useful for the company while shaping LSS strategy and developing LSS culture. It may be also a valuable guide to all researches and practitioners involved in continuous improvement actions.

3.1 Short Presentation of the Company

ACP produces surgical instruments and cardiovascular sets. It has a workforce of almost 2000 people. The company's production system has the following characteristics:

- high diversity of products – the company produces around 5 thousand different finished products using around 25 thousand semi-finished products and components,
- small lot production – individual orders for 10 to 500 pieces,

- large share of manual work in technological processes – wage cost accounts for around 40% of direct production costs;
- traditional machining techniques – a relatively high warehouse stocks levels of raw materials, semi-finished goods, production in progress, and finished products accounting for 25% of the annual manufacturing costs.

Until the mid-90s the company operated in a traditional way typical for a centrally planned economy. At the end of the 20th century, after passing through ownership changes, ACP started to implement some projects that only loosely referred to Lean Manufacturing assumptions. The initial improvement projects supported by Kaizen measures were geared towards reducing visible waste resulting from excessively high stock levels. The main focus of the new arrangements was on reducing inventories of work in progress and storages. The outcomes have been achieved by combining previously independent production cells into supermarkets operating as FIFO queues. In this way, the accumulation of excess inventories between production cells has been restricted, which translated into improved flow of materials between workstations.

However, the implemented projects failed to produce the anticipated outcome of 20–30% increase in labour productivity, as referred to in the literature. In order to determine the main reason behind this lack of progress, the operations of one of the machining departments have been investigated. The purpose of this investigation was to determine the following parameters of production cells:

- accumulated time of VA operations – to confirm whether the production cell operated according to the technological assumptions,
- accumulated time of NVA operations – in order to identify the main changeover operations taking place at workstations,
- accumulated time of operations classified as waste – machine downtime because the operator has left the workstation or is present, but is not performing the statutory duties.

The analysis revealed that:

- 37% of the working time in the analyzed area was classified as waste (NVA),
- 14% accounted for BVA operations that were essentially necessary,
- 49% accounted for VA operations.

To improve the situation measures were taken to eliminate BVA operations. As a result, following the introduction of 5S practices, the necessary information flow between production cells and the surrounding environment has improved. It has been assumed that 25% of these operations can be transformed into VA operations, which should translate into a 10% increase in labour productivity.

Four years after the implementation of LM solutions, it has been concluded that LM projects made it possible to reduce stocks and lead times, and to increase labour productivity by streamlining the flow of information and materials. The volume of cost savings has been estimated on the basis of the conducted analyses and concrete historical data from the previous years. In this respect, a clear upward trend in cost savings and increased profitability was demonstrated in the initial stage of the process, followed by stabilization of the respective values at a definite maximum level, and a downward trend in the final stage. LM projects are deemed to 'exhaust' their potential to become increasingly less effective over time. Also, problems with unstable production processes have been identified, manifested by high discrepancies/shortcomings levels, which was another side effect of the implemented solutions.

Taking all these observations into account, in 2010 the company decided to extend Lean Manufacturing to Six Sigma. In the first stage, the company focused on the professional education of its workforce. Between 2010 and 2017, over 60 employees attended training courses and graduated from post-graduate studies to obtain Green Belt qualifications. Moreover, six individuals obtained Black Belt qualifications.

From 2010 till 2017 about 90 Lean Six Sigma projects were implemented in ACP according to the DMAIC methodology. Projects performed brought financial benefits first of all. Some of them are at the level of up to 100 000 Euro per year. Part of the projects didn't bring financial benefits but improved work culture and work organisation.

The scope and purpose of projects were both defined by employees and managers (according to the bottom-up principle) with only discreet support of the company management. Therefore the Lean Six Sigma leaders decided to conduct a survey to identify possible regularities in the selection and running of projects. This kind of information can be useful for the company while shaping LSS strategy and developing LSS culture. It may be also a valuable guide to all researches and practitioners involved in continuous improvement actions.

3.2 Research Methodology

Organizing and Gathering and Data

Team of experts including people from ACP and from the Poznan University of Technology, analysed chosen improvement projects realized in ACP. 65 projects out of 90 performed in the years 2010–2018 were selected for the analysis. Reports of rejected projects were incomplete. Team of experts defined criteria to classify improvement projects. LSS projects were characterised by attributes, which were clustered into five groups:

G1 (Group 1) - baseline situation prior to the project
G1A1- disorder and disorganization - The dominating trend is difficult to determine. Examples of symptoms (chaos at individual workstations and in the surrounding environment, absence of working standards – operations are performed on a routine basis, lack of communication between workstations and the surrounding environment, visible stocks (in the production department and in warehouses, on desks), ineffective communication channels, blocked passageways)
G1A2 - hidden waste - The operations of the company are well organized, but the company struggles with long lead times and high stocks accumulated in the course of production. Bottlenecks in the flow of information, materials and works, large share of processes controlled manually. This condition can be manifested by accumulation of inventories of work in progress at work stations, lack of consistent principle of work assignment (employees themselves decide what to do, in what sequence, and when), delayed or irregular deliveries

G1A3 - unsatisfactory level of quality - In general, the operations of the company are well structured and no major delays occur, but the company experiences continuous problems by failing to meet the quality requirements: there are deficiencies and shortcomings; some of them are transferred to subsequent workstations or are even released from the company, resulting in high levels of quality-related complaints
G1A4 - New chances and opportunities - Everything is OK, but there is constant pressure to identify room for improvement to become even better

G2 (Group 2) - Project beneficiaries
G2A1 - marketing - Improved efficiency and effectiveness of marketing operations and/or of customer requirements and satisfaction surveys
G2A2 - planning - Effective and timely planning of demand for products and production resources (improved balance between needs and capabilities)
G2A3 - product design - Reduced period of product design and placement on the market, lower number of errors revealed in the manufacturing processes and/or during product use
G2A4 - design of manufacturing processes - Shorter design time, lower number of errors and corrections revealed in the manufacturing processes
G2A5 - maintenance - Improves accessibility and/or capabilities of machines, reduced failure rates of machines, reduced time of inspections/repairs, reduced changeover time
G2A6 - internal logistics - Shorter delivery time to workstations, better access to materials, reduced number of transport accidents, reduced storage space
G2A7 – procurement - Increased stock rotation, reduced stock levels, reduced storage space, lower number of complaints concerning the purchased materials and services
G2A8 - manufacturing - Lower number (share) of discrepancies/shortcomings, shorter changeover/cycle time, lower number of complaints concerning workmanship
G2A9 - audits - Improved effectiveness and efficiency of inspections
G2A10 - distribution - Improved sales efficiency, shorter complaint handling time concerning products/customer service
G2A11 - management streamlined decision-making, improved communication among employees and departments, improved employee motivation

G3 (Group 3) - the dominating improvement strategy
G3A1 - Lean Management/Lean Manufacturing - Measures aimed at streamlining production flow, balancing workload distributed across machines and workstations, destocking, reducing lead times
G3A2 - Six Sigma - Quality improvement measures (cutting down discrepancies, shortcomings, corrections), projects implemented according to the DMAIC methodology, focus on performance indicators and analysis (including statistical analysis)
G3A3 - Lean Six Sigma - Combined Lean and Six Sigma measures
G3A4 – Kaizen - Improvement of workstations, simple and low-cost solutions, short lead times
G3A5 –TQM - Awareness-raising among employees, workforce involvement in workstation improvement measures

G4 (Group 4) - Tools and methods of measurement and analysis
G4A1 - universal tools - Pareto, Ishikawa, relation diagram, etc.
G4A2 - statistical tools - Statistical tests, confidence intervals, Anova, etc.,
G4A3 - experiments - OFAT, factorial, regression function, correlation coefficients
G4A4 - tools and methods to plan and design - QFD, requirements and satisfaction surveys, FMEA, etc.
G5A5 -auditing tools and methods - Capacity indicators, control charts, statistical acceptance monitoring, MSA.

G5 (Group 5) - Type of measures
G5A1 - optimization of process parameters
G5A2 - training, awareness raising among employees
G5A3 - introduction of a standard
G5A4 - change of layout, flow
G5A5 - change of flow management
G5A6 - change of technology – construction
G5A7 - elimination of actions

The analysis was motivated by an observation, that baseline conditions (prior to project undertaking) are decisive all other attributes. In other words, from the five groups of attributes considered in this analysis, attributes depicting baseline conditions have the most comprehensive interaction. The aim of research was to find the strongest relations between attributes from G1 and attributes from G3, G4 and G5. Relation G1–G2 wasn't considered because it doesn't have a "cause and effect" nature. Attributes from these groups are independent.

Analysis of Attributes Distribution

Distribution of attributes in the five depicted groups is shown in Fig. 3a–3e.

Group 1 - Baseline situation - The baseline condition prevailing at ACP (Fig. 3a) was characterized by long lead times, or excessive stock levels, and quality-related problems. To a lesser extent projects were initiated from the need to organize work station or environment or to make improvement because new opportunities have arisen.

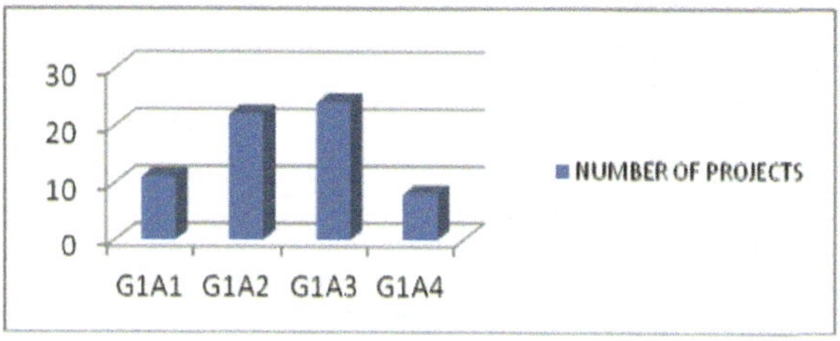

Fig. 3a. Baseline condition in the area subject to improvement.

The observed distribution of baseline attributes indicates implicitly that ACP is at the transitional stage on the way to excellent processes. The share of projects dealing with poor organisation and with looking for new opportunities as well is relatively low.

Group 2 - Project beneficiaries - In the analyzed 65 projects, 108 beneficiaries have been appointment. The majority of projects implemented (Fig. 3b) delivered positive results in the following areas: manufacturing (39 projects) and process design (22 projects).

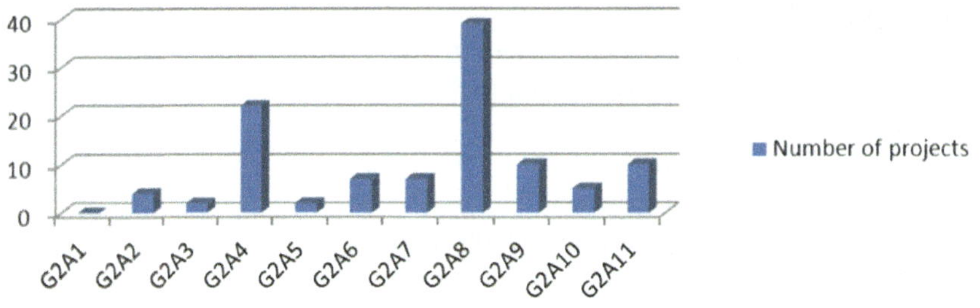

Fig. 3b. Beneficiaries of projects.

One of the reasons for such a distribution can be related to the nature of the company's operations. In ACP that sheer manufacturing processes are perceived as the basic processes. For this reason managers are convinced that they should have been improved in the first line. The auxiliary processes should be left - in their believes - for improvement in the second place. In addition manufacturing process are prevailing also in terms of quantity (over 90% employee are line workers).

Group 3 - Dominating concept - Majority of projects have been implemented in line with Six Sigma or Lean Manufacturing strategies (Fig. 3c).

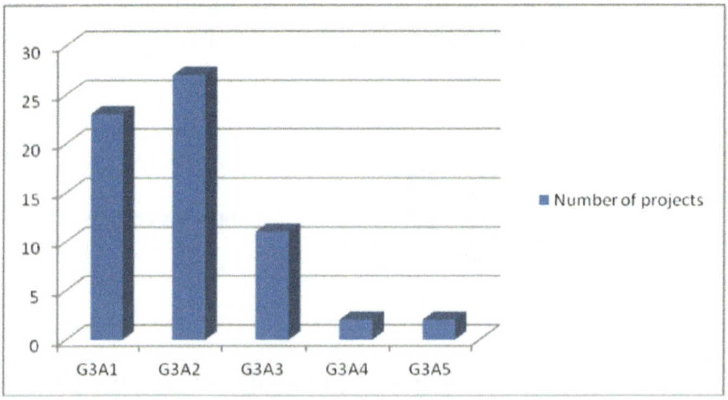

Fig. 3c. The dominating improvement strategy.

More in-depth project analysis could show that TQM was a leading concept only in projects associated with insufficient workers commitment. This explains why the number of TQM appointments is relatively small. But taking into account that TQM principles are universal it can be concluded that this concept was a background in all projects implemented. The same reflections refer to Kaizen. A deeper analysis showed that in some of them (about 20) Kaizen "mini-projects" was conducted.

Group 4 - Tools and methods - There were 87 groups of tools indicated in 65 projects. The diagram (Fig. 3d) indicates the domination of simple organizational tools, such as Ishikawa diagram, Pareto diagram, 5 Why, and ABCD analysis. Simple statistical tools and more advanced, like process control charts or MAS analysis have been used frequently as well.

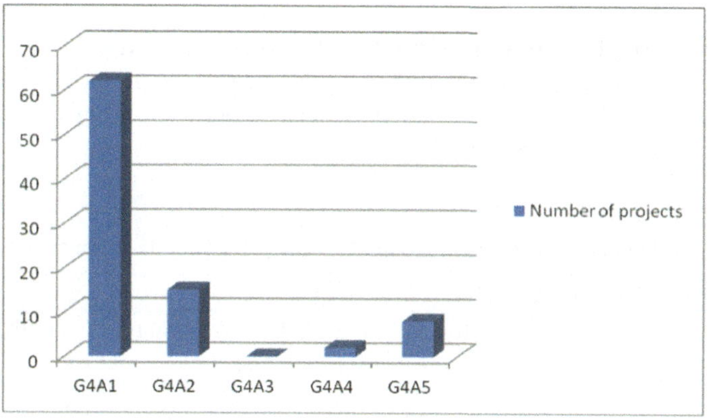

Fig. 3d. Tools and methods used

Relatively seldom experiments and methods for planning and designing products or processes were implemented. This may indicate an insufficient or even poor preparation of green - and black belts for use more advanced statistical tools.

Group 5 - Measures taken - The measures taken show with a strong bias in favour of standards implementation, followed by measures aimed at changing process parameters as well as technology or construction parameters (Fig. 3e).

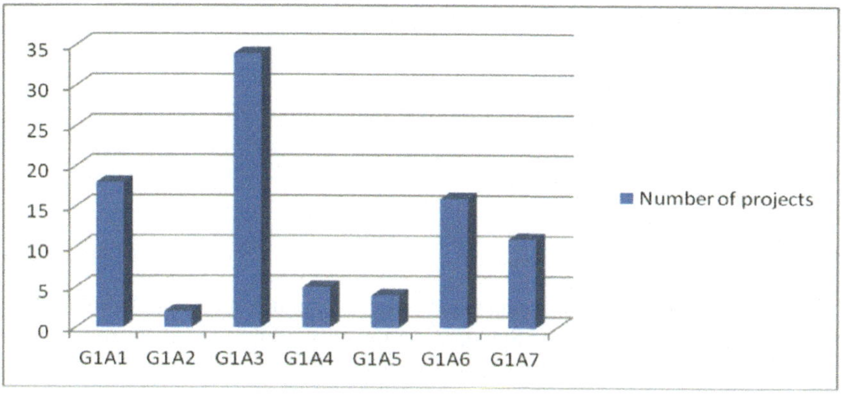

Fig. 3e. Type of measures

In summary, the results of analysis of 65 projects implemented lead us to the following conclusions:

1. The improvement opportunities mainly consisted in lead time reduction and in eliminating issues with quality.
2. The production department in the first place, followed by the process design area.
3. Six Sigma was the dominating improvement strategy, while Lean Manufacturing was only slightly less common.
4. Simple universal organizational techniques were favoured in the implementation of improvement projects.

Influence of Baseline Conditions on Other Projects Attributes

As already said relations between attributes belonging to the group G1 and attributes from the groups G3, G4 and G5 are of special importance. They are shown in a graphical form and in numbers (Table 1).

Table 1. Distribution of attributes in groups 2, 3 and 4 for various attributes from group 1.

Group	Attribute	Group G1			
		G1A1 disorder	G1A2 L/T	G1A3 quality	G1A4 OK!
G3	A1	64%	68%	0%	13%
	A2	9%	14%	83%	37%
	A3	27%	18%	13%	13%
	A4	0%	0%	0%	24%
	A5	0%	0%	4%	13%
G4	A1	85%	78%	65%	58%
	A2	15%	11%	22%	21%
	A3	0%	0%	0%	0%
	A4	0%	4%	0%	7%
	A5	0%	7%	13%	14%
G5	A1	7%	7%	36%	25%
	A2	7%	0%	0%	8%
	A3	58%	13%	36%	25%
	A4	7%	13%	0%	0%
	A5	7%	10%	0%	0%
	A6	7%	13%	24%	25%
	A7	7%	24%	4%	17%

Following observations are noticed:

- Group A1 (disorder and disorganization) - Lean Manufacturing and Lean Six Sigma were most commonly used. Standardization of working practices was the most common outcome of the improvement projects.

- Group A2 (hidden waste) - Lean Manufacturing was the preferred improvement strategy as well as universal organizational tools. Standardization of working practices and elimination of redundancies were the most common outcomes of the projects in the manufacturing, process design, and logistics (Fig. 3b).
- Group A3 (unsatisfactory quality) - Six Sigma was the preferred improvement strategy coupled with universal organizational and statistical tools. Implementation of standards and parameter modifications, mainly in the production area, were the most common outcomes of the projects.
- Group A4 (new chances and opportunities) - Six Sigma and Kaizen were the preferred improvement strategies, together with LM, LSS and TQM as well as universal organizational tools. As a result of the improvement projects, standards have been implemented, technology and parameters have been modified, and waste has been eliminated, predominantly in process design and manufacturing, as well as in the area of inspections and management.

In conclusion it can be said that LM and LSS strategies were demonstrated to be useful in areas which are considered immature in organizational terms, with visible chaos and excessive stock levels. Having reorganized the basic elements of the production system: workstations and the flow of resources, the next step is to improve quality parameters. In this case, the Six Sigma concept and a detailed analysis of a selected aspect or operation are recommended. At a more advanced stage of the company development, at the time when improvement involves exploiting the opportunities rather than problem solving, the strategies that support engagement of all employees, such as TQM, and aimed at increasing safety and comfort, such as Kaizen, in addition to SS, LM, and LSS, are more common than in the previous cases.

4 Summary

The analysis of improvement projects carried out in ACP over the last 15 years has shown their great diversity. It has been described in the article by five attributes (assigned to groups G1 to G5), referring to: motivation to undertake the project, beneficiaries of the project results, the dominating strategy of improvement, applied tools and methods of measurement and analysis and implemented improvement actions. The analysis results showed that motivation connected with the lack of order and proper organization of work as well as different forms of waste prevailed. It also revealed that projects referred mostly to the Lean Manufacturing approach and the relatively simple tools and methods of measurement were used. The results obtained are a valuable indication for ACP how to conduct the policy of improvement in the near and distant future. They suggest that despite good theoretical preparation, project leaders prefer using relatively simple tools and resources. Analysis of the project effects in ACP suggests that using of base/traditional organizational tools is overused (Fig. 4).

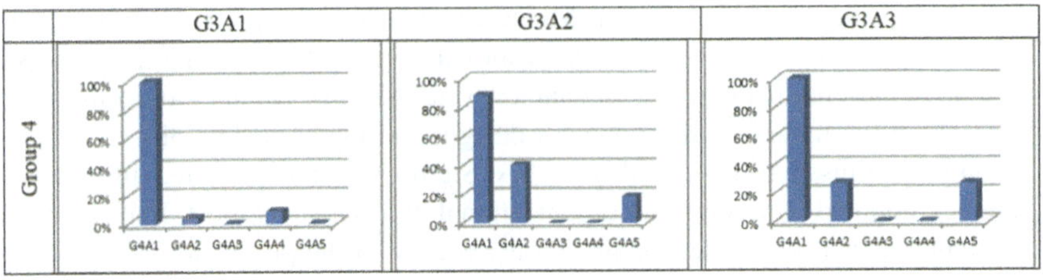

Fig. 4. Tools used according to the improvement strategy.

Some conclusion could be identified in shorter period if more sophisticated methods are implemented. It was found that project teams were not always properly prepared for the implementation of tasks within the selected CIS. Therefore, it is necessary to constantly inform employees and managers about the benefits already gained and to make them aware at the same time that only the expanding and deepening of their knowledge and skills in the scope of tools and methods of improvement allows the use of emerging opportunities. Planning of CI project could be used as a tool to organize training system in the company. For the planing of trainings should be taken into consideration dominant CIS according to base situation (Fig. 5).

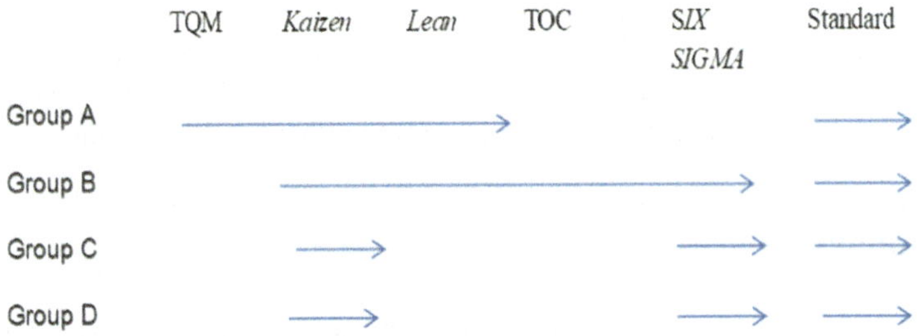

Fig. 5. A sequence of actions after the diagnosis of the base situation. Source [12].

Before a company declares its intention to create CIC, its management board and executives should identify and recognize the current status of the company in terms of organisational maturity of processes. In this context, four starting situations for the groups of projects can be appointed [12].

Group A – disorder and disorganization. The processes and operations of a company are disorganized and a feeling of general chaos prevails. The company lacks organizational culture that would foster conditions for effective business operations, and is missing a work organization appropriate for this purpose. This condition can be manifested by:

- disorder and chaos at workstations and in the surrounding environment,
- absence of documents defining working standards,
- visible stocks (in the production department and in warehouses),
- ineffective interpersonal communication channels,

Group B – hidden waste. A general feeling of order and structure, good working conditions, effective interpersonal communication. However, the lead times (Manufacturing Lead Time, Customer Lead Time) are long, with high stock levels persisting in the course of production. Many processes are controlled manually. This condition can be manifested by:

accumulation of inventories of work in progress at workstations,

lack of consistent principle of work assignment (employees themselves decide what to do, in what sequence, and when),

delayed or unrhythmic deliveries.

Group C – unsatisfactory level of quality. In general, there is order and no major delays occur, but the company experiences continuous problems by failing to meet the quality requirements:

there are deficiencies and shortcomings; some of them are transferred to subsequent workstations or even released from the company,

high levels of quality complaints are reported, filed by external and internal customers.

Group D – new chances and opportunities. No symptoms of incorrect operation of the company's subsystems are observed; however, there is still potential for improvement and development in specific business areas.

Obviously, it is difficult to assign a specific baseline condition to one of the discussed categories; still, some dominating features can be distinguished in the majority of cases. The assignment of the prevailing problem to one of the defined groups in a sense defines the underlying purpose of using a specific "sequence of improvement stages", which differ in terms of the dominating operation strategy of the company.

For problems in Group A concerning the general chaos across or in parts of system, CIP should be implemented by heavily relying on the Kaizen and Lean Management or Lean Manufacturing sequences, leading to the introduction of standards and good practices. The following methods and tools are recommended for a start: 5S, SMED, standardisation.

For problems with long lead times or bottlenecks at any point of a system (or process) – Group B – it seems more natural to initiate measures according to the principles of Lean Manufacturing stepped up, when needed, with the Kaizen tools, and more specifically: VSM, SMED, TPM, supermarkets, self-dependent maintenance, multitasking, Kanban boards, flow levelling and balancing, drum-buffer-rope mechanism, DoE, and quality capability indicators.

For problems assigned to Group C, it is recommended to use Six Sigma or Kaizen workflows. Kaizen is preferred for problems that can be solved intuitively, while Six Sigma works best for problems in which solutions are more deeply hidden. The following tools can be considered useful: brainstorming, QFD, CSI, flowchart, data pooling, Ishikawa diagram, Pareto diagram, ABC analysis, system flow diagram, 5Why diagram, histogram, capability indicators, control charts, DoE, and FMEA.

Kaizen and Six Sigma projects should prevail if starting point conditions belong to the Group D.

The tips provided are obviously a mere indication of possible ways in which to proceed. A creative and critical approach needs to be taken on a case-by-case basis. Lack of criticism or following the same old path can lead to dead ends.

After the identification and diagnosis of the base situation (e.g. disorder and disorganisation, hidden waste, unsatisfactory quality, new chances and opportunities), team knowledge about tools and methods should be checked and the plan of trainings for the project team should be establish. Different strategies involve different requirements in terms of staff training and the necessary changes in the organizational structure. For example LM requires that all employees of a company become involved in the implemented changes. In this context, the Six Sigma approach is different: a group of selected individuals in charge of the projects needs to hold the highest qualifications and attend specialist training.

The analyses and results discussed here provide evidence that there is no 'one-size-fits-all' approach to the implementation of the CI culture.

Each and every company has its own unique organizational culture and human resources profile, and is at a different stage of development. Therefore, the scope of activities at particular stages of the implementation of improvement projects should be adapted to the situation in the enterprise.

References

1. Timans, W., Ahaus, K., Solingen, R., Kumar, M., Antomy, J.: Implementation of continuous improvement based on Lean Six Sigma in small and medium enterprises. Total Qual. Manag. Bus. Excell. **27**(3), 234–309 (2016)
2. Aqulan, F., Al-Fandi, L.: Prioritizing process improvement initiatives in manufacturing environments. Int. J. Prod. Econ. **196**, 261–268 (2018)
3. Hamrol, A.: Quality Management With Examples. PWN, Warsaw (2007)
4. Kinney, M.R., Raiborn, C.A.: Cost Accounting: Foundations and Evolutions. Thomson South-Western, Mason (2009)
5. Grudzien, L., Hamrol, A.: Information quality in design process documentation of quality management systems. Int. J. Inf. Manage. **36**(4), 599–606 (2016)
6. Cox, J., Schleier, J.: Handbook: Theory of Constrains. McGraw Hill, New York (2010)
7. Zu, X., Frendendall, L., Douglas, T.J.: The evolving theory of quality management: the role of Six Sigma. J. Oper. Manag. **26**(5), 630–650 (2008)
8. Bicheno, J., Holweg, M.: The Lean Toolbox: Towards Fast, Flexible Flow. PICSIE Books, Buckingham (2004)
9. Starzyńska, B., Hamrol, A.: Excellence toolbox: decision support system for quality tools and techniques selection and application. Total Qual. Manag. Bus. Excell. **24**(5), 577–595 (2013)
10. Starzyńska, B., Grabowska, M., Hamrol A.: Effective management of practitioners' knowledge - development of a system for quality tool selection. In: 2016 3rd International Conference on Social Science (ICSS 2016), 9–11 December 2016, Shanghai, China, pp. 859–865 (2016)
11. Rewers, P., Trojanowska, J., Chabowski, P.: Analysis of the use of LM tools - research results, Logistics, no. 3 (2015)
12. Hamrol, A.: Strategies and Practices of Efficient Operation: Lean Six Sigma and other. PWN Scientific Publisher, Warsaw (2015)
13. Pero, M., Pozzi, R., Rossi, T.: Complexity reduction and kaizen events to balance manual assembly lines: an application in the field. Int. J. Prod. Res. **24**, 1–18 (2018)

14. Suri, M.: Quick Response Manufacturing: A Company Wide Approach to Reducing Lead Time. Productivity Press, Portland (1998)
15. Womack, J., Jones, T.: Lean Thinking. Free Press, New York (2003)
16. Marchwinski, C., Shook, J.: Lean Lexicon: A graphical glossary for Lean Thinkers. Lean Enterprise Institute, Brookline (2003)
17. Harry, M., Schroeder, R.: Six Sigma: Use of Quality Improvement to Improve Financial Results. Economic Office, Krakow (2001)
18. Hamrol, A.: A new look at some aspects of maintenance and improvement of production processes. Manag. Prod. Eng. Rev. **9**(1), 34–43 (2018)
19. Aruleswaran, A.: Changing with Lean Six Sigma, LSS Academy Sdn. Bhd. Malaysia (2009). ISBN 978-983-44582-0-1
20. Burton, T.T.: Accelerating Lean Six Sigma Results: How to Achieve Improvement Excellence in the New Economy. J.Ross Publishing, USA (2011). ISBN 978-1-60427-054-9
21. Yadav, G., Seth, D., Desai, T.N.: Application of hybrid framework to facilitate Lean Six Sigma implementation: a manufacturing company case experience. Prod. Plan. Control **29** (3), 185–201 (2018)
22. Ohno, T.: Toyota Production System: Beyond Large-Scale Production. Taylor & Francis Group, Boca Raton (1988)
23. Rother, M., Shook, J.: Learning to See. Lean Enterprise Institute, Brookline (2003)
24. Sundara, R., Balajib, A.N., SatheeshKumar, R.M.: A review on lean manufacturing implementation, techniques. Procedia Eng. **97**, 1875–1885 (2014)
25. Nielsen, P., Michna, Z., Sørensen, B.B., Dung, N.D.A.: Lean times - their behavior and the impact on planning and control in supply chains. Manag. Prod. Eng. Rev. **8**(2), 30–40 (2017)
26. Becker, Ch., Scholl, A.: A survey on problems and methods in generalized assembly line balancing. Eur. J. Oper. Res. **168**(3), 694–715 (2006)
27. Karwasz, A., Chabowski, P.: Productivity increase through reduced changeover time. J. Mach. Eng. **16**(2), 61–70 (2016)
28. Behrouzi, F., Wong, K.Y.: Lean performance evaluation of manufacturing systems: a dynamic and innovative approach. Procedia Comput. Sci. **3**, 388–395 (2011)
29. Parast, M.M.: The effect of Six Sigma projects on innovation and firm performance. Int. J. Project Manage. **29**(1), 45–55 (2011)
30. Zhen, H., Yuija, D., Min, Z., Xingxing, Z., Jiju, A.: An empirical investigation of the relationship between Six Sigma practises and organisational innovation. Total Qual. Manage. Bus. Excell. **28**(5), 459–480 (2017)
31. Shafer, S., Moeller, S.B.: The effects of Six Sigma on corporate performance: an empirical investigation. J. Oper. Manag. **30**, 521–532 (2012)
32. Swing, M., Jacobs, B.W.: Six Sigma adoption: operating performance impacts and contextual drivers of success. J. Oper. Manag. **30**(6), 437–453 (2012)
33. Trojanowska, J., Żywicki, K., Varela, M.L.R., Machado, J.M.: Improvement of production flexibility by shortening changeover time, Multiple Helix Ecosystems for Sustainable Competitiveness, part of the series Innovation, Technology, and Knowledge Management (2016)
34. Yong, L., Shihua, M., Li, Z.: Manufacturing strategies for time based competitive advantages. Ind. Manag. Data Syst. **112**(5), 729–747 (2012)
35. Liker, J.K.: Toyota Way: 14 Principles of Managing the World's Leading Production Company. MT Biznes Publishing House, Warsaw (2005)

36. Hong, C.: Implementing Lean Six Sigma to achieve inventory control in supply chain management. In: AIP Conference Proceedings, volume 1902, 7 November 2017, Article number 020037, 3rd International Materials, Industrial and Manufacturing Engineering Conference, MIMEC 2017, 6 December 2017, Miri, Malaysia; through 8 December 2017, Code 131755
37. Chaplin, L., O'Rourke, S.T.J.: Could lean and green be the driver to integrate business improvement throughout the organization? Int. J. Prod. Perform. Manag. **67**(1), 207–219 (2018)
38. Shanley, A.: Reinventing Lean Six Sigma for the pharmaceutical industry. Pharm. Technol. **41**(10), 76–80 (2017)

Total Innovation Management – Application in Large and Medium-Sized Manufacturing Enterprises in China

Mateusz Molasy(✉), Mariusz Cholewa, Maria Rosienkiewicz, and Joanna Helman

Faculty of Mechanical Engineering, Wroclaw University of Science and Technology, Wrocław, Poland
mateusz.molasy@pwr.edu.pl

Abstract. The paper introduces and describes the concept of Total Innovation Management (TIM) as the concept for more successful innovations management. In the paper the current status of innovation management in large and medium size manufacturing enterprises in China has been also presented as well as the characteristic elements for applying Total Innovation Management in these enterprises. It has been presented how the medium size manufacturing enterprises in China can improve their effectiveness by implementing the abovementioned, including a comparison of case studies.

Keywords: Innovation · Manufacturing · China · Total Innovation Management

1 Introduction

Large and medium size manufacturing enterprises play very important role in the economy of China. They are also one of the main drive forces of industrialization of the country. Currently the innovation ability of these enterprises is a key to China's further economic growth and competitiveness on international markets.

Manufacturing industry is the most important part of national economy in China. It contributes 40,5% of GDP [1] and its value is increasing: GDP from manufacturing in China increased to 222165 CNY HML in the third quarter of 2018 from 144656,40 CNY HML in the second quarter of 2018 [2]. However, many studies show that innovation capacity in China's large and medium-sized manufacturing enterprise is still relatively low [3]. Currently many enterprises realize that the level of innovation determinates the development of manufacturing and competitiveness. Some of them made innovation efforts but didn't achieve desired results – mostly due to lack of knowledge about innovation management. Currently applying advanced innovation management methods and techniques is necessary for improving capacity of innovation for China's manufacturing industry [4].

Therefore, the aim of this paper is to present the concept of Total Innovation Management (TIM) and present how to apply this innovation management concept to enterprises in China to support their growth of effectiveness and efficiency.

A. Hamrol et al. (Eds.): *Advances in Manufacturing II* - Volume 3, LNME, pp. 101–113, 2019.
https://doi.org/10.1007/978-3-030-17269-5_8

2 The Concept of Total Innovation Management

Over the years many definitions of innovation have been created. In 1911 Schumpeter [5] defined innovations as commercial or industrial application of something new: product, process, production methods; new market or sources of supply; a new form of running business. Schumpeter viewed innovation as a variable implementation with no standard constant of a set of combinations broken down into five cases:

- New product introduction – product that was not known by customers before.
- New method of production – technique that was never used in particular branch before.
- New market opening – starting up a business in area, where some branches of industry have not existed before.
- New source of raw material gaining – it could apply either to existing sources or creation of a new one.
- Reorganization of existing industry organization – creating or breaking monopolistic situation.

Following Schumpeter, many scholars shifted the focus of innovation research from economic growth at the macro level to enterprises innovation management at the micro level.

To the innovation concept Porter [6] incorporates technological improvements, better methods, ways of doing something. This can be revealed in product changes, process, new approaches to marketing, new forms of distribution. Author also implied the need to distinguish between invention and innovation. New processes, products and services must be put to a productive commercial use to make invention innovation. The point is that innovations are of practical use in providing new or improved products or services and/or enabling people and organizations to do things more effectively and/or efficiently.

Drucker [7] described innovation as a special tool for entrepreneurs, by which the changes make the opportunity to undertake a new business or provide a new service. Innovation is a specific tool of entrepreneurship - an activity that gives resources new opportunities for creation riches.

According to Kotler and Armstrong [8] innovation is connected with the good, the service and the idea that is perceived by someone as new. The idea could have existed before for a long time, but it is an innovation for the person who perceives it as new.

The issue if innovation management can be summarized as the process by which innovation is managed by affecting decisions, practices and actions, as a response to a vision or an opportunity – to achieve a certain target, like generation of idea, product or service. The issue has been studied for many years. Especially nowadays it can be observed that the number of researches focused on innovation management is growing rapidly [9–20].

Reviewing the main research on innovation management helps identify five phases [21]:

1. Research on individual innovation (1940s–1950s) - systematic research on innovation of enterprises at the micro level began to take place. In this beginning phase of innovation, the basic problems of innovation were still unsettled and research focused on the separate components. The outstanding characteristic of this phase of innovation theory research was the research philosophy of individual innovation management.
2. Research on organizational promotion (1960s–1970s) - research mainly studied the sources of innovation within organizations and focused on how to attain the objective of innovation and promote innovation in organizations through effective management of R&D activities.
3. Research on outsider involvement (1970s) - the third research phase emphasized interactive promotion of internal (R&D) and external sources (users).
4. Research on portfolio, integrated and systematic innovation (1980s–1990s) - theory further evolved into integrated innovation theory and systematic innovation theory.
5. Research on Total Innovation Management.

An overview of the key characteristics of generations of innovation framework models is presented in (Fig. 1).

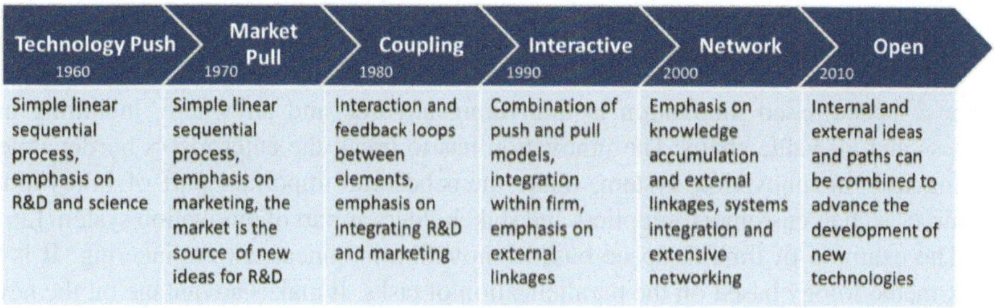

Fig. 1. Overview of the key characteristics of generations of innovation framework models.

Most of the early research has been focused on product innovations or process innovations. Dynamic market is clearly generating an impact for manufacturing enterprises to constantly improve their products. Great majority of the products is being designed to be safer, universal and functional. This adjustment can improve both: technical specifications, as well as the applied materials to improve products and services. This sort of innovation can be driven forward through newly acquired knowledge and technology or through new ideas applied in the using of previously known technology. The term is relevant within the concept regarding physical goods and services [22].

Another type of innovation widely interested by researchers and practitioners, that can appear in the enterprises, is a process innovation. Most common definition states that this approach is the implementation of new and improved or enhanced means of production or delivery [22]. Accordingly, it can be assumed that process innovation can be specified, anytime where there will be problem in the process. It can apply either to product quality, cost of the process, time of the process or many others.

Possible improvements in manufacturing enterprises might be applied not only to main production processes but also to logistics processes, new ways in human resources management, documentation flow and they might relate with introducing new machinery for main production process, reducing unit costs of finished goods or introducing new method of accepting deliveries as RFID identification.

The new paradigm of Total Innovation Management (TIM) is much wider than any other concept of innovations management, because it's not limited to combining different innovation factors. It points the innovation synergy between technology and non-technology areas in enterprise [23]. TIM emphasizes that enterprises need innovation not only in technology, but in all key parts of the business, e.g. strategy, marketing, business process innovation, organization innovation, institution innovation, and culture of innovation. Therefore, TIM requires that enterprise has to consider the innovation of total factors synergy, time and space-based innovation and all innovators. TIM also incorporates innovation by everyone, at all times and in all places [24].

TIM can be characterized as the synergy of two factors: 'time & space-based innovation' and 'all innovators'. First of them mean that innovations arise all the time and everywhere. The latter mean that everyone in the organization is the innovator. It's a kind of ecological system, and an effective platform to cultivate and enhance the core competency to win the sustainable competitive advantage [25].

2.1 Time & Space Based Innovation

Time & space-based innovation is innovating anytime and anywhere, including all process and all value chain. The innovation has to break the enterprise's border, integrating user in innovation system, make them become important part of innovation resource, such as customers, suppliers and stakeholders as part of innovation system [26].

The example of time & space-based innovation is concurrent engineering. It is a work methodology based on the parallelization of tasks. It makes advantage on the new product's development time, reduce the time required bring new product to market. It requires the coordination of market analysis, product design, process design, manufacturing and sale. Concurrent engineering makes the innovation time reduced greatly.

2.2 All Innovators

In order to innovate anytime and everywhere all people inside and outside of the organization should be involved in the process. People's creativity and skills are the greatest source of innovation. Paradigm of TIM requires the universal commitment and participation in innovation. Innovation is no longer function of R&D personnel alone, but all employee's behaviors. The top management should support this system to create the environment which promotes innovating and making employees realize that their creativity is valuable [27].

The application of 'all innovators' management is transformed into four phases, as follows [27]:

Phase 1. Experts to people. In the past the enterprises focused on experts in processed regarding innovations. The new challenge of innovation management is how to make everyone to participate the innovation processes. It is difficult to achieve enterprise value maximization only by experts participating in innovations. Aiming to increase the value of innovation management, enterprise must make all human resources involved in innovating and enhancing people's creativity in order to improve innovation efficiency of enterprises.

Phase 2. Passiveness to initiative. Activities around innovations are depending on conscious behavior of employees rather than coercive power from outside. Employees with greater autonomy, independence and responsibility to complete a task or innovation in new areas which their own interests, often can stimulate employee's innovation potential, to achieve greater innovation performance.

Phase 3. Organization development to the combine of organization and employees' development. Enterprise must focus on the combination of organizational and personal development and establishment of improving the policies of incentive system to encourage employees to continue to overcome and transcend themselves. Personal development is closely linked to the development of enterprises.

Phase 4. Individual innovation to innovation synergy. Instead of emphasizing individual creativity (which is not meaningless) more attention should be paid to corporate level of information and knowledge sharing between employees. Integrating all employees' innovation is possible by building teams, such as project innovation team, networked working team etc.

The phases have been presented in (Fig. 2).

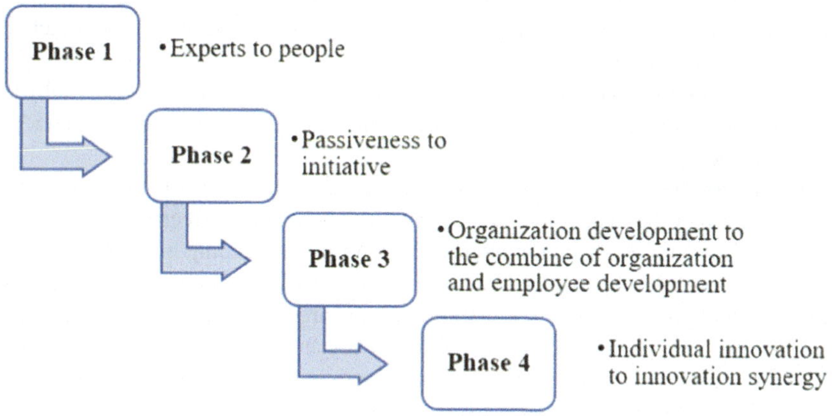

Fig. 2. Phases of transformation to 'All innovators'.

3 Total Innovation Management in Practice – A Study of Innovation Capability in Enterprises in China

China undoubtedly has been innovations leader for ages. The Chinese invented and implement to everyday use such inventions as gunpower, the compass, the waterwheel and paper money [28].

The national innovation system in China can be traced back to the early 1950s when the government embarked on the development of a modern economy under a planned economic system. At that time technological innovation was seen as an integral part of the modernization drive. In 1956 the Chinese government launched a campaign of 'advancing science' through the launch of the long-term technology development plan. The government was the primary source of investment in innovation and innovation was motivated by the perceived needs of the country's eco- nomic growth, social development and national security. In 1978 there have been radical change in the direction of economic development in China and the programme of reform and opening up to the outside world was introduced. It was proclaimed that science and technology are primary productive forces [29].

In recent years, China's economy has integrated with world's economy at an unprecedented pace. China's manufacturing industry is more and more widely integrated into the international division of labor system, the Chinese market has also become an important part of the international market. Government is focusing its efforts on advancing national technological innovation.

The innovativeness of Chinese large and medium sized enterprises is improving; however, several problems can be pointed:

- Innovation capacity remains weak. Large and medium manufacturing enterprises in China are still backward equipment, with lack of core technology, most technologies have been imported from abroad.
- The organizational structures remain inflexible. Enterprises remain multi-level management structures, easy to produce "internal bureaucracy" and management inefficiencies.
- Cultural attitudes hinder enterprises' innovation. Lack of atmosphere to encourage innovation and the lack of creative talents in whole society seems the issue.
- The application of information technologies for manufacturing enterprises in China are basically still in a scattered, isolated applications. At present, the knowledge management project based on integration of resource gathering, exchange and share has not been implemented, but it is the way for an enterprise to the innovative.
- Managers prefer short-term profits over long-time development, therefore innovation is ignored to make up time to address ongoing projects.

Comparing to Chinese enterprises the American enterprises seem to operate in totally opposite way. The U.S. enterprises always apply advance innovation management in faster manner than other countries. And they achieve better result. The American enterprises emphasize the need of synergistic combination of all major innovation factors with a dynamic adaptive innovation core. Also shows the culture of entrepreneurial spirit with openness, diversity and trust [30].

The features of Japanese enterprise are that enterprises have strong connection with extraneous innovation, have highly market-oriented support system for innovation, and top managers are usually involved to innovation process [31].

The recommendations for improvement in the system for managing innovations for Chinese enterprises have been built around 3 pillars reflecting the concept of Total Innovations Management:

- Improvements in technological innovations capacity
- Time & Space-based innovations
- Pattern for all innovators

The pillars have been presented in (Fig. 3).

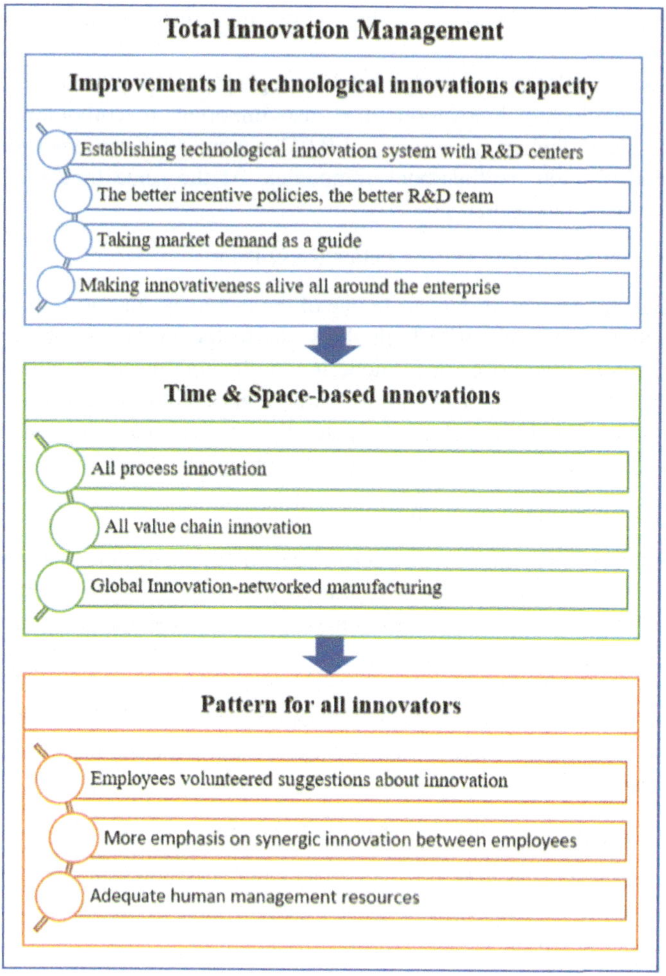

Fig. 3. The pillars of Total Innovation Management.

3.1 Improvements in Technological Innovations Capacity

Before applying TIM in Chinese enterprises, improvements in the technological innovation capacity should be done. It can be reached by few steps that has been identified:

1. Establishing technological innovation system with R&D centers.

Enterprises should have R&D centers to help improving the ability of developing technology. Also, enterprises should put more investment for technological innovation, increase expenses for research and development. Enterprise have to develop core technology with independent intellectual property right, according to it is own condition. That is important for establishing technological innovation system in enterprise. Technological innovation system should meet the requirement of market competition and enterprise development. For this reason, this system will help to improve the ability of quick response to market, the ability of resource distribution, also independent innovation ability.

Beside building R&D center and establishing system for enterprise itself, in the process of technological innovation, it is also important to corporate with outside stakeholders, such as universities, external R&D centers etc. Large enterprises can also cooperate with foreign multinational companies. At the same time the should pay attention to protect intellectual property.

2. The better incentive policies, the better R&D team.

Nowadays, the human resources are one of the factors in enterprise to be develop in China. The more employees with high skills, the better improve to ability of technological innovation. Enterprises need modern and effective incentive policies to attract talents, keep them and let them know how important their value for the enterprise is. Incentive systems should include both: material incentives and moral encouragement. Beside this, enterprises need to provide good training programs.

3. Taking market demand as a guide.

Technological innovations cost a lot – both, money and time. The result can be good but it also can be just a waste of resources. To avoid the latter, enterprises have to take market demand as a guide, do the technology research and development according to market. Therefore, before starting the process they need to deeply research about market demand. Enterprises should have departments responsible for market information research, strength ability of quick response and make decision when changing in domestic and foreign market, innovating product and technology according to market demand.

4. Making innovativeness alive all around the enterprise.

The important step has to be done not to lock the innovations in R&D centers. They knowledge need to be transferred to all departments and values of culture of sharing knowledge need to live in whole organization.

3.2 Time & Space-Based Innovations

In order to successfully implement Total Innovations Management Chinese enterprises must be prepared to manage according to Time & Space-based innovations. First of all, Chinese manufacturing enterprises must complete informatization. This will allow to build networks to share information and knowledge in anytime and everywhere. Networked manufacturing uses data flows to drive better communication, co-ordination, and control in and around manufacturing. Networked production will make manufacturing more decentralized and more flexible. It will emphasize the need for good infrastructure because it will require just-in-time deliveries, and it will require excellent communication between the plants.

To realize the steps of implementing Time & Space-based innovations three-steps process has been proposed, as follows:

Step 1. All process innovation. To easy achieve all process innovation, the innovation of logistic management is necessary. In manufacturing industry, the logistic management is throughout overall process of production. On one hand, logistic is the necessary condition that enterprises depend on for existence and development. Logistic will affect directly the essential operations of enterprise, such as planning, row materials and energy supply management, related information between the point of origin and the point of consumption, also the transport of final products. On the other hand, the continuity and equilibrity of production process depends on activities in logistics. Normally production process and logistic activities are tight hang together. Therefore, putting up a logistic information system becomes a general trend. This system not only is the carrier of information exchange between each department in enterprise, but also is the intercross and interpenetration between the manufacturing enterprise, supplier and customer. Based on information sharing, for better share market profile with supplier or another enterprise, will be further sharing the bigger amount of profile.

For manufacturing enterprise, the efficiency of supplying products and row material is limited and depend on the network construction of logistic. Therefore, the logistic department has responsibility for perfecting the network logistic network construction. In dynamic and competitive market environment, the classification of products, supply quantity, and requirement of manufacturing are changing all the time, therefore, it is necessary to adjust logistic network all the time to fit in this changing.

Logistic innovating needs to meet market demand, satisfying individualized demand of customer, establishing logistic management system which production and supply integration. According to trend of development of domestic and foreign logistic, building pattern logistic system is necessary. China's large and medium size manufacturing enterprise have to change traditional internal logistic system pattern, having clear consciousness of market economy, building long term partnership with supplier, establishing innovative interactive mechanisms of supply and demand.

Step 2. All value chain innovation. Modern and innovations-oriented supply chain management requires enterprise to focus on developing the core competence instead of paying attention to non-important business activities, because the enterprise resources are limited. Those non-important business activities can be outsourced to partners. This movement has been particularly evident in logistics, where the provision of transport, warehousing, and inventory control is increasingly subcontracted to specialists or logistics partners.

Step 3. Global Innovation-networked manufacturing. Nowadays, the information technologies effects on enterprise activities, such as manufacture, management, communication, transport, etc. They make manufacturing activities possible parallel in different places of the world. Enterprise's headquarters and manufacturing plants can be in different countries, as well as R&D centers.

Networked manufacturing is based on internal network in enterprise, integrates subsystems such as design, searching information about supply and demand in the market, manufacturing new product, building synergies of operations and finally selling the products. The advantage is quick response to market demand.

To successfully implement it top manager should take concept of innovation seriously, which means that enterprise should practice innovating all around and anytime, and make sure all employees understands the idea. Top management need to understand that effects appears not immediately. Moreover, certain financial support is needed for such investments as e.g. hardware and software.

3.3 Pattern for All Innovators

Based on the concept of All Innovators in TIM, if an enterprise wants to be better, bigger, and more innovative, it must explore the potential of everyone and make every staff member an innovator. The following steps need to be done in this area:

Step 1. Employees volunteered suggestions about innovation. The only way is to let employees all around the enterprise to propose changes, mostly in the processes they are familiar with. Organization needs respect employee's innovating behavior and their self-value, also need support from synergy of strategy, culture and organization, etc.

Step 2. More emphasis on synergic innovation between employees. Enterprise should not only emphasize that employee participate in innovation individually, also should more emphasize the corporation between employees, complementary advantages and also the knowledge sharing. Now, most of enterprises still encourage employees innovating by using method which set suggestion boxes. These methods are not focused on team work, its lead to innovation efficiency still low. Therefore, the concept requires that enterprise should provide a good working environment and encourage employees participate in innovating on corporate level with other departments. At the same time, enterprises are required integrate of all employees' innovative resource by build working team, such as innovation team, cross-functionally team, etc.

Step 3. Adequate human management resources. First of all, enterprises should recruit high skill workers and also offer them adequate training programs to improve their skill and knowledge. Secondly, establishes adequate incentive mechanism for all employees. It must provide support for employees that maximizes creativity. Providing proper training programs enhances abilities in the area of innovating.

Good strategy for innovation not only emphasizes entrepreneurial innovation but also emphasizes all employees involved in innovation. Employees not only are the participators but also the executors of strategies of innovation.

4 A Comparison of Case Studies

A detailed description of several case studies presented in [25] enables to assess the impact of TIM implementation on several key performance indicators of companies. Within this paper, the Authors compare the level of sales' increase in the four analyzed companies – namely: Sunyard System Engineering Co., Ltd. (Sunyard), Hangzhou Focused Photonics Inc. (FPI), Hangzhou Honyar Electric Appliance Co., Ltd. (Honyar), SUPCON Group Co., Ltd. (SUPCON). The results of an initial comparative analysis of these case studies are presented in Fig. 4.

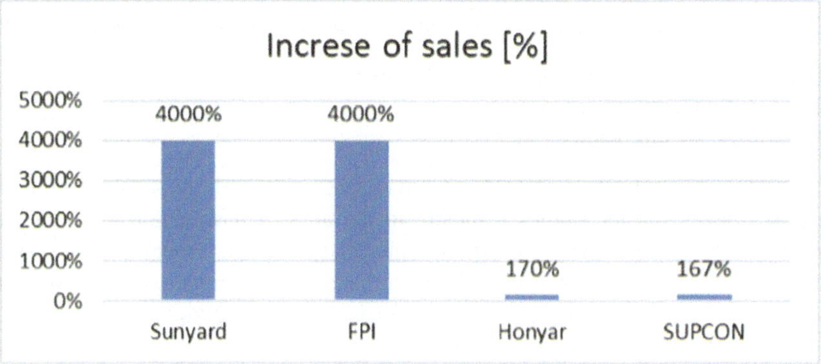

Fig. 4. Increse of sales after TIM implementation.

In Sunyard, which is a company developing R&D software and system integration products and operation services, the transition period including the Total Innovation Management implementation took approximately five years (2001–2005). Within this time, the increase of sales reached 4000%. The same result was achieved by FPI being a hi-tech enterprise, which i.a. develops, produces and sells the measurement instruments (transition period took 6 years). In terms of Honyar, which manufactures various types of building power distribution products, and SUPCON, which deals i.a. with development of innovative control and information technologies, the increase of sales was not that spectacular as in case of Sunyard and FPI, but still significant – respectively 170% and 167%. The transition period of both Honyar and SUPCON took approximately 3 years (2006–2008). The shorter time span of transition may justify the divergence between the achieved sales' increase of analyzed companies. It is also worth to mention another company – Uni-Power Guaranty in Zhejiang Province (UPG) – for which the annual net profit increased from 2006 to 2009 by 5742% due to the TIM implementation. The results of the initial comparative analysis show that implementation of Total Innovation Management is beneficial not only in terms of innovation performance, but also strictly in the monetary terms.

5 Summary

The manufacturing industry is the main source for employment in cities and towns of China and the major embodiment of international competitiveness of China. The major goal for China in the next 10 years is to become more innovative. The TIM concept has been presented to help Chinese enterprises to achieve it. China has no longer inhabited the road of traditional industrialization, it has to take the new path of industrialization driven by informatization. The new industrialization features include science and technology, good economic benefits and low energy consumption.

It was concluded that enterprises in China, to reach abovementioned goal, need to make deep changes in the management of innovations and these changes should be proceeded according to 3 pillars of the concept of Total Innovation Management:

- Improvements in technological innovations capacity
- Time & Space-based innovations
- Pattern for all innovators

The issues of application of TIM in China's large and medium-sized manufacturing enterprises needs further studies – mostly in the area of individualization and better fitting the ways and methods to specific enterprises. Further research may show the practical issues of implementations of TIM and analysis of limitations. Moreover, they may point out exact patterns for shaping the usage of the method.

References

1. National Bureau of Statistics of China Homepage. http://data.stats.gov.cn. Accessed 14 Nov 2017
2. Trading Economics. https://tradingeconomics.com/china/gdp-from-manufacturing. Accessed 2 Dec 2018
3. Bradsher, K.: 'Social Risk' Test Ordered by China for Big Projects, New York Times, 12 November 2012
4. Tianhu, S.: Correct understanding of China's manufacturing industry. China Mach. Electr. Ind., 25–28 (2014)
5. Schumpeter, J.: The Theory of Economic Development: An Inquiry into Profits, Capital, Credit, Interest, and the Business Cycle. Transaction Publishers, New Brunswick (1934)
6. Porter, M.E.: The Competitive Advantage of Nations. Free Press, New York (1990)
7. Drucker, P.: Innovation and Entrepreneurship. Routledge, London (2007)
8. Kotler, P.T., Armstrong, G.: Principles of Marketing. Pearson, Boston (2019)
9. Chen, J., Yin, X., Mei, L.: Holistic innovation: an emerging innovation paradigm. Int. J. Innov. Stud. **2**, 1–13 (2018)
10. Frishammar, J., Richtnér, A., Brattström, A., Magnusson, M., Björk, J.: Opportunities and challenges in the new innovation landscape: implications for innovation auditing and innovation management. Eur. Manag. J. (2018)
11. Haneda, S., Ito, K.: Organizational and human resource management and innovation: which management practices are linked to product and/or process innovation? Res. Policy **47**, 194–208 (2018)

12. Hernandez-Vivanco, A., Bernardo, M., Cruz-Cázares, C.: Sustainable innovation through management systems integration. J. Clean. Prod. **196**, 1176–1187 (2018)
13. Hernandez-Vivanco, A., Cruz-Cázares, C., Bernardo, M.: Openness and management systems integration: pursuing innovation benefits. J. Eng. Technol. Manag. **49**, 76–90 (2018)
14. Kusterer, D.J., Schmitz, P.W.: The management of innovation: experimental evidence. Games Econ. Behav. **104**, 706–725 (2017)
15. Lendel, V., Hittmar, S., Siantova, E.: Identification of the main levels in the management of innovation processes. Procedia Econ. Finance **26**, 1108–1112 (2015)
16. Lendel, V., Hittmár, Š., Siantová, E.: Management of innovation processes in company. Procedia Econ. Finance **23**, 861–866 (2015)
17. Lin, H.F., Su, J.Q., Higgins, A.: How dynamic capabilities affect adoption of management innovations. J. Bus. Res. **69**, 862–876 (2016)
18. Maier, D., Vadastreanu, A.M., Keppler, T., Eidenmuller, T., Maier, A.: Innovation as a part of an existing integrated management system. Procedia Econ. Finance **26**, 1060–1067 (2015)
19. Mir, M., Casadesús, M., Petnji, L.H.: The impact of standardized innovation management systems on innovation capability and business performance: an empirical study. J. Eng. Technol. Manag. **41**, 26–44 (2016)
20. Roehrich, J.K., Davies, A., Frederiksen, L., Sergeeeva, N.: Management innovation in complex products and systems: the case of integrated project teams. Ind. Mark. Manag. (2018)
21. Xu, Q., Chen, J., Xie, Z., Liu, J., Gang, Z., Wang, Y.: Total innovation management: a novel paradigm of innovation management in the 21st century. J. Technol. Transfer **32**, 9–25 (2007)
22. European Commission, Oslo Manual, Third Edition OECD, 47–52 (2005)
23. Yang, Z., Gang, Z., Xie, Z., Bao, G.: Total innovation management: a new emerging paradigm of innovation management, pp. 261–262. IEEE (2003)
24. Menke, M., Xu, Q., Gu, L.: An Analysis of the Universality, Flexibility, and Agility of Total Innovation Management: A Case Study of Hewlett–Packard, pp. 50–52. Springer Science & Business Media, LLC (2006)
25. Xu, Q., Chen, J., Shou, Y., Liu, J.: Leverage Innovation Capability. Application of Total Innovation Management in China's SMEs' Study. World Scientific Publishing, Singapore (2012)
26. Xu, Q., Xin-Ru L., Gang, Z.: The evolvement of innovation management - based on the opinion of innovation degeneration. J. China Univ. Geosci. (Soc. Sci. Ed.) **4**(5) (2004)
27. Xu, Q.R., Jia, F.H., Zhangshu, X. and Gang, Z.: Everyone involved in innovation on total innovation management. Stud. Sci. Sci. **21** (2003)
28. Abrami, R.M., Kirby, W.C., McFarlan, F.W.: Why China can't innovate. Harv. Bus. Rev. **92** (3), 107–111 (2014)
29. Ding, X., Li, J.: Incentives for Innovation in China. Building an Innovative Economy. Routledge, New York (2015)
30. Xu, Q.-R., Gu, L.-F.: Comparison of enterprises' TIM pattern between China and US. Stud. Sci. Sci. **22**(6), 658–662 (2004)
31. Lam, A.C.: Training and innovation, the Japanese way. Eng. Manag. J. **3**(6), 263–268 (2010)

Implementation of EPM Methodology in Production Plants

Andrzej Mróz(✉)

Department of Production Management and Logistics,
Faculty of Management and Production Engineering,
Lodz University of Technology, ul. Wolczanska 215, 90-924 Lodz, Poland
andrzej.mroz@p.lodz.pl, andrzej.mroz@ame.org.pl

Abstract. The aim of this article is to present the results after the implementation of Early Product Management methodology within the World Class Manufacturing system in a worldwide agriculture company. EPM is an advanced project management method, which is concentrated on new product launches and requires high interdisciplinary skills and team spirit cooperation. Its main goal is to bring the project launch cost savings and high performance of the product from the initial concept stage to the final design release. By utilization of the complex combination of project participants' experience, knowledge and motivation, the final product release should meet the project objectives, timing and cost score. In this paper, the specific stages of the project are described whereas the general KPIs and KAIs targets are presented. The review presents the case study based on the successful project implementation of an agriculture machine within years 2015–2018. The research methods include the elements of experiment, direct observation and system simulation, which naturally appear within the project flow. EPM methodology, as the universal approach, can be used in various project types and management applications. For this reason, it is especially recommended for project managers and project team members (e.g. manufacturing and design engineers, logistics and quality specialists).

Keywords: Early Product Management · World Class Manufacturing · Product management · Cost deployment · Advanced Manufacturing Engineering

1 Introduction

Nowadays, in times of a rapid technological development on one hand, the globalization and strong market competition on the other, modern firms must resolutely concentrate on innovation and loss reduction to survive. Furthermore, many branches of the economy are directly correlated with the development of these organizations on the national scale, e.g. developed sector of services. The meaningful importance of manufacturing processes optimization in European countries on micro and macroeconomy was emphasized in the Bruegell think-tank report [1].

A. Hamrol et al. (Eds.): *Advances in Manufacturing II* - Volume 3, LNME, pp. 114–143, 2019.
https://doi.org/10.1007/978-3-030-17269-5_9

Fig. 1. Modern logo of WCM.

WCM - World Class Manufacturing (Fig. 1) is a vivid concept of continuous improvement philosophy throughout manufacturing plants, which adopts multiple process inputs. These originate from specific areas - logistics, manufacturing, quality- and concentrate on the maximization of final profit with the corresponding respect to safety and environmental standards.

Originally, WCM focuses on continuous improvement of organizations and requires commitment of all company structures for final success [2]. Once implemented in the 80s by R. J. Schonberger, *WCM* has become a world-renowned ideology, which is especially common in automotive and machinery companies.

The sourcing of this methodology is linked to Toyota Production System (TPS) and its goal - highest quality at lowest cost and in shortest lead-time. This can be achieved by the optimization of process methods, standardized work, minimization of required resources, detection and elimination of abnormalities.

The literature presents the description of *WCM* as a system directed at the customer satisfaction by collaboration of Total Quality Control (TQC), Computer Integrated Manufacturing (CIM) and Just-in-Time production (JIT) [2]. Some researchers recognize WCM as a form of Total Quality Management (TQM), Total Productive Maintenance (TPM), Total Industrial Engineering (TIE), Just in Time (JiT)/Lean Management (LM) [3, 4]. Indeed, all these statements are correct, however too general and definitely must be analyzed more thoroughly.

First, specific numerology is closely connected with WCM methodology:

- 0 - zero:
  zero defects, zero waste, zero breakdown, zero inventory, zero accidents, etc.
- 7 - seven:
  seven steps of approaching improvement.
- 10 - ten:
  ten technical pillars, ten managerial pillars.

On its own, *WCM* is a very complex concept, and was originally composed of 10 technical and 10 managerial pillars. The 10 technical pillars formed a technical temple of WCM, which was presented in (Fig. 2). Each of them, treated separately represented specific area and specific targets.

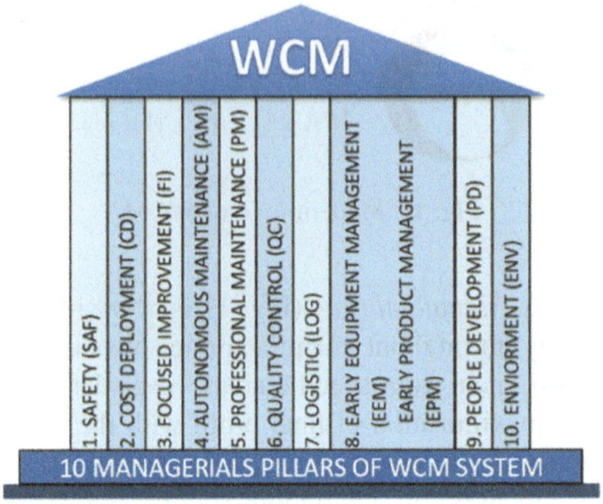

Fig. 2. Technical & managerial temple of WCM system [5].

By definition, the introduction of *WCM system* is not only change in the business excellence strategy only. More important is the change in a way of people's thinking, motivation and dedication to their work. It is based on a model area, which once optimized, in principle should expand on other areas, operative units, plant and all company sites (Fig. 3). At some point, it resembles Gemba, known from Lean Management.

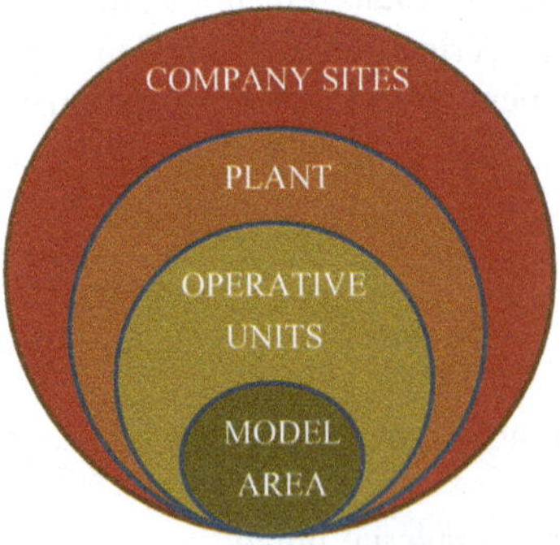

Fig. 3. Model area has expansive character within the organization structure.

The implementation of WCM system into a manufacturing plant is one of the most important strategic decisions for the company. However the key to success is not only the top management involvement, but also popularization of this concept among every worker within the organization.

The author encourages the reader to investigate the literature, in case of a deeper interest in *WCM* methodology - its main assumptions, implementation and methods of verification status. As a reference, positions [6–13] are highlighted. Some articles related to WCM are presented below.

Practical implementation of WCM methodology in an automotive company was described by Stanek et al. in their paper [4]. The authors indicated general tools, which are ubiquitously used in production system optimization:

- S-TAG (Safety Tag).
- AM-TAG (Autonomous Maintenance Tag).
- POKA-YOKE.
- 5S.
- 5Why.
- 4M.
- OPL (One Point Lesson).
- EWO (Emergency Work Order).

In conclusion, the originators excerpted benefits from WCM system implementation in production plant, which are:

- Number of breakdowns reduction - by focusing on conceptual preventions.
- Product quality improvement - by recognizing and eliminating sources of defects.
- Safe working conditions - by disposal of risk factors.
- Increase productivity rates - by process optimization.

Sagwan and Digalwar conducted a study on the evaluation of the WCM systems implementation based on Indian automotive industry in an article [14]. The authors used PVA (Performance Value Analysis) model in order to determine critical success factors and performance variables used for successful implementation of WCM. Most of them include:

- Top management commitment,
- Knowledge management,
- Employee training,
- Innovation and technology,
- Vendor management,
- Production planning and control,
- Quality,
- Flexibility,
- Cost.

The general assumptions of WCM system and a case study in logistics and customer service area were analyzed by Felice and Petrillo in their work [15]. The authors identified losses in the production line, which were covered in Pareto diagram. Hence, inventory reduction and increase of customer satisfaction was achieved by synchronization of production with sales and the reduction of material handling. The conducted study led to implemented solutions in logistics system resulting in firm's economic score.

More detailed overview of WCM system was performed by the same authors in their work [16]. The paper was separated into three main sections: starting from literature overview of WCM history, a list of general tools and an industrial case study. In practical part of their publication, the authors concentrated on WO (Workplace Organization) pillar activities supported by CD (Cost Deployment) and LOG

(Logistics) inputs. Following PDCA approach and by using main WCM tools - Muda/NVAA analysis (job stratification, spaghetti chart, Kaizen standard), material classification and ergonomic simulation (Jack software), significant productivity growth and ergonomics improvement were achieved.

Although the literature on WCM system in general is widely available, the author finds essential need to present each specific methodology in more detail. In this paper, one of the most important technical pillars in WCM system - Early Product Management is analyzed.

2 Early Product Management (EPM)

Originally, Early Product Management (EPM) and Early Equipment Management (EEM) formed a complete pillar of WCM temple. The company experience shows that though in some areas they infiltrate each other, due to their elaborate performance, must be treated separately. Hence, in recent elaborations, modernized WCM temple is presented as below (Fig. 4).

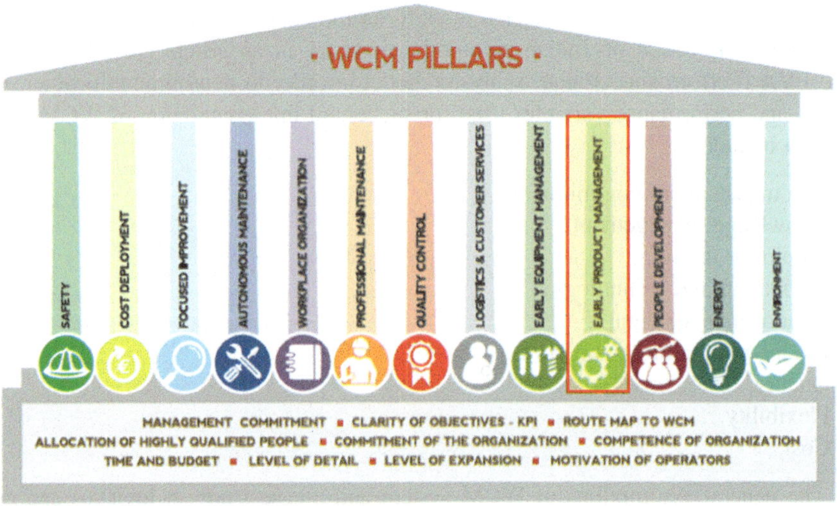

Fig. 4. Modernized technical temple of WCM and EPM pillar.

Early Product Management is a philosophy concentrated on new product- its implementation, from concept-design stage to first line run and delivery to customer. It covers multiple aspects of product lifecycle, however especially quality, cost deployment, logistics and workplace organization pillars are involved. The main goals of EPM can be noted by Q, C, D:

- *Q* - *QUALITY* of the new product should meet all targets.
- *C* - *COST* of the new product tracked at every project stage milestone, not exceeding project target.
- *D* - *DELIVERY* date to customer as scheduled in the project plan.

As a reference, EPM competence cross section was presented in (Fig. 5).

LEGEND	
SYMBOL	PILLAR NAME
SAF	SAFETY
CD	COST DEPLOYMENT
FI	FOCUSED IMPROVEMENT
QC	QUALITY CONTROL
LOG	LOGISTICS
EEM	EARLY EQUIPMENT MANAGEMENT
EPM	EARLY PRODUCT MANAGEMENT
ENV	ENVIORMENT
WO	WORKPLACE ORGANIZATION

Fig. 5. EPM methodology as the main drive among other WCM technical pillars.

EPM pillar team consists of a leader (usually Launch Manager of a new product) and competent members e.g. Advanced Manufacturing Engineer, Design Engineer, Quality Specialist, Safety Inspector, Logistics Engineer, Value Analyst. The purpose of this multidisciplinary cooperation is to analyze the influence of a new product release on the specific area/process and if needed - suggest technical changes to design engineering and board management.

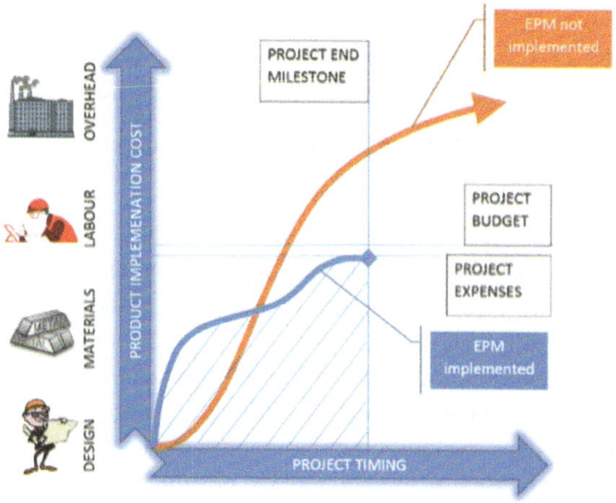

Fig. 6. Project timing versus product implementation cost with and without EPM methodology.

The main aim of EPM pillar activities is to reduce cost and time of implementation of new products [5]. To visualize this, (Fig. 6) presents cost distribution of the new launches.

In situation, when EPM is not implemented (i.e. design is not thoroughly validated), firms generate waste on scrap material, extra labor due to reworks and general overhead. In these circumstances, effective implementation of the product becomes doubtful. This concerns not only project timing milestones but also overall investments.

Furthermore, a delayed delivery of a product to the market can result in market share decline and far-reaching financial losses, which can be difficult to recover in future. Success of the project in this case is a matter of favorable coincidence and contingency.

EPM methodology in line with GPD procedure requires specific activities and their recurring verification in seven steps. This approach allows to control the budget expenses and planning of the activities in order to meet the project milestones.

It is estimated, that proper design analysis and prototyping prior to line release can bring approximately 50% cost saving at project release stage. This can be achieved by elimination of past issues, bringing technological innovation and design optimization, simultaneously meeting technical, ergonomic or safety requirements.

Taking into consideration the cost share in project expenses, the visible differences are also pronounced. The conceptual graphs are presented in (Fig. 7).

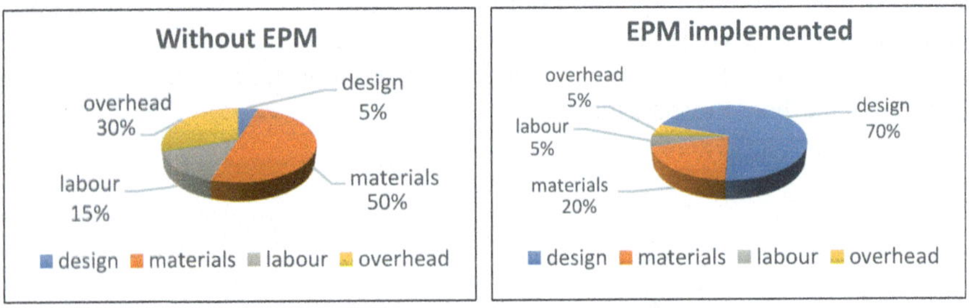

Fig. 7. New product launch cost distribution.

Projects, which are not following EPM methodology, generate major expenses during first series production on the line caused by longer assembly time including explicit NVAAs, quality issues and scrap of material. Hence, high overhead cost is observed.

In line with EPM methodology, the strongest emphasis is on product optimization at design and prototyping stage. Based on past knowledge and proactive approach, many defective aspects are excluded herein.

Although, more expenses are generated by design engineering, prototyping and testing (hence higher design and materials share in total), significant lower implementation cost on the line is achieved. In the light of the above, the main rule of EPM can be formulated:

EPM methodology in project management within WCM system involves all activities, with particular emphasis on proactive and knowledge management approach, in order to deliver products meeting the quality, cost and functional (technical) requirements within project schedule.

2.1 Main EPM Nomenclature

HPV - hours per vehicle
Includes manufacturing operation time to produce a single part, system key structure or a complete product.

DFM - Design for Manufacturing
Method of manufacturing process optimization or unification depending on available technology, equipment, resources, materials, etc.

DFA - Design for Assembly
Method of assembly process improvement by elimination of NVAAs, access issues, secondary operations, introduction of Poka-Yoke and Error-Proof solutions, etc.

EPM checklist
Collection of questions/checkpoints included in EPM-info cards, which should be verified during new product launches.

B/C - Benefit to Cost ratio
Coefficient recognized as a quotient of benefits to investments, calculated in specific return period. B/C coefficient is dependent on the reduced loss value, running cost, amount of investments and corresponding return period:

- 1 year - minor or moderate investments required, calculated according to (1).

$$\frac{B}{C} = \frac{Benefit}{Cost} = \frac{Reduced\,loss - Running\,cost}{Amount\,of\,investment} \tag{1}$$

- 3 years - major investments required e.g. advanced equipment, infrastructure modification, building construction, etc., calculated according to (2).

$$\frac{B}{C} = \frac{Benefit}{Cost} = \frac{3x(Reduced\,loss - Running\,cost)}{Amount\,of\,investment} \tag{2}$$

CD - Cost Deployment
One of the main technical pillars in WCM temple responsible for identification of production losses and cost reduction. For more details, refer to positions [17–19].

CA - Cost avoidance

Includes all theoretical cost, which could be generated unless antecedent solution was implemented. Because of its "imaginary" nature, plays secondary role in the plant financial analysis.

EPM Info - EPM information

Official document prepared by a competent personnel, including recognized issues/problems or improvement suggestions. The document is presented to design engineering or product management board. It highlights HPV and material cost impact analysis, before and after the design change. As a driving factor, coefficient B/C is calculated. An example of EPM Info Template is presented in (Fig. 8).

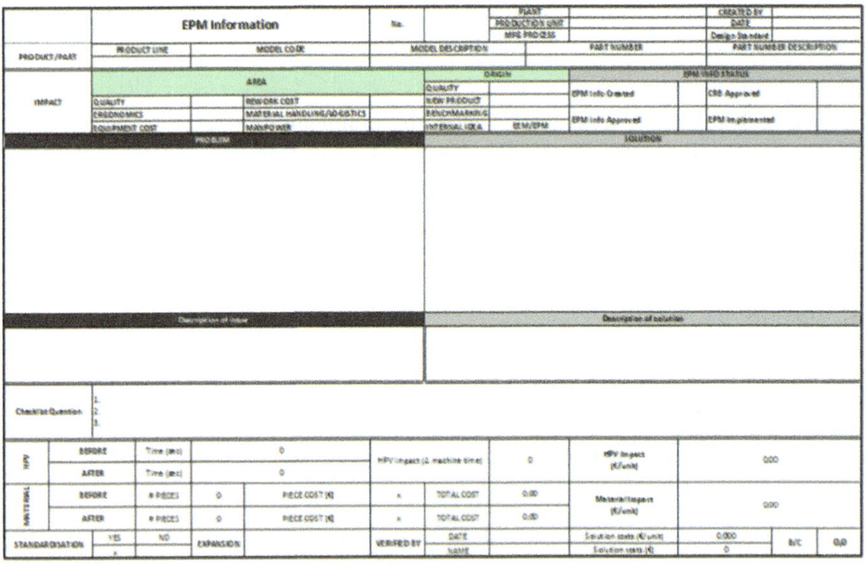

Fig. 8. EPM info template.

EPV test - Engineering Product Validation Test

A test of product performed by the engineering department at early design & prototyping stage. The aim of validation is to check the "physical" product with regard to technical efficiency, functionality and design coherence.

CQA - Customer Quality Audit

Internal audit performed in-house by certified inspectors in terms of product performance from customer's point of view - technical parameters, assembly completeness and visual aspect.

DMU - Digital Mock-up

Computer software (generally CAD) used for model validation at earliest product design stage. An example of design analysis in Teamcenter Visualization Mockup was presented in (Fig. 9).

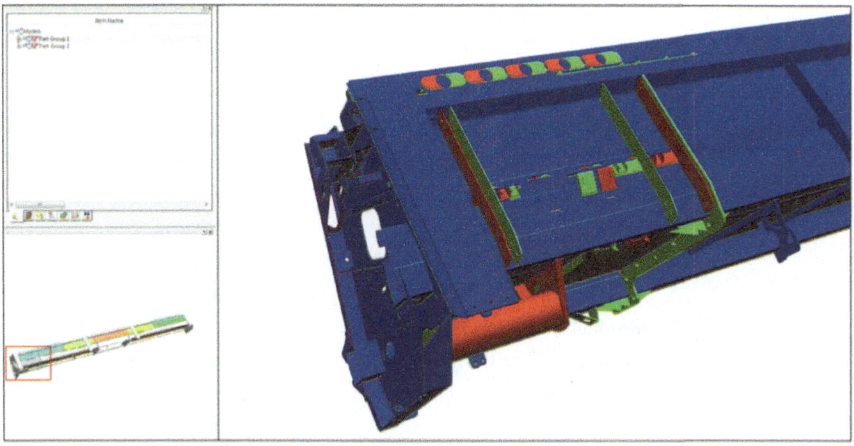

Fig. 9. Header design update under detailed study.

2.2 30 EPM Tools

Early Product Management utilizes 30 tools, which are divided by the phase character (Table 1):

Table 1. List of EPM tools.

No.	Description	Phase
1.	5G	Reactive
2.	5Why1How/5Why	Reactive
3.	7 WCM Tools	Reactive
4.	7 Steps of EPM	Reactive
5.	GPD Process	Reactive
6.	Project selection methodology for CD	Reactive
7.	EPM Cost Deployment	Reactive
8.	EPM Info template	Reactive
9.	EPM info process& database structure	Reactive
10.	Change Review Board (CRB)	Reactive
11.	DR System	Preventive
12.	Design Standards	Preventive
13.	Checklist process for new products	Preventive
14.	Front loading	Preventive
15.	7 Steps of QM	Preventive
16.	Preventive QA Matrix	Preventive
17.	Safety risk assessment	Preventive
18.	DFMEA methodology	Preventive

(continued)

Table 1. *(continued)*

No.	Description	Phase
19.	3M Analysis & Optimization	Preventive
20.	Benchmarking	Preventive
21.	Part/Tool interference analysis	Proactive
22.	Virtual manufacturing	Proactive
23.	WPI	Proactive
24.	Proactive QA matrix	Proactive
25.	8 stages of incoming material	Proactive
26.	Product and Process FMEA	Proactive
27.	Intrinsic safety design	Proactive
28.	Maintainability design	Proactive
29.	QFD	Proactive
30.	Concurrent engineering	Proactive

The sites, which have implemented EPM methodology at high level, should exhibit mostly proactive approach (the graphical interpretation was presented in (Fig. 10). Only this guarantees meeting scheduled program objectives and smooth delivery of a product to the market.

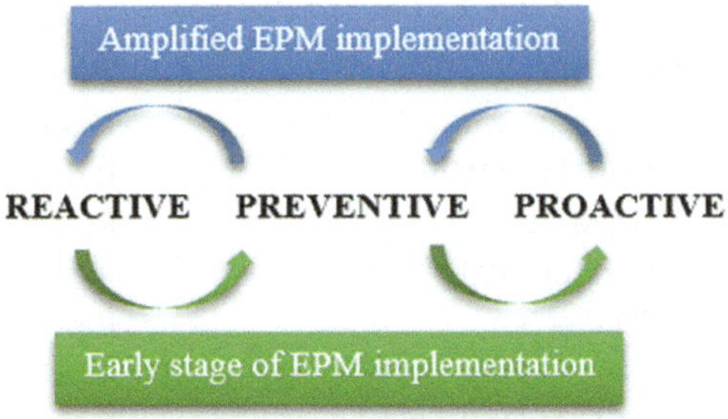

Fig. 10. EPM approach for new launches.

The reactive EPM tools are based on past & current product experience, which in WCM system create *Knowledge Management*. By their usage, KM is expanded onto other products and involved processes, which plays vital role in the development of organizations and their strong market position [20–24].

2.3 7 Steps of EPM

All technical pillars in WCM consist of seven steps, which are determined by product development stages. These are defined by the Global Product Development (GPD) procedure, which includes best practices and practical experience from past product implementations and improving projects. Accordingly, Early Product Management consists of seven steps, which are related to new product launch (see Table 2).

Table 2. 7 Steps of EPM

Step	Name	Detailed description
0.	Knowledge management	Includes all knowledge collected in steps 1–7
1.	Planning	Planning, Mission, Targets, Team organization
2.	Develop concept	Line and cycle Macro-planning, Product concept alignment
3.	Design feasibility	Workstation concept, Pillars Proactive Analysis (1st)
4.	Design optimization	Workstation detailed design, Pillars Proactive Analysis (2nd)
5.	Process implementation	Workstation Off-line Test, Workstation installation
6.	Process verification	Production Trial, Workstation Certification
7.	Optimization & Startup	Production Startup, Optimization

The correlation between EPM steps and GPD program implementation flow was presented in (Fig. 11).

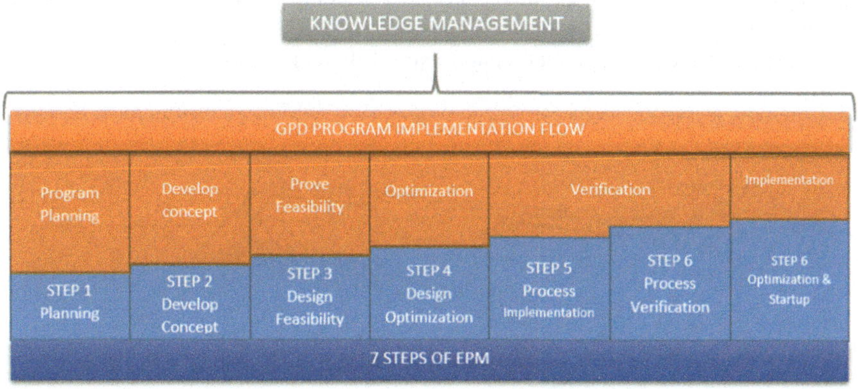

Fig. 11. A Correlation between 7 Steps of EPM and GPD Procedure.

2.4 EPM KPIs and KAIs

Successful project implementation must be accompanied by constant monitoring of KPI/KAI indexes. In terms of EPM approach, these are closely correlated with cost and quality management. Therefore, main KPIs include:

- HPV index.
- Material cost.
- CQA (Customer Quality Audit) score.
- FTQ (First Time Quality) index.
- Approved EPM Info documents.
- Number of design modifications after Design Release stage.

The corresponding KAIs:

- Number of created EPM info cards.
- Number of design remarks and comments.
- Number of checkpoints added to EPM checklist.
- Number of design reviews and following design changes.

3 Description of the Project

A review presented in this paragraph is based on the author's professional experience as Advanced Manufacturing Engineer and leader of EPM project team in Agriculture Company. As an example, a project of a cornheader row unit is analyzed. Combine cornheader is presented in (Fig. 12) and row unit in (Fig. 13).

The market study conducted by the company revealed a need from customers for a new product - narrow row unit. Based on the research, main project targets were set:

(1) Weight reduction versus currently offered product.
(2) Depletion of the frame geometry and size of components to enable narrow corn spacing.
(3) The cost of the new product cannot exceed the current one.

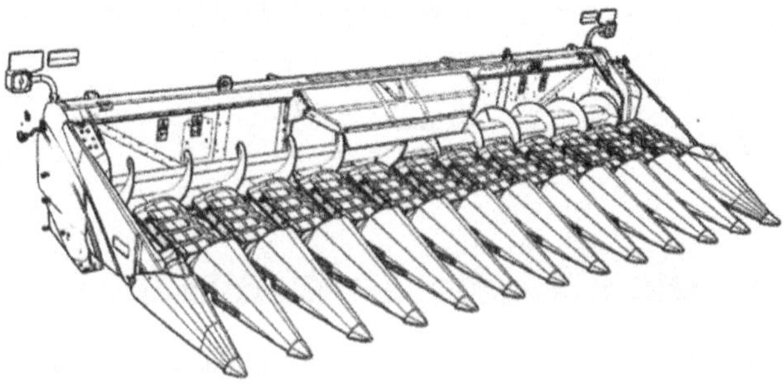

Fig. 12. Combine cornheader.

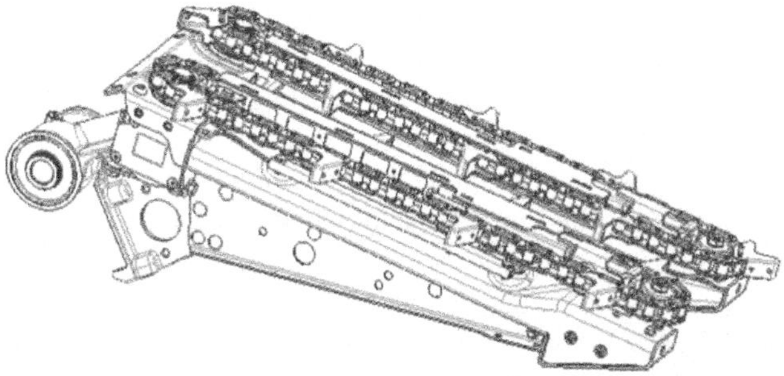

Fig. 13. Narrow row unit.

3.1 EPM Methodology Flow

STEP 0: KNOWLEDGE MANAGEMENT

As long as the production is running and new implementations are taking place, the company is in continuous process of learning and gaining experience. In WCM system, this process is recognized as Knowledge Management.

It is essential, that all these activities are properly documented so that the related information could be utilized in new product launches. In this situation, EPM Info card plays a special role. Its main sources come from:

- DMU analysis.
- Drawing reviews and comments.
- CQA reports.
- Quality matrix.
- Nonconformities reports.
- Kaizen network.
- Benchmarking.
- DFMEA & PFMEA sessions.
- Product validation & testing reports.
- Company know-how, Best-practices, patents.

EPM Info cards are formal documents, which include:

- Specific impact area (pillar affiliation).
- Detailed description of a problem and solution.
- Specific questions, which are collected in EPM checklist database.
- Cost avoidance per unit analysis (HPV, Material saving).
- B/C factor.

An example of EPM Info card was presented in (Fig. 14). The problem, which was revealed at a prototyping stage of a product, concerned time-consuming usage of eyebolts for transport purposes. A proposal of adding technological holes, which had been suggested to design engineering, was accepted. As a result, a dedicated load leveler was designed and implemented on the workstation. The transportation time was reduced by 80% which resulted in B/C factor equal to 12,4.

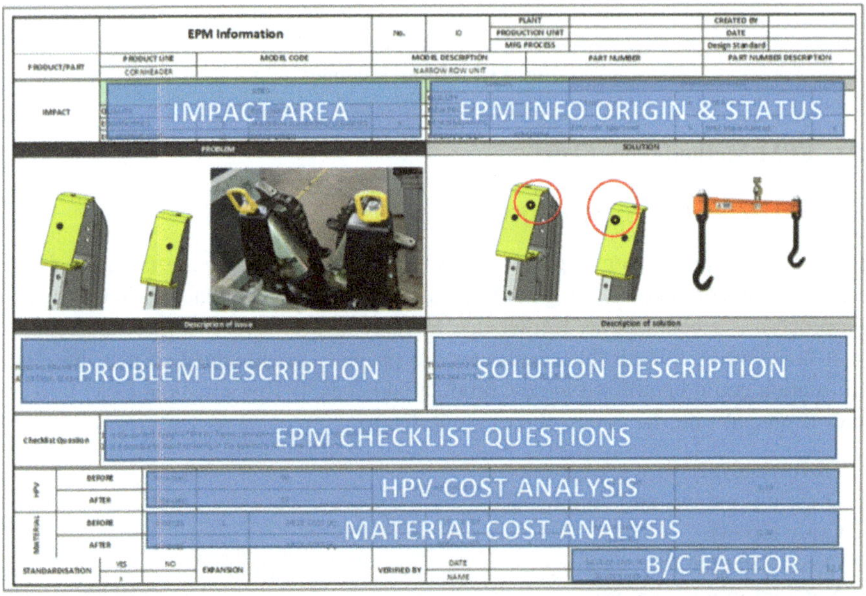

Fig. 14. An example of EPM Info card.

The cards are collected in EPM Info database and are easily accessible to every worker in organization. Number of EPM Info cards and corresponding trend, track the activity of EPM Pillar during specific time range. EPM Info trend graph that was created based on EPM Info database was presented in (Fig. 15), whereas its distribution per pillar in (Fig. 16) and per sourcing in (Fig. 17). In a similar way, number of added checkpoints in EPM checklist database is tracked.

At every new project launch, the team should verify the existing newly added checkpoints and cards, in order to avoid past errors repetition and effective financial loss. This input is essential for proper DFMEA analysis, which indicates optimal routes for designers, advanced manufacturing engineers, and plays a driving role in further project implementation.

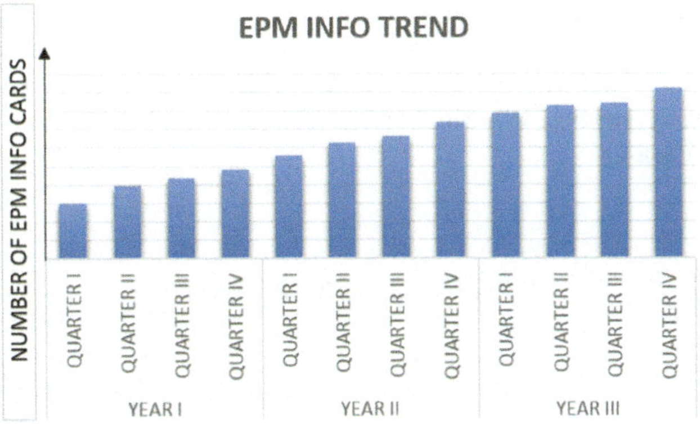

Fig. 15. EPM Info trend.

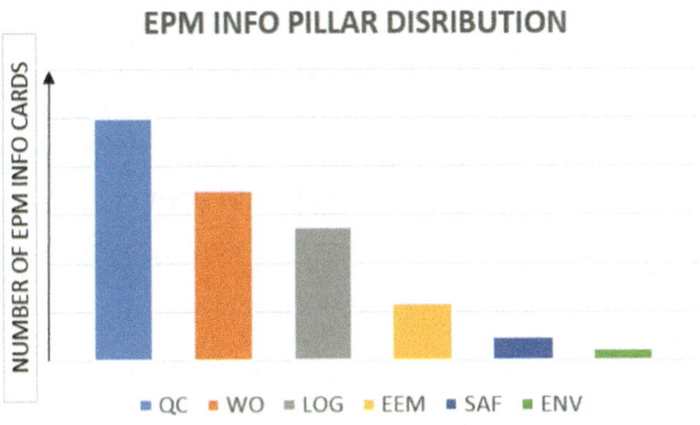

Fig. 16. EPM Info card graph - distribution by pillar.

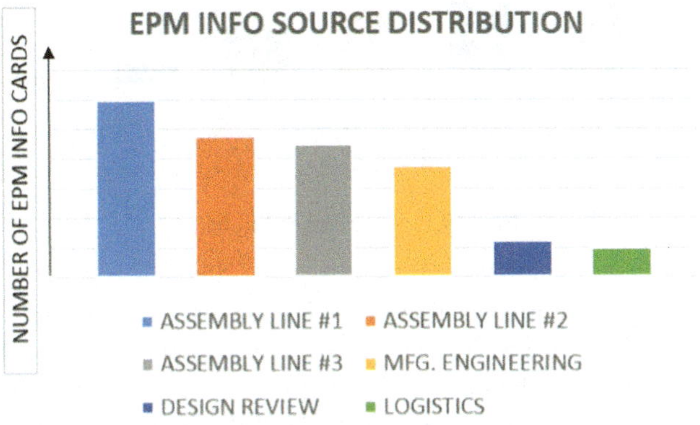

Fig. 17. EPM Info card graph - distribution by sourcing.

Another important aspect in EPM methodology and KAI at step 0 is a number of checked drawings and added comments. It can be presented as split between reviewed drawings without comments and with comments (Fig. 18) or more detailed, depending on the comment type (Fig. 19). Similarly, to EPM Info cards, comments are collected in common database.

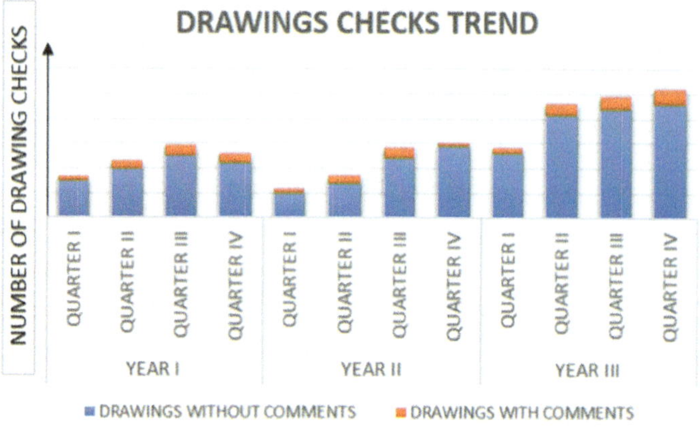

Fig. 18. Drawings checks trend graph.

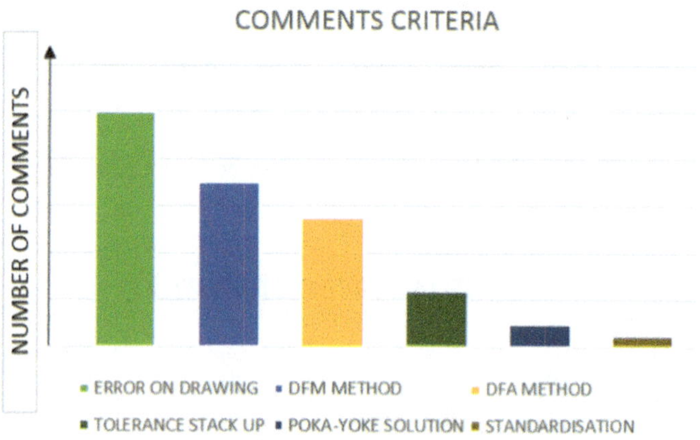

Fig. 19. Comments split chart.

Cost aspect plays a significant role in EPM approach. In most cases, it represents *cost avoidance* type and means hypothetical loss elusion in either project launch budget (e.g. launch hours reduction by precise activity planning) or product life cycle (e.g. part cost reduction by design modification). As a result of the above, EPM info cards are additionally evaluated by their average B/C ratio (Fig. 20) or cost avoidance per product (Fig. 21).

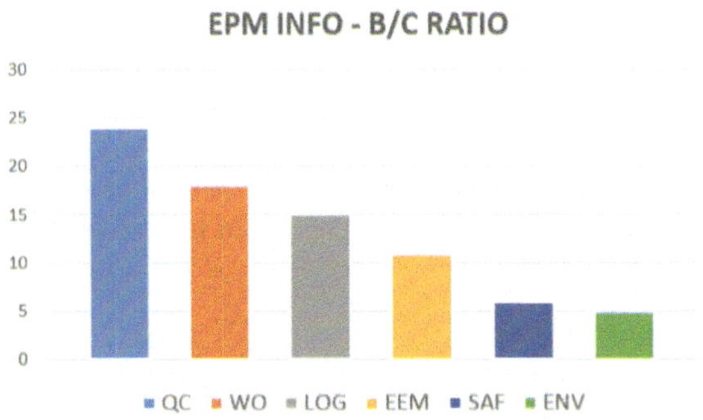

Fig. 20. EPM Info B/C ratio.

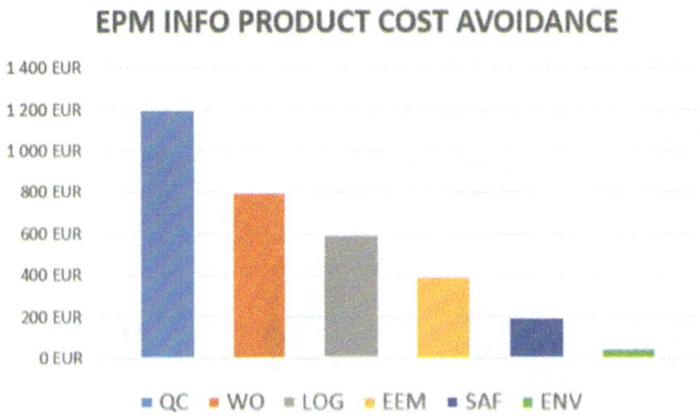

Fig. 21. EPM Info product cost avoidance.

Knowledge management is a continuous process, which is based on the gathered experience and exhibits rather reactive and preventive approach to product management. The proper action listing and standardization at Step 0 are essential for effective KM use and create the fundamental basis for EPM development into proactive approach.

STEP 1: PROGRAM PLANNING

At this stage of EPM Methodology, initial data on specific product was collected. In this example, the narrow row unit was the developed concept of the existing product (since 2012). The average yearly volumes exceed 10000 units, which in agriculture industry can be described in terms of serial production.

Due to its continuous and repeatable production character, quality of the current product has been well tracked. The units have been submitted to Customer Quality Audits (CQA) and regularly controlled in terms of identified defects. All nonconformities have been collected in quality matrix whereas improvements in Kaizen and EPM info documents.

The areas for improvement are also indicated by the current product cost deployment matrix (C Matrix). This is a data collection, which utilizes knowledge about identified losses on the current product and stratifies them in terms of the project stage.

According to the matrix presented in (Fig. 22), Loss A, Loss B and Loss C generated significant waste in product launch budget, and should be avoided in future implementations. In the titled project, the three main losses were identified:

- Design complexity.
- Late design release milestone.
- Tooling issue.

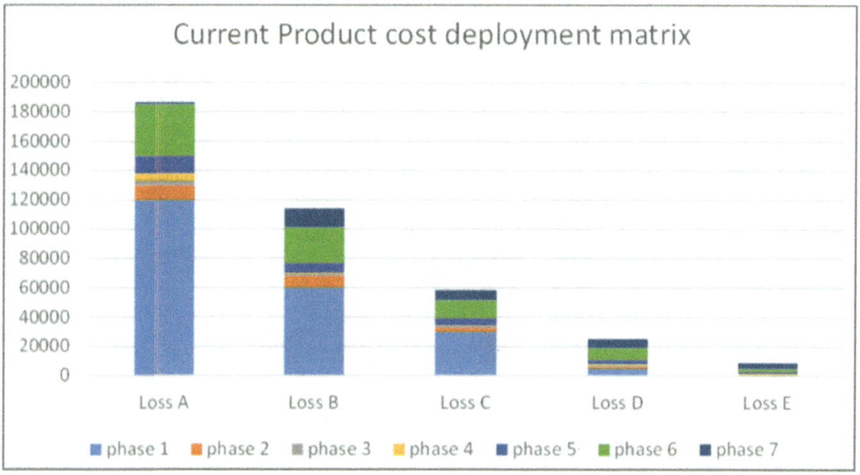

Fig. 22. Example of cost deployment matrix.

The above tools had been initially used in *Front-loading* of activities. Based on Knowledge Management approach, critical items have been identified at the beginning of the new project. These can be presented as below:

- Quality of the frame weldments must perform complete penetration between joined components without visible cracks.
- Weld spatters must be reduced.
- Frame geometry must conform to technical documentation.
- Number of welded and assembled components must be reduced.
- Automation of assembly processes is required.
- Facilitation of Poka-Yoke solutions.
- NVAA elimination.

Some of the identified issues were presented in (Fig. 23).

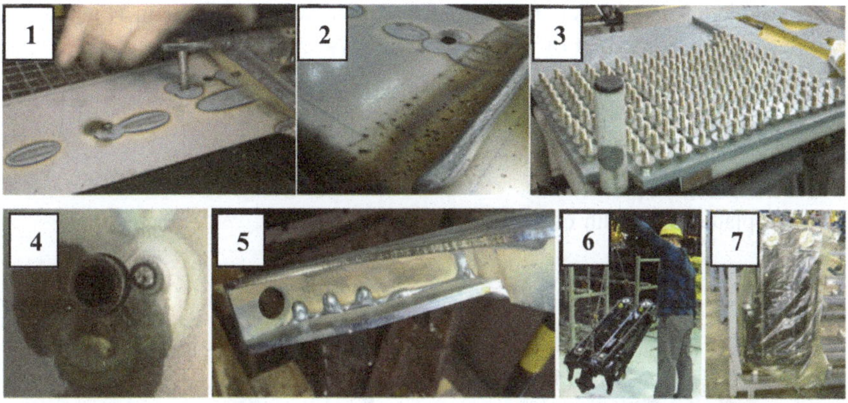

Fig. 23. Examples of identified issues on current product (Frontloading process): (1) Rethreading of welded nuts due to the spatters. (2) Large number of the weld spatters. (3) Excessive NVAA due to the preassembly of bolt and two washers. (4) Misposition of the weld nuts. (5) Deformation of parts and weld defects. (6) Problematic handling of the parts. (7) Problematic anti-corrosion foil wrapping.

Concurrent to past issues analysis and to facilitate the best in class new design, a benchmarking study had been conducted (Fig. 24). A group of design engineers analyzed several products of competitive brands and specified technical requirements and engineering solutions for the new project.

Fig. 24. Benchmarking of competitive products (external) and company best practice (internal).

The above sources had been used in the new concept design. Due to its strictly "technical" character, close cooperation between manufacturing and design engineering was established. The project team members have played active roles in the information exchange and the conceptual work. The effect of this collaboration resulted in initial product concept breakthrough.

A new model and its detailed virtual teardown was presented in (Fig. 25). The initial design was composed of several system keys:

(1) Separated welded frame - consisting of LH&RH legs, pedestals and top plate.
(2) Gearbox.
(3) Gathering chains & tensioning system.
(4) Pinching Stalk rolls.
(5) Deck-plates and width control system.

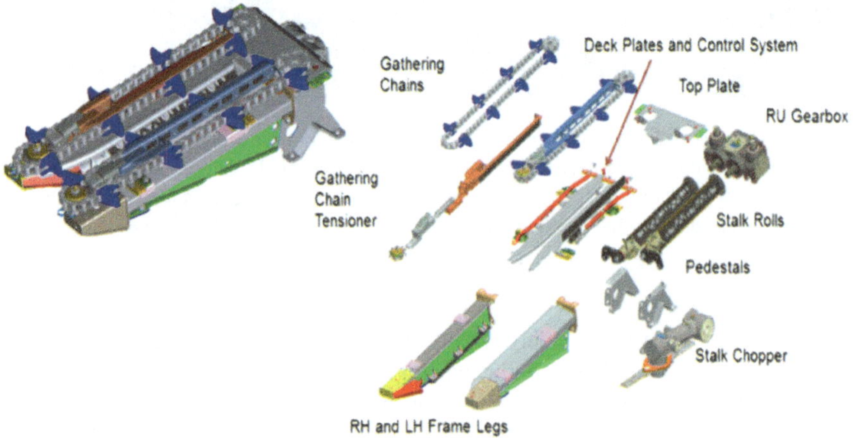

Fig. 25. Product concept and virtual teardown of a row unit.

Based on initial documentation and product BOM, the components were grouped by their sourcing. The visualization of this share was presented in (Fig. 26).

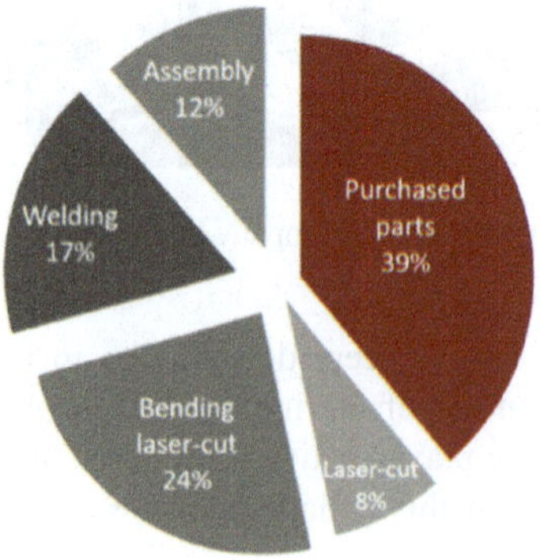

In-house processes included:	**63**
Laser-cut	8
Bending, laser	25
Welding	18
Assembly	12
Outsourcing:	**40**
Machining	24
Forging	5
Plastic Forming	11
Total	**103**

Fig. 26. Components' grouping by sourcing in concept design.

STEP 2: DEVELOP CONCEPT

At this stage of EPM Methodology, DFA, DFM and virtual manufacturing techniques are used for model optimization. An example of a model change resulting from Design for Assembly approach was presented in (Fig. 27).

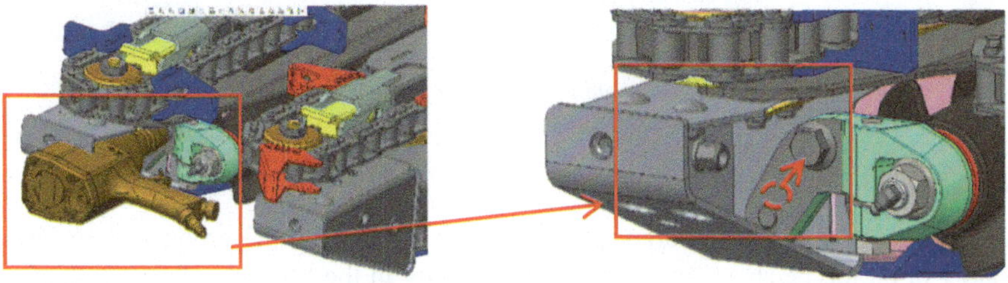

Fig. 27. Tooling access issue.

The indicated bolt's location was problematic due to a limited tooling access. In case of prototype builds, application of required torque would be impossible. The model was studied with the virtual wrench resulting in the bolt position modification.

Another modification in the design concerned deck-plate tensioning system. In order to reduce number of components, rotating arms replaced the sliding system (visualized in Fig. 28). In this way, seven components were eliminated, which brought reduction in material cost and HPV index. The B/C factor was equal to 35.

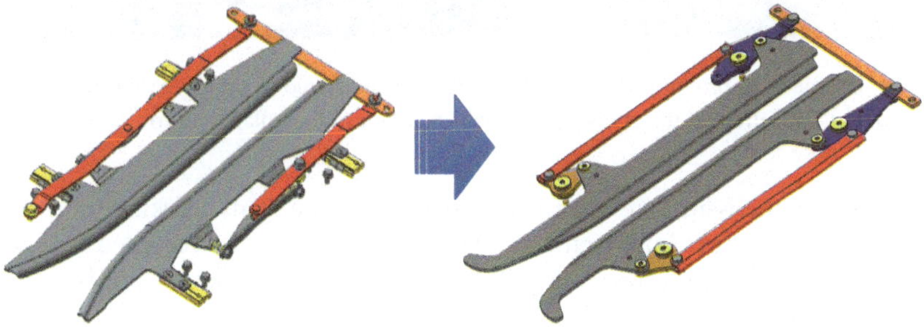

Fig. 28. Deckplate tensioning system modification.

In order to avoid large NVAA caused by the preassembly of components (see Fig. 29), a flange bolt with patch lock replaced a set of a bolt with two washers. In this way, ineffective handling of parts was eliminated which resulted in HPV equal to −100% and material cost reduction by 8%. The overall B/C coefficient was equal to 46.

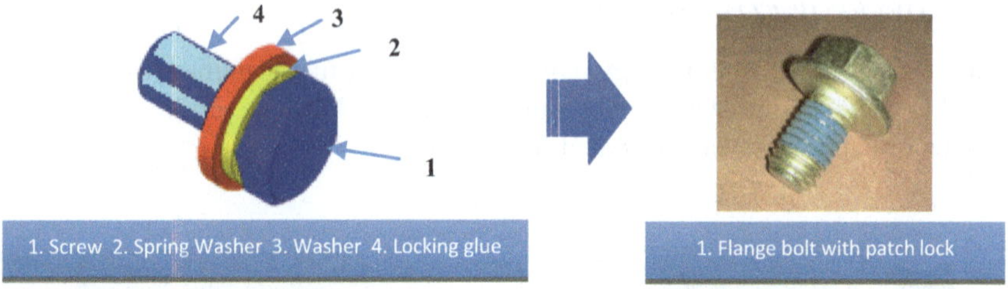

Fig. 29. Implementation of flange bolts with patch lock.

STEP 3: DESIGN FEASIBILITY

At this step of EPM Methodology, a prototype is built by the product engineering, usually in prototype shop. This is the crucial phase in the lifecycle of the product and first physical verification if all technical assumptions are achievable. Manufacturing Engineering role in prototyping is very essential, as prior to floor production, there is a unique chance to verify the product design in terms of applied technology, process flow and essential equipment. The major activities related to narrow row unit prototype builds were presented in (Fig. 30).

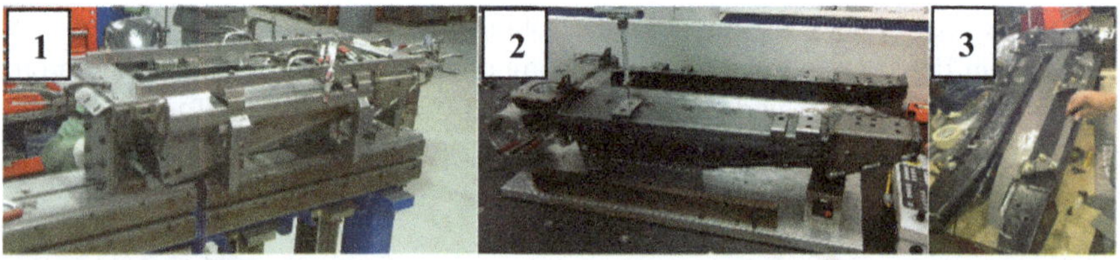

Fig. 30. Engineering prototype build: (1) Welding of the frame in prototype fixture. (2) CMM geometrical verification of the frame. (3) A test of rotating deck-plate control system.

As soon as the issues are identified, the project team starts working on their resolution and brings new improvements. All these initiatives are recorded by specific Kaizen projects and corresponding EPM-Info cards. Below, some further improvements are presented.

By using virtual manufacturing approach, number of frame-welded components was reduced as long as the welds were shortened. In (Fig. 31) a new frame model composed of deep-drawn parts was presented. Despite high level of tooling investment, this change resulted in HPV and material cost reduction by almost 35% and B/C factor equal to 6.

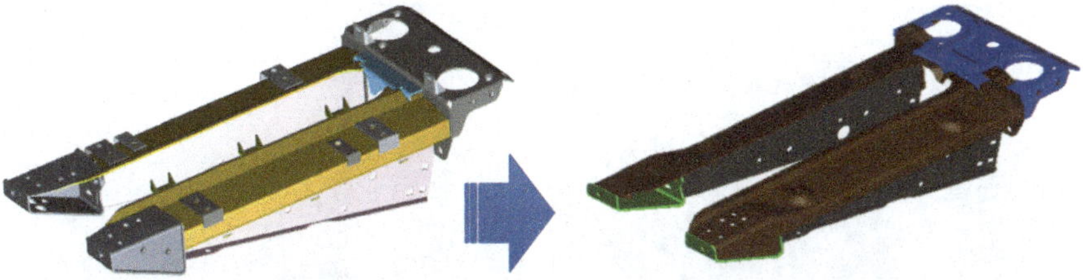

Fig. 31. Reduction of welded components by implementation of deepdrawn legs and top plate.

Use of weld-nuts generated numerous quality issues including spatters after welding or paint leaks on thread surface. In order to repair these defects, all nuts had to be rethreaded (Fig. 32). To eliminate these past problems (Fig. 23 issues 1 and 4), implementation of rivet nuts was suggested. Prior to the insert implementation, a cost analysis was done indicating HPV equal to −100%, material saving −59% and B/C index equal to 10.

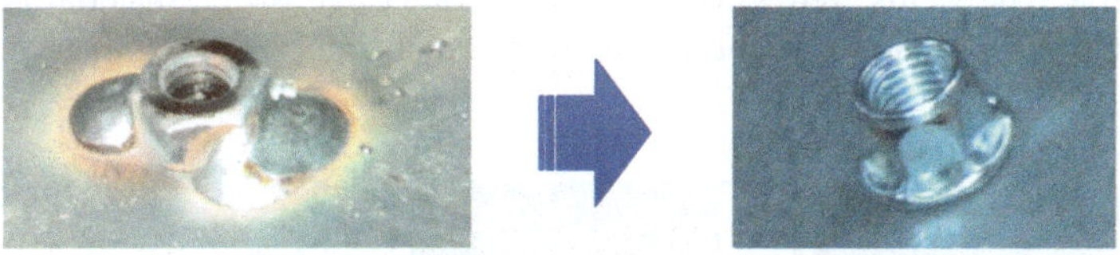

Fig. 32. Steel inserts replaced the welded nuts.

STEP 4: DESIGN OPTIMIZATION

In fourth step of EPM Methodology, development build of a product is performed. The main responsible sites are still product engineering and manufacturing, however due to high maturity of design and processes, other pillars (especially quality, logistics, and safety) should take active part in this activity. This collaboration should expand during DFMEA, PFMEA, 3M, Risk assessment, Inventory Flow sessions.

In the titled project, the conducted activities led to the following conclusions - welding quality and the efficiency of the product's test-run station must be improved. To meet these objectives, in cooperation with EEM and WO Pillar the solutions presented below were proposed:

- Welding quality - introduction of low spatter technology CMT (Cold Metal Transfer) - issues 1,2 and 5 in (Fig. 23).
- Test-run station - a run- through concept with integrated assembly trolleys which reduces handling-issues.

The main advantages of CMT welding technology are its low-spatter character (even 99% less than in standard short-circuit welding), high speed (+50% versus standard MAG technology) and low heat emission (approx. −50% less than in short circuit). The comparison between MAG and CMT methods was presented in (Fig. 33).

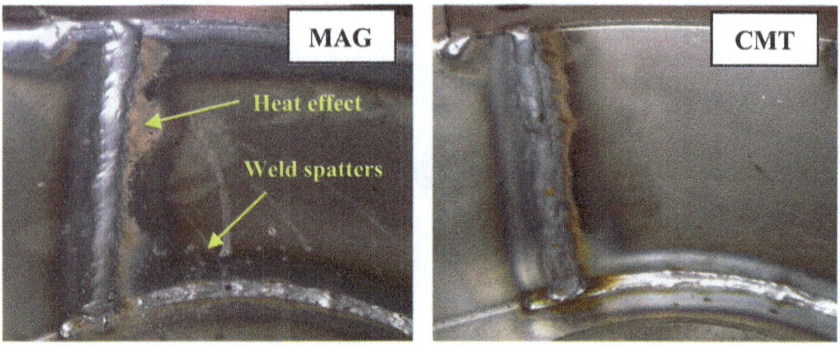

Fig. 33. Comparison between MAG and CMT welds.

The above CMT characteristics were definitely meeting the requirements for the product manufacturing however, the material thickness of the frame was greater than the maximum value suggested by the tooling supplier. For this reason, technical tests were performed, in which proper welding parameters and proper protective gas type were chosen. The welds were checked in terms of visual aspect, material penetration and external cracks (Fig. 34). The results were satisfactory.

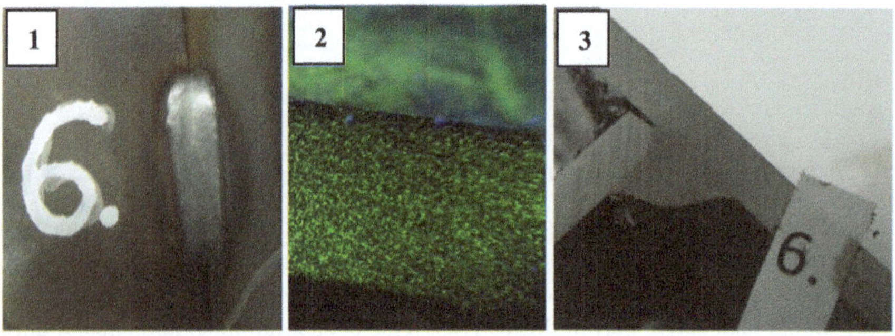

Fig. 34. Weld quality control: (1) External look of the weld- no spatters and porosity. (2) Marker crack verification - no crack detected. (3) Penetration test - material fully penetrated.

Another important case at this stage of the project was the design of new test-run station and resulting assembly line layout. As a starting point, current product assembly line was analyzed.

Below in (Fig. 35) a problematic area was marked. Between the last assembly station (5) and the test run station (6), the products were transported using hoist, which was a time consuming operation. After the test, the row units were manually rotated using dedicated tooling and placed into transport stillage.

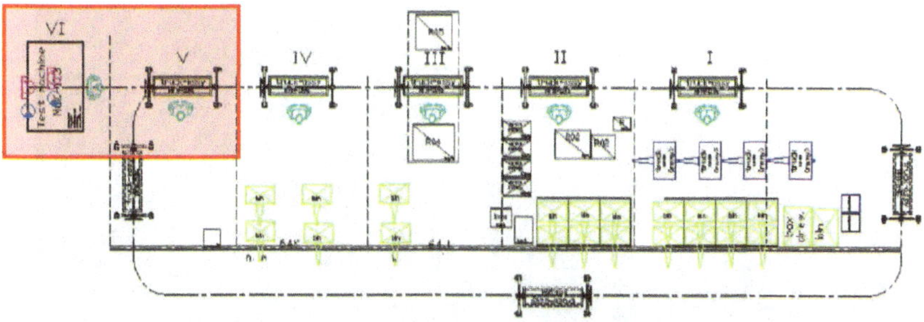

Fig. 35. Current product assembly line station.

The project team suggested an innovative solution, which included an automated run-through test station with fully integrated assembly trolley. Additionally, to increase operator's safety, the row unit rotation was planned to be performed inside the testing machine in an automated way. In order to meet these objectives, assembly line layout had to be modified according to the concept presented in (Fig. 36).

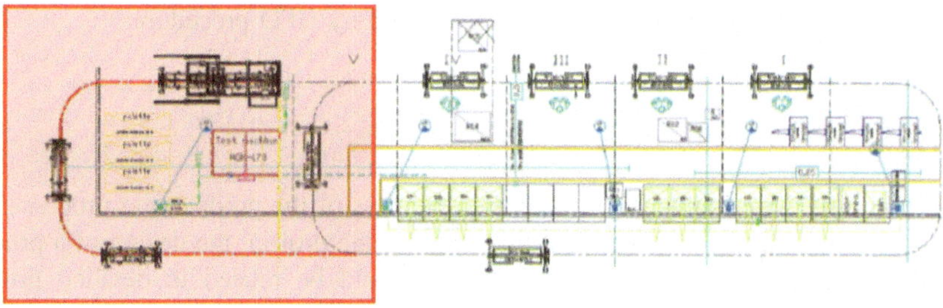

Fig. 36. Modified concept of the row unit assembly line layout with run-through test station.

The effects of the works on design of the new test-run station are presented in (Fig. 37) and its physical prototype in (Fig. 38).

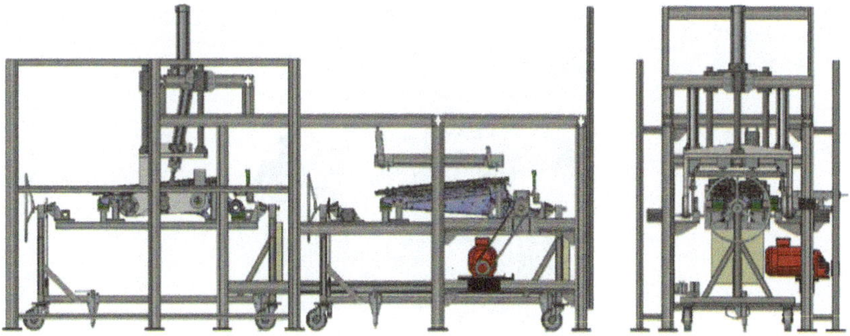

Fig. 37. Visualization of the run-through test station.

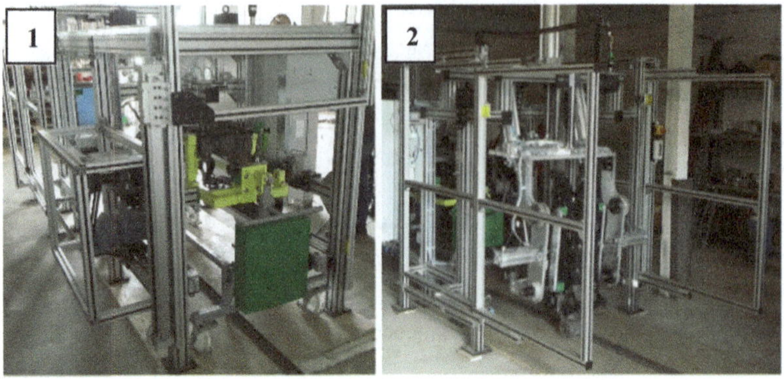

Fig. 38. Prototype build of the run-through test station: (1) Loading gate. (2) Protected rotating & unloading area.

STEP 5–STEP 7: PROCESS VERIFICATION AND OPTIMIZATION

The last 3 steps of EPM Methodology are concentrated on manufacturing process optimization through validation and implementation on the production line. This is achieved through major project milestones, defined by GPD procedure:

- Prebuild.
- 4P Build.
- JOB1.

The aim of these builds is to prove the readiness of the product and processes for final production launch. Implicitly, at this stage of the project, the design of a product should be frozen and most of the activities should be related to detailed process improvements using same techniques as in step 4. Some engineering changes can be introduced due to major cost savings or technical-field test results however due to high level of implementation on the production stations, the changes must be thoroughly discussed with manufacturing site.

Quality policy plays important aspect in this moment of project maturity. Hence, the detection of all nonconformities on the line is very important. To facilitate this approach, a concept of visual control run-through station was proposed (Fig. 39).

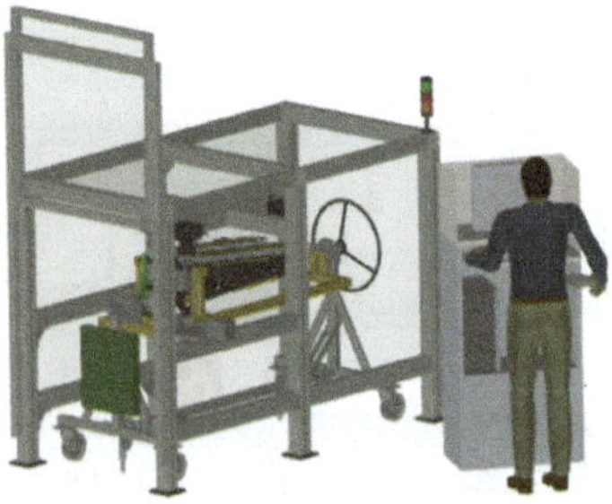

Fig. 39. Concept of visual run-through test station.

Critical assembly parts have been identified which include snap-ring installation. Trial verification tests were performed and presented in (Fig. 40).

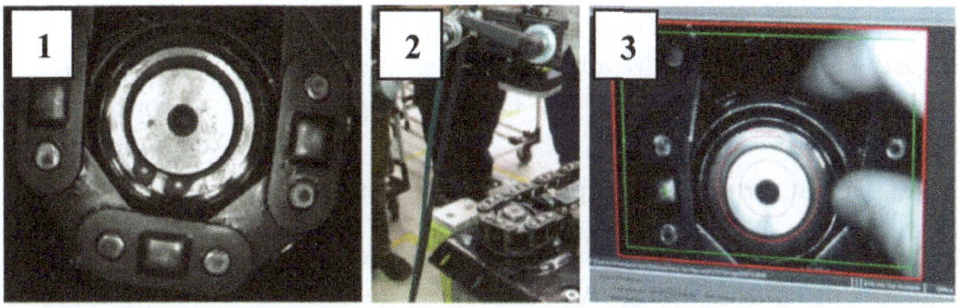

Fig. 40. Visual inspection trials: (1) Properly installed snap ring. (2) Visual control system. (3) Error detection - snap ring missing.

4 Conclusions

In this paper, the general description of the successful EPM methodology implementation within WCM system was presented. The actions taken in terms of the EPM methodology led to a significant reduction in the costs of implementing the new product. As a result of the applied tools, nearly 80 improvements were developed, bringing significant savings in the project budget. Additionally, due to the fact that the cost of the product was verified at each stage of the project implementation, the final cost of the product was 5% lower than the initially accepted one. The cost deployment analysis made it possible to identify further areas requiring improvement, among which it was determined:

- Tooling issue.
- Supplier quality issue.
- Late test/validation failure.

The above areas will be designated for correction during the next implementation project. In the analyzed case study, all project targets i.e. technical, cost and timing aims were achieved. The effects of the highlighted activities on the assembly line area were presented in (Fig. 41).

Fig. 41. Effects of EPM activities on the shop floor: (1) Assembly line area. (2) Run-through test station.

Due to the brief character of this publication, only basic assumptions were highlighted. The author will develop EPM concept in a more detailed overview in specially dedicated publications, which should emphasize the aspects below:

- Importance of cost deployment in EPM targets.
- Cost aspect in terms of EPM approach.
- WPI (Workplace Integration) in terms of proactive approach within EPM Methodology.
- Quality aspect in terms of new product launches.

In conclusion, EPM is a complex methodology, which includes a list of specific activities in specific project maturity. Due to its "innovative" character, it is constantly striving for constant improvement- in order to reach the best in class- a world class level.

References

1. Veugelers, R.: Manufacturing matters in the EU. MWP Advanced Manufacturing, p. 16 (2013)
2. De Felice, F., Petrillo, A.: Key success factors for organizational innovation in the fashion industry. Int. J. Eng. Bus. Manage. **5**, 47–57 (2013)
3. Piasecka-Głuszak, A.: Implementation of World Class Manufacturing in a production company on the Polish market. Economics of the 21st century, vol. 4, no. 16, pp. 52–65 (2017). (in polish)
4. Stanek, K., Czech, P., Barcik, J.: Methodology of World Class Manufacturing (WCM) in Fiat Auto Poland S.A., Scientific Papers of the Silesian Technical University, Series: Transportation, vol. 71, pp. 65–72 (2011). (in polish)
5. Mróz, A.: About some aspects of Advanced Manufacturing Engineering Department in WCM-oriented production plants. Manage. Prod. Eng. Rev. **9**(4) (2018)
6. Ishikawa, K.: What is Total Quality Control: The Japanese Way. Prentice-Hall, Inc., Englewood Cliffs (1985)
7. Schonberger, R.J.: World Class Manufacturing: The Lessons of Simplicity Applied. Simon & Schuster, New York (1986)
8. Ishikawa, K.: Introduction to Quality Control. Chapman & Hall, Tokyo (1990)
9. Womack, J.P., Jones, D.T., Ross, D.: The Machine that Changed the World. Rawson Associates, New York (1990)
10. Taguchi, G.: Taguchi on Robust Technology Development: Bringing Quality Engineering Upstream. ASME Press, New York (1993)
11. Womack, J.P., Jones, D.T.: Lean Thinking-Banish Waste and Create Wealth in Your Corporation. Simon & Schuster, New York (1996)
12. Bicheno, J.: The Lean Games and Simulations. Picsie Books, Buckingham (2015)
13. Gunn, T.G.: Manufacturing for Competitive Advantage: Becoming a World Class Manufacturer, pp. 28–29. Ballinger Pub, Cambridge (1987)
14. Sangwan, K.S., Digalwar, A.K.: Evaluation of world-class manufacturing systems: a case of Indian automotive industries. Int. J. Serv. Oper. Manage. **4**(6), 687–708 (2008)
15. De Felice, F., Petrillo, A.: Optimization of manufacturing system through world class manufacturing. IFAC-PapersOnLine **48**(3), 741–746 (2015)

16. De Felice, F., Petrillo, A., Monfreda, S.: Improving operations performance with world class manufacturing technique: a case in automotive industry, operations management, IntechOpen, pp. 1–30 (2013). http://dx.doi.org/10.5772/54450. Accessed 27 Sep 2018
17. Yamashina, H., Kubo, T.: Manufacturing cost deployment. Int. J. Prod. Res. **40**(16), 4077–4091 (2002)
18. Framinan, J.M., Ruiz, R.: Guidelines for the deployment and implementation of manufacturing scheduling systems. Int. J. Prod. Res. **50**, 1799–1812 (2011)
19. Silva, L.C.S., Kovaleski, J.L., Gaia, S., Garcia, M., de Andrade Júnior, P.P.: Cost deployment tool for technological innovation of world class manufacturing. J. Transp. Technol. **3**, 17–23 (2013)
20. Digalwar, A., Sangwan, K.S.: Role of Knowledge Management in World Class Manufacturing: an Empirical Investigation, pp. 415–419. IEE (2011)
21. Dalmarco, G., Maehler, A.E., Trevisan, M., Schiavini, J.M.: The use of knowledge management practices by Brazilian startup companies. RAI Revista de Administração e Inovação **14**, 226–234 (2017)
22. Nowacki, R., Bachnik, K.: Innovations within knowledge management. J. Bus. Res. **69**, 1577–1581 (2016)
23. Al-Emran, M., Mezhuyev, V., Kamaludin, A., Shaalan, K.: The impact of knowledge management processes on information systems: a systematic review. Int. J. Inf. Manage. **43**, 173–187 (2018)
24. Mardani, A., Nikoosokhan, S., Moradi, M., Doustar, M.: The relationship between knowledge management and innovation performance. J. High Technol. Manage. Res. **29**, 12–26 (2018)

Approaches to Design for Six Sigma. A Confusing Redundancy

Adam Hamrol and Matthew Barsalou(✉)

Poznań University of Technology, Poznań, Poland
matthew.barsalou@gmail.com

Abstract. There are various approaches available for designing new products or services and Design for Six Sigma (DFSS) is one of them. Unfortunately, there is a confusing redundancy of options available for organizations seeking to implement DFSS so this research searched eight journals to identify possible approaches used for DFSS. Four data databases were then searched using the previously determined acronyms to identify the most common approaches to DFSS. This paper provides a comparison of the various approaches and concludes with recommendations for those seeking to implement DFSS.

Keywords: DFSS · Six Sigma · Quality · Product development

1 Introduction

The design process is the process of creating a new product or service [1]. Organizations usually use a stage-gate design process with multiple gates used as control points. Typical stage-gate steps include generating ideas, scoping, creating a business case, developing the product or process, evaluating the product or process, and launching the product or process [2]. An alternative and more efficient approach to design is concurrent engineering, which uses multifunctional teams that design the various aspects of a product in parallel [3]. Stage-gates tend to be a linear process while concurrent engineering reduces development time due to the use of parallel operations.

A specific approach to design as introduced as a part of Six Sigma. The Six Sigma methodology was created at Motorola in the 1980s as a method to be competitive [4] against high quality Japanese products. Six Sigma then spread to companies such as General Electric by the 1990s [5] and many others around the world.

Design for Six Sigma (DFSS) "is a process to define, design and deliver innovative products that provide competitively attractive value to customers in a manner that achieves the critical-to-quality characteristics for all the significant functions" [6]. DFSS is a variation on Six Sigma; however, the basic Six Sigma approach suggests following one set of project phases, called DMAIC (Define, Measure, Analyze, Improve and Control), as an approach to quality improvement. [7]. The define phase is where a project charter is written, goals are stated, a problem statement is created, and project planning steps including scheduling, budgeting, and milestones are established [8]. The measurement system is assessed through the use of a measurement system analysis (MSA) and measurements are taken to establish baseline performance [9]. The

A. Hamrol et al. (Eds.): *Advances in Manufacturing II* - Volume 3, LNME, pp. 144–154, 2019.
https://doi.org/10.1007/978-3-030-17269-5_10

analyze phase is where the sources of variation in a process are analyzed using statistical [10]. Once sources of variability are understood, improvements are implemented in the improvement phase. Often, these improvements are identified through the use of design of experiments (DoE) [11]. The final Six Sigma phase is the control phase where improvements are maintained. Typically, statistical process control (SPC) is implemented to monitor process performance after improvements have been implemented [12].

There are many possibilities when using DFSS. In the last forty years, you can observe in the literature such design cycles as DMADV, IIDOV, I2DOV, IDOVI, IDOIV, IDDOV, DMEDI, DCOV, DCCDI, and CDOV.

GE used the phases Define, Measure, Analyze, Design, and Verify and called the cycle DMADV. The first phase of DMADV is define. Here, the business case for the project is established, customer and market data is collected and a project budget is created. The customer and market data is then evaluated during the measure phase. Critical to Quality (CTQ) factors are also identified at this point. The analyze phase is where data from the previous phases are translated into product requirements. The design phase is where the actual design is created and evaluated. The final phase is where the design is verified [13].

The first phases of IDDOV (Identify, Define, Develop, Optimize, and Verify) is where the customer's needs are identified and a business case is created. Then the product requirements are defined and concepts and design are developed. The design is then optimized to achieve robustness and finally verified to ensure it meets requirements [14].

With IDOV (Identify, Design, Optimize, Verify), a business case is made, critical to quality factors are identified and optimized, and the final design is validated [15].

When using IDEAS (Identify, Design, Evaluate, Assure, and Scale-up), the best product or service concept is selected based on the voice of the customer and then critical to quality characteristics are identified and a design is created. The design is then optimized checked to ensure it meets requirements [16].

IDOVI (Identify, Define, Optimize, Validate, and Incorporate) is used as a roadmap for Sarbanes-Oxley compliance through a DFSS driven creation of an accounting system [17].

The first two phases of I2DOV (Invention, Innovation, Develop, Optimize, and Verify) are where business goals, strategies and the mission are determined and markets and the market segment are defined. A technology roadmap is create and then product is developed, optimized, and then verified [18].

IIDOV (Invent, Innovate, Develop, Optimize, Verify) is a slight variation on I2DOV and IIDOV and I2DOV have been used interchangeably [18]. Here, customers' needs and technological tendencies are identified, needs are then classified by importance and converted into technical requirements, and alternatives solutions are considered. Systems, sub-systems, and components are then designed, optimized, and verification testing is performed [19].

The first step of CDOV (Concept Development, Design Development, Optimization, and Verify Certification) is identifying customer needs and classifying them by importance. These requirements are then translated into technical requirements and the system and components are designed inconsideration of the inputs between each. Robust design is then used to optimize the resulting designs and the final design is then verified using acceptance tests [19].

When using DCCDI (Define, Customer, Concept, Design, and Implementation), the first phases is where the customer requirements are defined and critical to quality metrics are established. A risk review is also performed and then a design concept is created and evaluated. The design is then refined and implemented in production [20].

DCOV (Define, Characterize, Optimize, and Verify) is a variant from Ford Motor Company and consists of defining customer requirements, characterizing the requirements, creating and optimizing the design, and then verifying the design [21].

DMEDI (Define, Measure, Explore, Develop and Optimize, and Implement) is used by Caterpillar Financial Services Corp [22] and PriceWaterhouseCoopers [23].

When using ICOV (Identify, Characterize, Optimize, and Validate), the first step is identifying requirements and translating them into something measurable. The required resources and possible risks are also identified during the first phase. During the second phase the design concept is created, the measurement system is validated, concept designs are assessed and the risks of each concept are evaluated. The optimize phase is where Key Product/Process Output Variables are identified (KPOVS), transfer functions between KPOVs and Key Product Input Variables (KPIVs) are developed, performance criteria are established and risks are again assessed. The final phase is validate, where the design is tested and validated, control and action plans are established and overall risks are evaluated [24]. There is disagreement amongst authors as to what each abbreviation stands for with ICOV, which is also explained as identify, characterize, optimize, and verify [23].

There is also a second version of ICOV where the validate phase has been replaced by verify. Here, the identify phase is where reliability requirements considering the operating environment are identified. A design is then created based on the requirements and optimized to ensure robust and reliable performance. Optimization is then performed to ensure reliability [25].

Another version of DFSS is DVDOV (Define, Concept, Design, Optimize, and Verify), where customer's and their needs are determined, solutions are considered, designed, optimized, and verified [26].

Such a large number of different phases and their sequences raises the question of whether it is a matter of a deliberate action or a desire to demonstrate an original, unique approach? That's why the authors decided to determine which versions of the DFSS phases are the most prevalent and answer the question why.

2 Literature Research

To answer the question posed in the previous chapter various publications were searched using the search phrase "DFSS" with the resulting articles read to determine which variations of DFSS, if any, were used.

Scientific journals searched were International Journal of Quality and Reliability Engineering International Journal of Lean Six Sigma, Total Quality Management and Business Excellence, The TQM Journal and Quality Engineering. Peer-reviewed practitioner journals were also used; these were Quality Progress (Quality Progress), and Six Sigma Forum Magazine (Six Sigma Forum Magazine). The online publications iSixSigma (iSixSigma) and Quality Digest (Quality Digest) were also used. All journals were searched in June 2018 with no limits on how far back the results went, except: International Journal of Quality and Reliability Engineering (March 1999 till June 2017), International Journal of Lean Six Sigma (2010 till 2016), Total Quality Management and Business Excellence (January 2003 till December 2016), The TQM Journal (2008 till 2016).

Tables 1 and 2 list the results for the publications searched and Table 2 lists the various types of DFSS phases identified. (Figure 1) shows the results for the DFSS phases identified.

Table 1. Publications and search results.

International Journal of Quality and Reliability Engineering		iSixSigma		International Journal of Lean Six Sigma		Quality Digest		Quality Engineering	
Phases	Total	Phases	Total	Phases	Total	Phases	Total	Phases	Total
CDOV	0	CDOV	0	CDOV	0	CDOV	0	CDOV	0
DCCDI	0	DCCDI	1	DCCDI	0	DCCDI	0	DCCDI	0
DCDOV	0	DCDOV	0	DCDOV	0	DCDOV	0	DCDOV	0
DCOV	2	DCOV	0	DCOV	0	DCOV	0	DCOV	1
DMADV	6	DMADV	9	DMADV	10	DMADV	6	DMADV	3
DMEDI	0	DMEDI	2	DMEDI	2	DMEDI	0	DMEDI	0
I2DOV	0	I2DOV	0	I2DOV		I2DOV	0	I2DOV	0
ICOV	0	ICOV	1	ICOV	1	ICOV	0	ICOV	0
IDDOV	0	IDDOV	0	IDDOV	1	IDDOV	1	IDDOV	0
IDEAS	0	IDEAS	0	IDEAS	0	IDEAS	0	IDEAS	0
IDOV	2	IDOV	6	IDOV	2	IDOV	0	IDOV	1
IDOVI	0	IDOVI	1	IDOVI	0	IDOVI	0	IDOVI	0
IIDOV	0	IIDOV	0	IIDOV	0	IIDOV	0	IIDOV	0

Table 2. Publications and search results continued.

Quality progress		Six Sigma Forum Magazine		The TOM Journal		Total Quality Management and Business Excellence		Total Occurrences	
Phases	Total	Phases	Total	Phases	Total	Phases	Total	Phases	Total
CDOV	0	CDOV	1	CDOV	0	CDOV	0	CDOV	1
DCCDI	0	DCCDI	0	DCCDI	0	DCCDI	0	DCCDI	1
DCDOV	0	DCDOV	1	DCDOV	0	DCDOV	0	DCDOV	1
DCOV	0	DCOV	2	DCOV	0	DCOV	1	DCOV	4
DMADV	8	DMADV	9	DMADV	3	DMADV	2	DMADV	57
DMEDI	0	DMEDI	0	DMEDI	0	DMEDI	0	DMEDI	4
I2DOV	0	I2DOV	1	I2DOV	0	I2DOV	0	I2DOV	1
ICOV	1	ICOV	2	ICOV	0	ICOV	0	ICOV	5
IDDOV	0	IDDOV	0	IDDOV	0	IDDOV	1	IDDOV	4
IDEAS	0	IDEAS	1	IDEAS	0	IDEAS	0	IDEAS	1
IDOV	1	IDOV	2	IDOV	3	IDOV	3	IDOV	19
IDOVI	0	IDOVI	0	IDOVI	0	IDOVI	0	IDOVI	1
IIDOV	0	IIDOV	2	IIDOV	0	IIDOV	0	IIDOV	2

Four large databases were then searched using the search term "DFSS" and the abbreviations for the various phases. DFSS was included to narrow results down to DFSS related results to avoid any occurrence where the aberrations may have had use in a non-SixSigma field. The databases were from Taylor & Francis (Taylor & Francis), American Society for Quality (ASQ) (American Society for Quality), Science Direct (Science Direct), and EBSCO (EBSCO) and the results are listed in (Table 3).

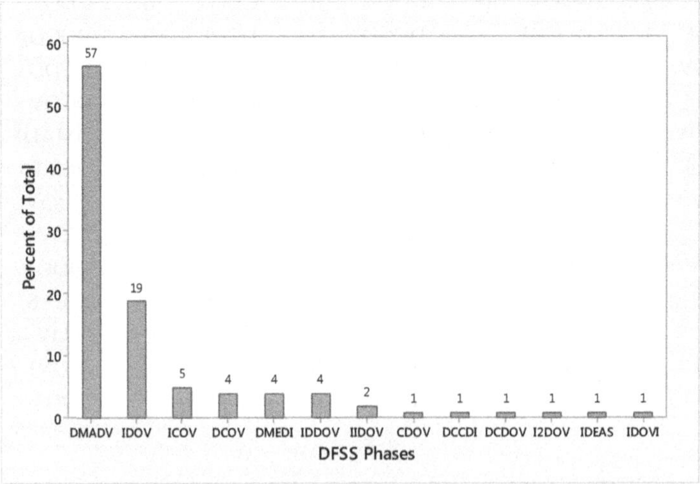

Fig. 1. Graph of publication totals by phases.

Table 3. Database results and totals by phases.

Phases	ASQ	Science Direct	Taylor & Francis	Total Occurrences
CDOV	1	0	1	2
DCCDI	1	0	1	2
DCDOV	1	0	5	6
DCOV	3	0	5	8
DMADV	49	17	28	94
DMEDI	1	0	3	4
I2DOV	2	0	1	3
ICOV	4	1	1	6
IDDOV	3	0	5	8
IDEAS	132	93	105	330
IDOV	6	9	7	22
IDOVI	0	0	0	0
IIDOV	1	0	0	1

It is easy to see that there was only one occurrence of IDEAS (Identify, Design, Evaluate, Assure, and Scale-up) in the publication search, but 767 occurrences in the database. This may be explained as occurrences of the word "idea" and not the DFSS methodology. For example, an article containing the sentence "Organizations transitioning to level five must again change their ideas of management and empower employees" [27, p. 41] was located using the keywords DFSS and IDEAs, but the only reference to ideas was in a non-DFSS connotation. The IDEAS methodology was then excluded from further consideration. Totals, not including IDEAS, are shown in (Fig. 2).

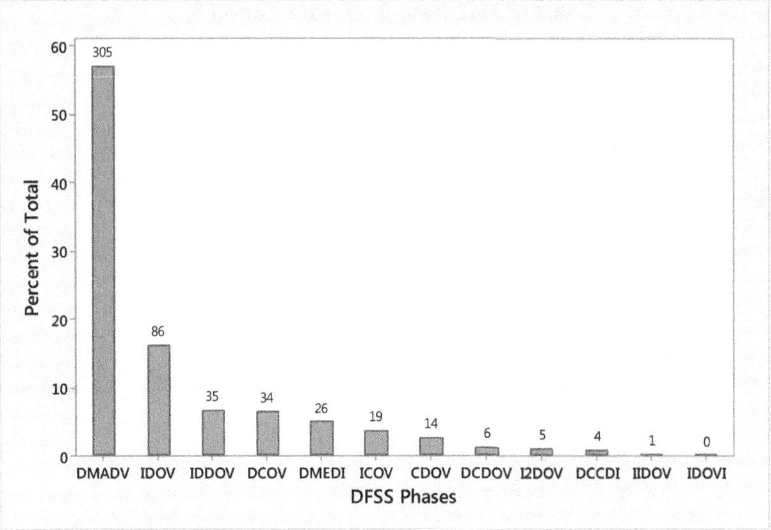

Fig. 2. Graph of database results and totals by phases.

The most common result for DFSS phases was DMADV. There were 305 database results for DMADV. The high number of occurrences for DMADV is potentially because this was the original DFSS methodology. The second most common occurrence IDOV with 86 database occurrences. IDDOV came in third for databases results with 35 occurrences with DCOV slightly behind with 34 occurrences. These four phases made up 86% of all results as shown in the Pareto chart in (Fig. 3).

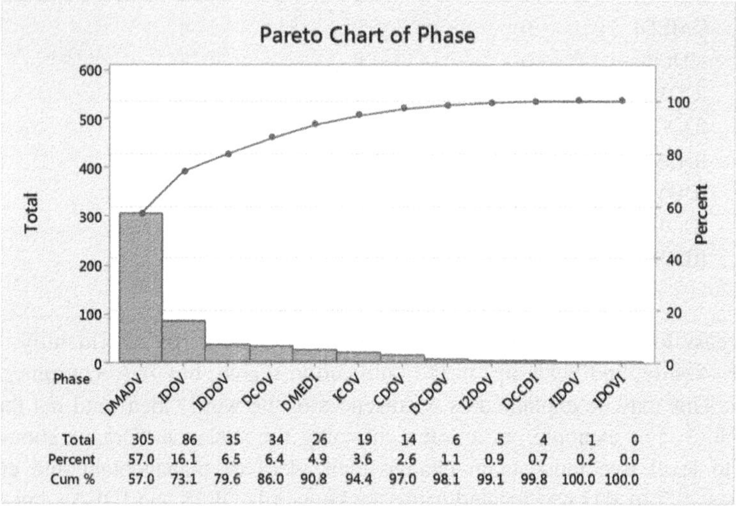

Fig. 3. Pareto chart of database results.

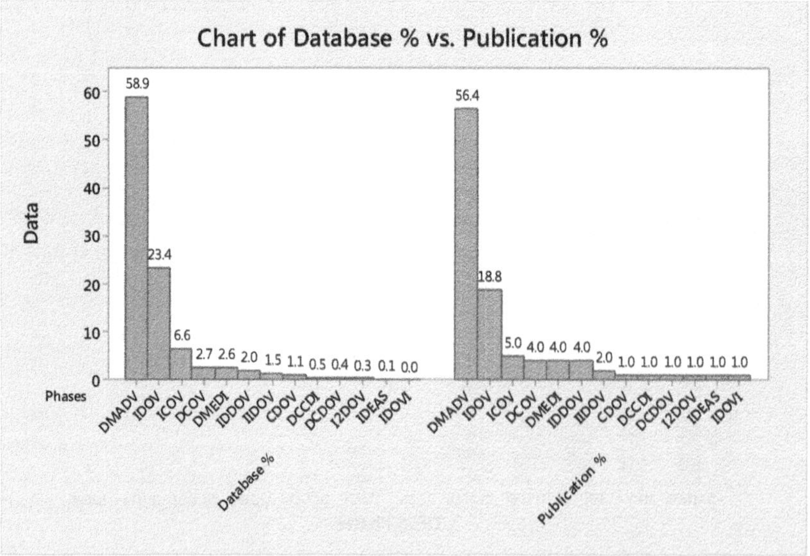

Fig. 4. Comparison of database versus publication results.

Both the databases and publication searches resulted in DMADOV, IDOV, and ICOV as the results that occurred the most. In both cases DMADIV and IDOV were by far the most common results as shown in (Fig. 4).

DMADV was DFSS result that occurred the most so this was compared to DMAIC, which is used in regular Six Sigma projects. The journals were again searched using "DMADV and "DMAIC" and the total number of results for each journal is shown in Tables 4 and 5. There were far more results for DMAIC with DMADV occurring in only 11% of all results.

Table 4. DMAIV versus DMADV.

International Journal of Quality and Reliability Engineering		iSixSigma		International Journal of Lean Six Sigma		Quality Digest		Quality Engineering	
Phases	Total	Phases	Total	Phases	Total	Phases	Total	Phases	Total
DMAIC	3	DMAIC	Appr. 2,650	DMAIC	497	DMAIC	372	DMAIC	100
DMADV	13	DMADV	406	DMADV	21	DMADAV	40	DMADV	8

Table 5. DMAIV versus DMADV continued.

Quality progress		Six Sigma Forum Magazine		The TOM Journal		Total Quality Management and Business Excellence		Total Occurrences	
Phases	Total	Phases	Total	Phases	Total	Phases	Total	Phases	Total
DMAIC	242	DMAIC	164	DMAIC	497	DMAIC	90	DMAIC	Appr. 4,615
DMADV	37	DMADV	22	DMADV	9	DMADV	13	DMADV	569

3 Discussion

Special design phases are used in DFSS. Their names and sequences are very diverse. The research presented in this article showed that three sequences dominate in the literature:

- DMADV (Define, Measure, Analyze, Design, and Verify)
- IDOV (Identify, Design, Optimize, Verify)
- ICOV (Identify, Characterize, Optimize and Validate)

If you look closely at the names of phases used in different cycles, you can show that the differences are apparent. (Figure 5) lists the variations of actions that occur during the DMADV phases.

DMADV Phases	
Define	• Project selection • Establish project charter with budget and deliverables • Establish the business case • Project PERT chart • Select project team members
Measure	• Voice of the Customer (VoC) • Kano Model • Critical to Quality (CtQ) • Quality Function Deployment (QFD) • Process and value stream maps • Measurement System Analysis (MSA) • SIPOC
Analyze	• Concept development • Preliminary designs • Design Failure Modes and Effects Analysis (DFMEA) • Identify tolerances • TRIZ analysis • Pugh matrix • Design of Experiments (DoE) • Reliability testing
Design	• Final design • Prototype build and evaluation • Establish Lean processes • Pilot production • Poke Yoke in production • Process capability study
Verify	• Process Failure Modes and Effects Analysis (PFMEA) • Control plan • Statistical Process Control (SPC) • Review complaints and customer surveys

Fig. 5. DMADV phases.

Table 6 shows how the various DFSS phases compared to the more common DMADV approach. Although there are many possible names for DFSS phases, the tools used and actions that take place during the phases are comparable to the tools used and the actions that take place in DMADV phases. All these many labels can be described using Define, Measure, Analyze, Design and Verify.

Table 6. DMADV compared to other approaches

DMADV	Define	Measure	Analyze	Design	Verify
CDOV	Concept Development			Design Development	Optimization/Verify Certification
DCCDI	Define	Customer		Concept/Design	Implementation
DCDOV	Define	Concept		Design/Optimize	Verify
DCOV	Define	Characterize		Optimize	Verify
DMEDI	Define	Measure	Explore	Develop and Optimize	Implement
I2DOV	Invention	Innovation		Develop/Optimize	Verify
ICOV	Identify	Characterize		Optimize	Validate
IDDOV	Identify	Define		Develop/Optimize	Verify
IDEAS	Identity			Design/Evaluate	Assure/Scale-up
IDOV	Identity			Design/Optimize	Verify
IDOVI	Identity		Define		Optimize/Validate/Incorporate
IIDOV	Invent	Innovate		Develop/Optimize	Verify

4 Conclusions

An organization attempting to implement DFSS could face confusion with so many potential approaches to DFSS. The original DFSS approached, DMADV, started at General Electric [16] and the many variants are due to various companies customizing the processes to fit their own needs such as DCOV at Ford [21].

Whether using DMADV, ICOV or any of the other variations what truly matters is the content of the methodology. The customer's needs and wants must be determined, understood and translated into measurable quantities. Preliminary designs that meet the requirements must be created and then evaluated. The selected design then needs to be verified. Using DMADV will achieve these goals and has the advantage of a larger body of knowledge available to support its use during a DFSS project. Somebody new to DFSS will have access to far more resources if they use DMADAV. Furthermore, the plethora of DFSS methodologies results in an unnecessary redundancy and can cause unnecessary confusion for organizations that are new to DFSS and they don't offer advantages over the better known DMADV methodology.

References

1. Stevenson, W.J.: Production/Operations Management, 6th edn. McGraw Hill, New York (1999)
2. Cooper, R.G.: Wining at New Products: Creating Value Through Innovation. Basic Books, New York (2017)
3. Anderson, D.M.: Design for Manufacturability: How to use Concurrent Engineering to Rapidly Develop Low-Cost, High-Quality Products for Lean Production. CRC Press, Boca Rotan (2014)

4. Carnell, M.: 3.4 million: Six Sigma in everything we do? Qual. Prog. 2003 **40**(1), 67–68 (2007)
5. Pepper, M.P.J., Spedding, T.A.: The evolution of Lean Six Sigma. Int. J. Qual. Reliab. Manag. **27**(2), 138–155 (2010)
6. Watson, G.H., DeYong, C.F.: Design for Six Sigma: caveat emptor. Int. J. Lean Six Sigma **1**, 66–84 (2010)
7. Kumar, M., Jiju, A., Madu, C.N., Montgomery, D.C., Park, S.H.: Common myths of six sigma demystified. Int. J. Qual. Reliab. Manag. **25**(8), 878–895 (2008)
8. George, M.L., Rowlands, D., Price, M., Maxey, J.: The Lean Six Sigma Pocket Tool Book. McGraw-Hill, New York (2005)
9. Breyfogle III, F.W.: Implementing Six Sigma: Smarter Solutions Using Statistical Methods, 2nd edn. Wiley, Hoboken (2003)
10. Gryna, F.M.: Quality Planning and Analysis, 4th edn. McGraw-Hill, New York (2001)
11. Breyfogle, F.: Integrated Enterprise Excellence Volume III - Improvement Project Execution: A Management and Black Belt Guide for Going Beyond Lean Six Sigma and the Balanced Scorecard. Citius Publishing Inc., Austin (2008)
12. Benbow, D.W., Kubiak, T.M.: The Certified Six Sigma Black Belt Handbook. ASQ Quality Press, Milwaukee (2009)
13. Watson, G.H.: An approach to software design. Six Sigma Forum Mag. **3**(3), 22–26 (2004)
14. Chowdhury, S.: Design for Six Sigma: The Revolutionary Process for Achieving Extrodanary Profits. IL Dearborn Trade Publishing, Chicago (2005)
15. Munro, R.A., Maio, M.J., Nawaz, M.B., Ramu, G., Zrymiak, D.J.: The Certified Six Sigma Green Belt Handbook. Quality Press, Milwaukee (2008)
16. Brue, G., Launsby, R.G.: Design for Six Sigma. McGraw-Hill, New York (2003)
17. Kowansky, E., Josyiula, R., Martin, J.W.: DFSS can help ensure sarbanes-oxley act compliance, iSixSigma. https://www.isixsigma.com/operations/finance/dfss-can-help-ensure-sarbanes-oxley-act-compliance/. Accessed 17 June 2018
18. Creveling, C.M., Slutsky, J.L., Antis Jr., D.: Six Sigma in Technology and Product Development. Prentice Hall, Upper Saddle River (2003)
19. Echeveste, M.E., Rozenfeld, H., Sonego, M.: Potential application of Six Sigma tool in the integrated product development process. J. Braz. Soc. Mech. Sci. Eng. **38**(8), 2499–2511 (2016)
20. Tennant, G.: Design for Six Sigma: Launching New Products and Services Without Failure. Gower Publishing Company, Burlington (2002)
21. Ferryanto, L.: DFSS: lessons learned. Six Sigma Forum Mag. **4**(2), 24–28 (2005)
22. Daniels, S.: At your service. Qual. Prog. **37**(6), 58–65 (2004)
23. Yang, K., El-Haik, B.S.: Design for Six Sigma: A Roadmap for Product Development, 2nd edn. McGraw Hill, New York (2009)
24. Mader, D.P.: Frontiers of quality: design for six sigma and your current design process. Qual. Prog. **36**, 92–94 (2002)
25. Hu, M.: Reediness plan put into action. Six Sigma Forum Mag. **13**(1), 17–91 (2013)
26. Berkshire Hathaway: The Process Design Process: Wading Upstream to the Source. More Steam Blog (2001). https://www.moresteam.com/blog/index.cfm?blog_id=58. Accessed 28 Oct 2018
27. Duffy, G.L.: Leveling up. Qual. Prog. **49**(6), 38–44 (2016)

Cooperation of Education and Enterprises in Improving Professional Competences - Analysis of Needs

Maciej Szafrański(✉), Marek Goliński,
Magdalena Graczyk-Kucharska, and Małgorzata Spychała

Poznan University of Technology, Strzelecka 11, 60-965 Poznań, Poland
maciej.szafranski@put.poznan.pl

Abstract. Enterprises achieve success in the labour market thanks to the implementation and use of adequate strategies, but also thanks to adequate resources management and proper relationships with the surrounding. The main resources currently most searched for by employers are human resources. In the market situation with scarcity of employees, where their quantitative and qualitative structure is insufficient for the enterprises to realise their tasks, the key problem for businesses is to find ways to search for suitable employees. In the situation of insufficient resources of suitable employees, enterprises should develop, in a planned and systemic manner, methods of obtaining and preventive provision of necessary human resources in connection with the business development plans. Within the developed personnel management strategies, the enterprises should also define the subjects to cooperate with in this area, as well as plan the rules of this cooperation. This paper discusses the problem of evaluating forms of supporting schools by enterprises, i.e. which tools and means of communications should be best used to obtain employees with best suited competences. It also ponders the question which forms of cooperation should schools develop in order to improve the quality of education and to increase the probability of matching students to the needs of current labour market. Research verification included the analysis of the number of businesses offering a given form of support and the average evaluation of importance of this form.

Keywords: Vocational education and training · Labour market · Competence management

1 Introduction

Current development trends, both social and economic, are a vital justification for vocational education and training and the development of human capital. A better-qualified employee is the initial condition for sustained development [1–3], which requires continuous improvement of qualifications [4]. One of the main sources of economic growth is human capital. Knowledge-based economy relies its growth on minimum secondary and tertiary qualifications, frequently with a narrow professional profile. In this area not only good vocational education is necessary, but also its social

A. Hamrol et al. (Eds.): *Advances in Manufacturing II* - Volume 3, LNME, pp. 155–168, 2019.
https://doi.org/10.1007/978-3-030-17269-5_11

popularisation and a close cooperation with modern branches of industry [5]. Employment opportunities are open to the employees who meet the requirements of enterprises regarding currently possessed competences, and vocational education and training are dependent on certain social and economic factors [6].

The issue of scarcity of qualified employees and the related mismatch of skills requires vocational improvement that will facilitate the transfer from learning to work. Numerous publications observe and point to the aspect of education quality and too big a gap between education and the labour market (too theoretical knowledge, inadequate cooperation with employers in the development and acceptance of syllabuses, curricula unrelated to obtaining soft skills that are required in the labour market) [7–9].

The need to use an innovative approach in vocational education and training, building a relationship with the surrounding and a systemic approach to solving this problem is discussed in many publications [10, 11].

The area of practice and the possibility of continuous verification of skills gained are of great significance. Experience gained in the 'chain of knowledge production' should be verified in the loop feedback [12, 13].

While analysing the quality of education, effects achieved should be translated both into the results of participants of education, and into system benefits for the region or country. In case of education, the quality expressed with the level of skills that has an influence on economic growth of GDP [14] is of greater importance than the number of persons learning professions not connected with economic support. Another factor lowering the quality of education is the inadequate education in competences recommended by enterprises. Among competences searched for by the economy are: innovation, teamwork, leadership skills, openness to social needs [15–17].

Commonly repeated theses of the necessity of effective education for the needs of economic growth, supported with numerous examples quoted by source literature, call for specific activities which will strengthen the relationship between an enterprise and school, or – from a wider perspective – labour market – education.

The goal of this paper is to identify and evaluate the forms of support of the educational process, understood as cooperation between at least two subjects. In case of the research presented, identification and evaluation of the support forms were done from the perspective of the unit educating future employees, i.e. technical school, and from the perspective of the requirements of the labour market, i.e. representatives of enterprises.

From the variety of research methods used over several years, the final choice was to use a direct survey with a questionnaire, carried out in 38 technical schools and 40 enterprises. Research results quoted in this paper were preceded by projects realised from 2010, aimed at improving the quality of education in vocational schools on both secondary and tertiary level [18, 19], by identification, development and implementation of effective forms of supporting the educational process, while taking into consideration the demand for competences in the labour market of Wielkopolska region.

A deepened case study carried out at the stage of research realisation and verification of research results facilitated the improvement of the relationships between educational institutions and enterprises. The detailed scope of the paper concerned:

(a) Which forms of support are present in the relationship between enterprises and schools?
(b) How important are individual forms of support in schools and enterprises?
(c) Is the importance of individual forms of support convergent from the point of view of schools and enterprises?
(d) Is the size of the business (expressed in the number of employees) related to the number of forms of support offered to schools?

2 Selected Forms of Cooperation Between Entrepreneurs and Vocational Schools

The history of cooperation between vocational education institutions and entrepreneurs is as long as vocational education itself [20]. Success in the labour market is difficult to achieve if professional career is started with theoretical background only. For this reason schools initiate contacts with enterprises in order to help students gain practical skills in their profession. The most popular form of gaining skills is temporary traineeship in the enterprise. Time spent with the employer is proportional to the education level and varied depending on the profession. Internship is an extension of traineeship, and it involves the candidate in a more challenging way, and frequently results in a job offer. Employers are increasingly more involved in opinion making and helping schools prepare curricula [21]. Entrepreneurs also offer schools support in preparing qualifying vocational courses, they may facilitate access to modern techniques and technologies and realise common activities in the area of career advising [22].

In the current situation in the labour market, entrepreneurs perceive schools as a source of candidates for work. Traineeship and internship programmes enable recognition of the candidate's potential and streamline the recruitment process. However, it should be remembered that employers must realise a set of goals, among them are: orientation to product quality, profit increase, improvement of competitiveness in the market.

Clusters/regional networks are of a greater importance, they serve as support for local organisations. Contacts with the local industrial environment are of key importance in innovation processes, also for many SMEs; regional resources include contextual knowledge, both open and hidden, specific for the area [23].

The enterprises studied belong to Wielkopolska Education and Economic Network, which facilitated the research. It is also an example of benefits resulting from building relationships within the network of cooperation.

3 Research Results

The research described in this paper was realised in the education area and for that reason there were epistemological bases related to the character of research questions and methods used [24]. Due to the objective of the research, it is located in the area of management and parametrise procedures of personnel management, and in a more specific context – competences management [25].

While realising the research described in this article, the authors made an effort to follow the principles of correctness of the choice of research methods and tools, adapting them to the research topic. It was also necessary to avoid errors described as a crisis of educational research and related to the division between research of practice and theory, and to avoid proving theses ahead of research [26–28].

The scope of research presented was identification and evaluation of support forms of the educational process, understood as cooperation between at least two subjects. The condition for cooperation was a mutual benefit of the partners. Enterprises have their own goals of cooperation, and so do schools. Based on these goals, many possible forms of cooperation between enterprises and schools have been developed. They have become the subject of research in order to broaden knowledge of research problems defined above. The research procedure is presented in (Fig. 1).

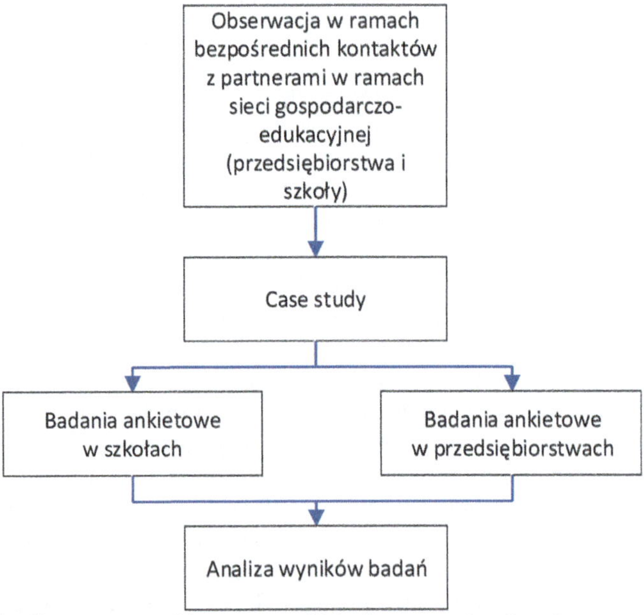

Observation in direct contact with partners within the economic-education network (enterprises and schools)
Survey in schools
Survey in enterprises
Research analysis

Fig. 1. Research procedure used to evaluate the convergence of needs of enterprises and vocational schools in the aspect of supporting schools by enterprises Source: Own elaboration.

In research up to 2017 (observation and case study), 25 types of forms of supporting schools offered by enterprises have been identified. They are presented in Table 1. There were two stages of identification of these forms of support. Authors of the article are part of a team of Poznan University of Technology, which realises a long-standing project 'Time for Professionals BIS – Professional Great Poland' – CZBIS (1st edition 2012–2015, 2nd edition 2015–2022). The project is financed by the EU, with Regional Government of Wielkopolska Voivodeship as the leading partner. This Project aims at creating a steady, functioning educational-economic network [29, 30]. In order to bring the model network closer to reality, activities were limited to one region. It is Wielkopolska Voivodeship (Poland), with more than 3.5 million inhabitants. Contacts between enterprises and schools showed that there are various forms of supporting schools, with several of them repeated frequently. Information concerning these forms of support was organised and tentatively verified in 2017 together with cooperating enterprises. Based on a case study, models of cooperation between enterprises and schools were developed [31–33].

Table 1. Identified forms of supporting schools used by enterprises. Source: Own elaboration.

Item	Direct relationships
1	Traineeship organised for students
2	Internship organised for students
3	Dual education
4	Patronage over the school
5	Co-organising patron classes in the school
6	Participation in meetings with students in the school
7	Inviting students to visit enterprises (trips, workshops, etc. – short time)
8	Transfer of teaching materials to schools and teachers
9	Cooperation with the school in preparing examination of vocational qualification
10	Organising vocational teaching
11	Running a school centre/ science park/ open laboratories
12	Participation in local job fairs
13	Participation in meetings regarding local labour market
14	Organising competitions for school students
15	Advertising job offers
16	Cooperation with job career advisors
17	Accepting students of vocational schools from other countries
18	Cooperation in the area of vocational education with technical universities
19	Cooperation in development and implementation of pilot curricula of vocational education
20	Cooperation in preparing and running conferences and seminars
21	Co-organization of dedicated conferences and training for the regional labour market
22	Funding scholarships for students
23	Funding internship/scholarships for teachers
24	Meetings with parents
25	Equipping laboratories, workshops, etc.

In the third stage of the research, enterprises and schools were asked to evaluate the importance of identified forms of support. The scale used was from 1 (the least important) to 5 (the most important). Already existing relationships in the regional educational-economic network were used for the research. The invitation was accepted by 38 schools and 40 enterprises with a total number of nearly 30,000 employees. Among the enterprises studied there were 5 small, 13 medium and 22 big ones (including 6 very big, i.e. with more than 1,000 employees). Nonprobability sampling was used for the research. Small enterprises were excluded from the analysis. The character of cooperation of these enterprises with schools is significantly different from that of medium and big ones.

Two forms of support from (Table 1) were excluded from the evaluation:

(a) meetings with parents (item no. 24) – since that form of support was identified after carrying out research in schools, which made it impossible to compare evaluation of its importance
(b) equipping laboratories, workshops, etc. (item no. 25) – since in the survey for enterprises that form of support was omitted, which should be deemed as a research error,

In enterprises, HR employees – most often managers of these organisational units – evaluated the importance of these forms of supporting schools, regarding benefits that enterprises may obtain by helping schools. They count mainly on benefits expressed in the number of employed school graduates and the quality of their competences.

Schools that encounter a continuous lack of finance to buy equipment and to pay high-class specialists, look for all kinds of support in order to improve the quality of costly vocational education. Their opinions are influenced by other premises than those in enterprises.

3.1 Evaluation of the Importance of Forms of Support by Schools and Enterprises

Average evaluation of the importance of forms of support made by enterprises and schools is presented (Fig. 2). Additionally, grey bars show differences in the evaluation of individual forms of support. Bars above the x-axis mean that that particular form of support is more important in enterprises than in schools studied. Numbers of individual forms of support refer to item numbers from (Table 1).

None of the forms of support are completely unimportant for the enterprises or schools. Average evaluation of enterprises vary between <2.60; 4.49> (gap RB = 1.89), and average evaluation of schools vary between <3.00; 4.84> (gap RS = 1.84). Both parties valued especially highly organising traineeship and internship for students, and schools valued also: meetings with parents and equipping laboratories, workshops, etc.

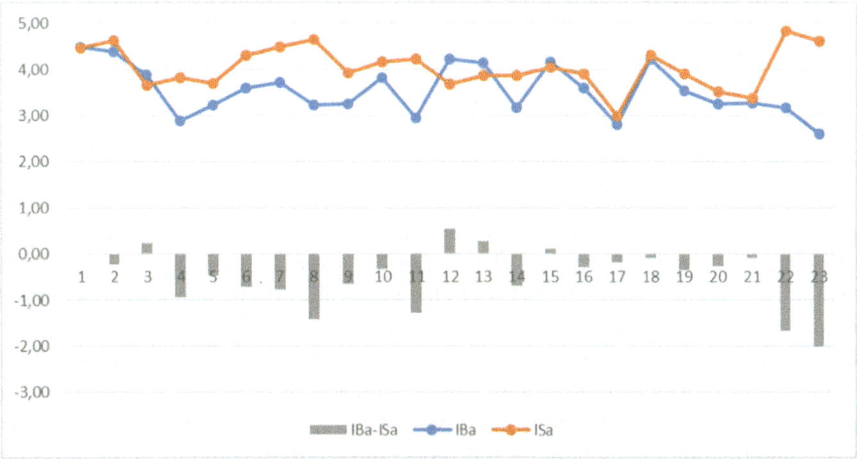

Fig. 2. Comparison of average evaluation of importance of forms of supporting schools done by enterprises and schools cooperating within the network (B = 35, S = 38). Legend: B – businesses, S – schools, IBa – average importance from the perspective of businesses, ISa – average importance from the perspective of schools Source: Own elaboration.

3.2 Convergence in Evaluation of Importance of Forms of Supporting Schools and the Areas of Divergence of Evaluation

Evaluation of enterprises and schools must be deemed as largely convergent. Analysis of Fig. 1 shows that most forms of support are valued higher by schools than by businesses. Equally highly (values above 4) are valued those forms of support as traineeship, internship, advertising job offers, cooperation in vocational education with technical universities.

The first 3 forms of support are traditional methods of cooperation of the subjects studied. They are recognisable, supported by the system of formal education, and accepted. The fourth form was popularised in schools and enterprises studied in the research by the Regional Government of Wielkopolska Voivodeship and Poznan University of Technology within projects realised in Wielkopolska Voivodeship since 2010. A similar score may not be achieved in regions where the technical university is not involved in cooperation with the regional government and the vocational education environment. Evaluation of the form ‘cooperation in the area of vocational education with technical universities’ shows the high quality of such cooperation and the developed model of cooperation with schools and entrepreneurs, which is satisfactory for every party studied. For the authors such high values were surprising.

Significantly more important forms of support from the perspective of schools are: patronage over the school, inviting students to visit enterprises (trips, workshops, etc. – short time), cooperation with the school in preparing examination of vocational qualification and organising competitions for school students, although some of them did not score the highest, e.g. patronage over the school, co-organisation of patron classes in the school.

The following forms of support were evaluated very highly, and definitely much higher than in enterprises, by the schools:

- transfer of teaching materials to schools and teachers,
- funding scholarships for students,
- funding internships/scholarships for teachers.

The research was carried out in Polish schools, hence a high demand for teaching materials is not surprising, since schools in that country are underequipped and they cannot actually run basic practical classes [34]. It must be stressed, however, that the situation is gradually improving thanks to the use of European Union funding aimed at purchasing equipment. However, the EU funding is used to purchase basic equipment, which is worn-out after several years. Therefore it is necessary to look for other forms of supporting schools than just transferring equipment. It must also be added that this equipment is mostly insufficient to prepare students for work in enterprises located in the neighbourhood of the school. A new model of activity is currently being developed at Poznan University of Technology in cooperation with Wielkopolska Voivodeship, schools and businesses within the afore-mentioned CZ BIS project.

Schools are greatly interested in scholarships for students and teachers. The scholarship is understood as financial support for persons with an exceptional level of competences. It should be treated analogically to sponsoring good athletes. From the perspective of enterprises, such activity should be treated as CSR rather than PR. As research shows, businesses are not really interested in these forms of support. In individual conversations, some entrepreneurs point to the fact that financing students is not profitable, since it does not guarantee that they will come to work at the enterprise. As to financing teachers and improving teachers' competences, entrepreneurs do not see any benefits for the business. Other research carried out within the CZ BIS project shows that teachers are not interested in unpaid internship in enterprises.

The diagnosis of needs of 79 schools participating in the project, carried out in 2017, showed that only 15 teachers were interested in internship, 318 were interested in training courses, 25 in postgraduate studies, and 17 in qualification courses. Internship covered 40 unpaid working hours, with travel costs reimbursed.

Some entrepreneurs are very interested in having teachers in internship. However, they are interested in longer periods (at least one month that a teacher would spend in the enterprise). In addition, entrepreneurs are of the opinion that the value they give to the school is the time devoted and the know-how transferred. In return, they expect a higher level of education, related to the specific character of the enterprise that accepts the teacher in internship. On the other hand, teachers look for forms of support that will accelerate their career progression, which is defined in legal acts.

Entrepreneurs look for ways to communicate their needs in the area of employment and want to have an influence on shaping the local labour market. Hence, a high score given to participation in job fairs and in meetings regarding the local labour market. From the point of view of these particular benefits for schools, they are not the highest-valued forms of support.

3.3 Convergence of Importance of Individual Forms of Support for Enterprises

While analysing the research results, the character of enterprises and its influence on chosen forms of support were taken into consideration. From among cross-dependencies, taking into account the size of the population studied and representativeness of the research, the size of the enterprises (number of employees) and the frequency of the chosen forms of support and its value for the enterprise were analysed.

In the research, data collected in very big, big and medium-size enterprises were taken into account. It was decided to check tentatively if there are any significant differences between scores given to forms of support between these two different types of enterprises. The results are shown in (Fig. 3). Samples of enterprises were relatively small, therefore it may only be tentatively concluded that there is no big divergence between the value of importance of the forms of support in big and medium-size enterprises. The only significant difference was in the evaluation of accepting students from vocational schools from other countries. Although in both groups of enterprises the average scores were not the highest, the difference between them is clear. For medium-size enterprises, it is 3.46 and for big one it is only 2.41. Reasons for such divergence may be the subject of further research following a new questionnaire on a bigger sample.

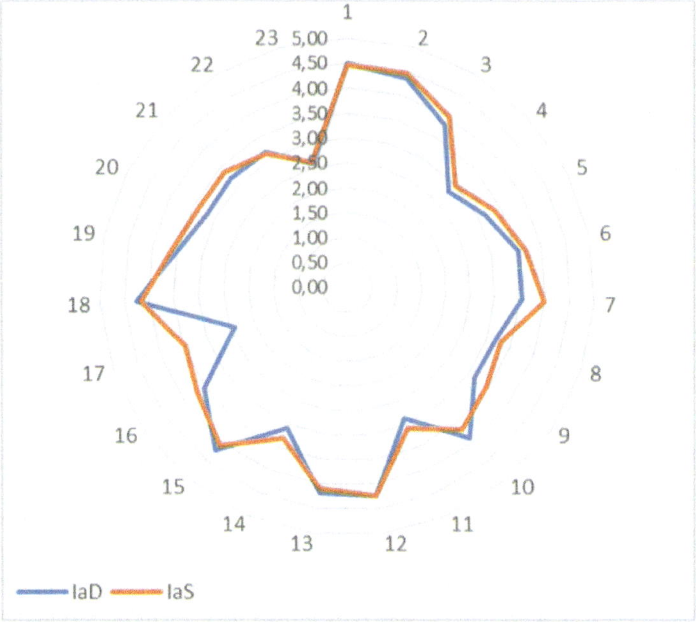

Fig. 3. Divergence in the evaluation of forms of supporting schools by big and medium-size enterprises. Legend: IaD – average importance in big enterprises, IaS – average importance in medium-size enterprises Source: Own elaboration.

Interdependency between the number of subjects using a given form of support and the average importance of this form was also analysed. The analysis shows that there is a clear correlation between the average score of the form of support and the number of enterprises that use it (Fig. 4). It was checked by calculating the coefficient of Pearson correlation, which is 0.834, for B = 30, N = 2 and $\alpha = 0.05$, where B – number of studied enterprises is 30 and not 35, since in 5 of the enterprises not all 23 identified forms of support were evaluated, N – number of degrees of freedom, α – statistical significance. It may be added that the correlation is present also for the statistical significance of $\alpha = 0.001$. Model variance, through coefficient of determination R^2 was on the level of 0.756, which allows for the opinion that it is a good matching.

The higher number of businesses use a certain form of support, the higher score it gets. Businesses value highly those forms of support that they use. It does not mean automatically that this form of support must bring expected results for the enterprise, related to its business goals. If that is the case – it should be verified, in other, much deeper research. Time, work and costs it would require are so high that first it would be best to carry out a case study or a multiple case study.

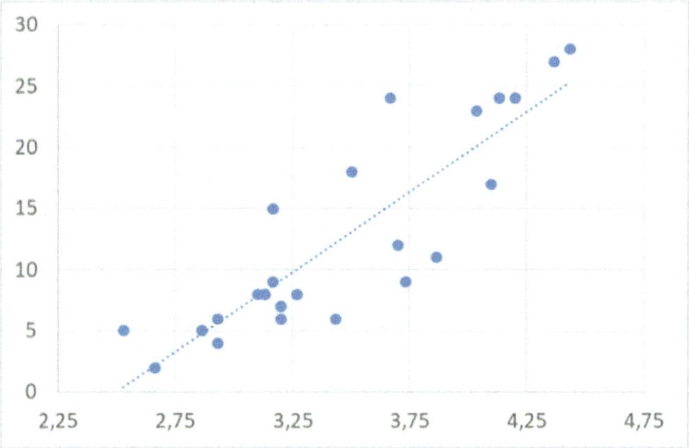

Fig. 4. Average importance of the form of support and the number of enterprises using it. Own elaboration.

Interdependency between the size of the enterprise (based on the number of employees) and the number of forms of supporting schools was also examined. As a result of the examination of correlation between the number of employees and the number of forms of support it must be stated that there is none. For example, there is an enterprise with about 6,000 employees which uses 5 forms of supporting schools, and another company, with about 100 employees, which uses 14 forms of support. Further research is necessary to explain if the scope and intensity of support has an influence on the employment efficiency, for example. This score may also be connected with a different policy of personnel management used by big enterprises that operate in international corporations.

The research also showed indications of individual, additional forms of support. In the research more forms of support were identified, which from the perspective of enterprises are significant, and the use of which should be considered. Employers indicated the following forms of supporting schools:

- cooperation between the school, office, enterprises – a network of cooperation,
- support for study groups in schools (e.g. provision of equipment),
- support for foreign trips of students to companies collaborating with the entrepreneur,
- running classes by representatives of enterprises,
- one-day workshops in schools, with participation of employees, during which students would realise projects,
- workshops on how to write a CV.

A bigger number of studied enterprises would allow for the analysis of the correlation, e.g. between the type of business activity, the recruitment policy, how many years the business is on the market, and pro-social orientation visible in CSR activities. During the research, it was also noted that there is a varied way of involving managerial staff and HR employees in the recruitment policy. This aspect is also worth developing in further qualitative research on the choice and importance of forms of cooperation between an enterprise and a school.

4 Discussion and Conclusions

This paper presented results of research on forms of supporting schools by enterprises. Like in any cooperation, both parties are motivated by achieving their own goals. In case of schools, it is connected with realisation of the curriculum approved and with the pursuit of an improved level of education. In case of businesses, it is obtaining employees and a long-term, preventive activity, by active influence on the education system.

The research realised referred, partly through the research area, to the educational part, but because of the effect of the relationship examined, it referred to the area of management. One of the most important conclusions resulting from the analysis of research is the superior role of economy over education system. Enterprises in the described partnership should play the creative and initiating role. As a consequence, it is necessary to cover higher costs and accept higher risk connected with the failure of realised activities.

However, while analysing current economic situation connected with the dynamic development and an unsatisfied demand for employees (including adequate competences), improvement of the relationship between an enterprise and the educational surrounding is the key problem of human resources management, if not of management understood as the set of activities allowing a business achieve the goals.

The results of the research are of both pragmatic and theoretical dimension. The practical significance of realised activities was observed as early as at the research stage, which caused persons responsible for human resources management to carry out an audit of benefits resulting from cooperation with schools. Schools also began to choose carefully forms and subjects to enter into cooperation with, knowing that candidates and students are increasingly more aware when building their career path, and population decline forces schools to improve the educational offer.

Basic conclusions resulting from observation and research confirm the following theses:

- employer must look for employees in an active, systematic way and long in advance,
- enterprises must build strong relationships with the subjects surrounding them, with special attention paid to cooperation with schools,
- methods of employer-school cooperation and support forms resulting from it should be equally beneficial for both parties,
- audit of benefits resulting from employer-school cooperation should be carried out regularly and frequently, forms of support should be continuously improved,
- forms of realised support should be adapted to individual subjects and take into account their specific character (size, branch, etc.),
- situation in the labour market, where the final decision belongs to the employee, forces businesses to change their approach to the problem of human resources management.

Even though research presented in this article has already found practical use in the improvement of the relationship between enterprises and schools, as well as in the modified forms of recruiting employees, it is necessary to further analyse and improve the form of a relationship between an enterprises and a school.

Varied importance of forms of support seen by the two parties of the relationship may be confirmation of divergence in expectations regarding benefits, which in turn is a premise to have frequent and detailed conversations on the subject of cooperation. Exchanging opinions and experience connected to the relationship economy – education should be inspired not only by those two parties, but it should be long-term and of a systemic character. For this reason the aspects of forms of cooperation between schools and companies must be the subject of activities performed by local self-governments and other entities responsible for the education system and economy. Initiatives creating networks for the improvement of the education quality mentioned in this paper are also an example of good practices in the creation of a relationship between an enterprise and a school.

References

1. Myszewski, J.M.: Sustainable improvement in education. Bus. Process. Manag. J. **24**(6, 5), 1381–1392 (2018)
2. Anand, C.K., Bisaillon, V., Webster, A., Amor, B.: Integration of sustainable development in higher education – a regional initiative. J. Clean. Prod. **108**(1), 916–923 (2015)
3. Pike, P.: Missing - the people's voice: organisation for economic co-operation and development. Policy Futur. Educ. **12**(6), 753–757 (2014)
4. Morozkina, A., Sabelnikova, E.: Education systems and multilateral development banks: international practices and perspectives. Int. Organ. Res. J. **12**(1), 24–42 (2017)

5. Wallenborn, M.: Vocational Education and Training and Human Capital Development: current practice and future options. Eur. J. Educ. **45**(2), 181–198 (2010)
6. McCoshan, A., Drozd, A., Nelissen, E., Nevala, A.M.: Beyond the Maastricht communique: developments in the opening up of VET pathways and the role of VET in labour market integration: consolidated final report. ECOTEC Research and Consulting, Brussels, Belgium, European Commission, Directorate General for Education and Culture (2008)
7. Donlevy, V., Curtarelli, M., McCoshan, A., Meierkord, A.: Study on obstacles to recognition of skills and qualifications. Final Report, Directorate-General for Employment, Social Affairs and Inclusion. Publications Office of the European Union, Luxembourg (2016)
8. Asheim, B.T., Isaksen, A.: Regional innovation systems: the integration of local 'sticky', and global 'ubiquitous' knowledge. J. Technol. Transf. **27**(1), 77–86 (2002)
9. Likar, B., Macur, M., Trunk-Širca, N.: Systemic approach for innovative education process. Kybernetes Int. J. Syst. Cybern. **35**(7/8), 1071–1086 (2006)
10. Halvarsson, L.A., Gustavsson, M.: Conditions for employee learning and innovation – interweaving competence development activities provided by a workplace development programme with everyday work activities in SMEs. Vocat. Learn. **11**(1), 45–63 (2018)
11. Gessler, M.: Educational transfer as transformation: a case study about the emergence and implementation of dual apprenticeship structures in a German automotive transplant in the united states. Vocat. Learn. **10**(1), 71–99 (2017)
12. Bauer, K., Fisher, F.: The education research–practice interface revisited: a scripting perspective. Educ. Res. Eval. **13**(3), 221–236 (2007)
13. Brunet, I., Rodríguez-Soler, J.: Vocational training and innovation: research on innovation transfer between VET schools and enterprises. Revista de Educación **365**, 177–201 (2014)
14. Valente, A.C., Salavisa, I., Lagoa, S.: Education quality and economic performance in Europe. In: Proceedings of the European Conference on Knowledge Management, ECKM, no. 3, pp. 1028–1036 (2014)
15. Musyarofah, S.: The need for new paradigm of sustainability reporting in higher education. Int. J. Econ. Policy Emerg. Econ. **5**(4), 342–352 (2012)
16. Chydenius, T., Gaisch, M.: Work-life interaction skills: an exploration of definitional and functional perspectives within the Austrian and Finnish ICT. Bus. Perspect. Res. **4**(2), 169–181 (2016)
17. Brown, L.A.: The course valuation model and 10 steps to increase course value: the business communication course. J. Educ. Bus. **90**(6), 340–346 (2015)
18. Szafrański, M., Goliński, M.: Interactive Laboratory of Technical Knowledge Acceleration in Program of Acceleration of Knowledge of Technology, Mathematics and Natural Science in Poland. Publishing House of Poznan University of Technology (2010)
19. Szafrański, M., Goliński, M.: System for Professionals – monitoring employers' demands for key competences in Wielkopolska. In: Recent Advances in Computer Science, Proceedings of the 19th International Conference on Computers, 19th International Conference on Circuits, Systems, Communications and Computers, pp. 184–199 (2015)
20. Simmons, R.: Civilising the natives? Liberal studies in further education revisited. Br. J. Educ. Stud. **63**(1), 85–101 (2015)
21. Employers' cooperation with vocational schools: The Present and Future. National Center for Supporting Vocational and Continuing Education, Warsaw (2013)
22. A detailed description of the Priority Axes of the Operational Program Knowledge Education Development 2014–2020. Ver 12, Warsaw (2018)
23. Mas-Verdú, F., Ortiz-Miranda, D., García-Álvarez-Coque, J.M.: Examining organizational innovations in different regional settings. J. Bus. Res. **69**(11), 5324–5329 (2016)
24. Furlong, J.: New Labour and teacher education: the end of an era. Oxf. Rev. Educ. **31**(1), 119–134 (2005)

25. Vanderlinde, R., Van Braak, J.: The gap between educational research and practice: views of teachers, school leaders, intermediaries and researchers. Br. Educ. Res. J. **36**(2), 299–316 (2010)
26. Badley, G.: The crisis in educational research: a pragmatic approach. Eur. Educ. Res. J. **2**(2), 296–308 (2003)
27. Mc, I.D.: Bridging the gap between research and practice. Camb. J. Educ. Teachers' Good Pract. Res. **35**(3), 357–382 (2005)
28. Biesta, G.: Bridging the gap between educational research and educational practice: the need for critical distance. Educ. Res. Eval. Int. J. Theory Pract. **13**(3), 295–301 (2007)
29. Perrotta, C., Bailey, C., Garside, C.: Culture, technology and local networks: towards a sociology of 'making' in education. Camb. J. Educ. **48**(5, 3), 553–569 (2018)
30. Bohne, C., Eicker, F., Haseloff, G.: Competence-based vocational education and training (VET): an approach of shaping and networking. Eur. J. Train. Dev. **41**(1), 28–38 (2017)
31. Szafrański, M.: Models of businesses' support for technical knowledge development in Wielkopolska region – a qualitology approach. In: Conference Proceedings, Engines of Urban and Regional Development, 6th Central European Conference in Regional Science, Slovakia, pp. 128–137 (2017)
32. Szafrański, M., Goliński, M., Wiecek-Janka, E., Graczyk-Kucharska, M., Spychała, M.: Nonformal activities for the improvement of the development of vocational competences – the polish example. In: Book of Abstracts of the 1ST International Scientific Conference, Teaching Methods for Economics and Business Sciences, p. 18 (2017)
33. Graczyk-Kucharska, M., Szafrański, M., Goliński, M., Spychala, M., Borsekova, K.: Model of competency management in the network of production enterprises in industry 4.0—assumptions. In: Hamrol, A., Ciszak, O., Legutko, S., Jurczyk, M. (eds.) Advances in Manufacturing. LNME, pp. 195–204. Springer, Cham (2018). https://doi.org/10.1007/978-3-319-68619-6_19
34. Education at a Glance 2018: OECD Indicators. OECD Publishing, Paris (2018)

Computer Modeling and Simulation in Engineering Education: Intended Learning Outcomes Development

Paweł Litwin(✉) and Dorota Stadnicka

Rzeszow University of Technology, Rzeszow, Poland
plitwin@prz.edu.pl

Abstract. Engineers need to know a wide spectrum of modern techniques and tools to be able to apply them in their professional work. Modeling and simulation are examples of the activities that are commonly undertaken by modern engineers, especially in the context of Industry 4.0 concept. Therefore, it is important that in the process of engineers' education the issues related to modeling and simulation should be strongly emphasized. This article attempts to show the benefits of using modeling and simulation techniques in the education of engineers. In addition, it proposes corresponding intended learning outcomes consistent with the Conali ontology and the Bloom's taxonomy. It also presents examples of modeling and simulation applications in engineering courses and shows the structure of possible tasks to be performed in the practical implementation of simulation.

Keywords: Modeling and simulation · Engineering education · Intended Learning Outcomes · Conali ontology

1 Introduction

Simulations allow to present the real world with the use of a simplified model, which represents actions undertaken in a process or the phenomena occurring in the process.

Students, in many cases, have no possibility to observe how a real process functions. Therefore, simulations can be used to analyze the behavior of the process. Moreover, companies are interested in process simulations performance, for example in order to check the consequences of the changes proposed in the process.

Many examples of process simulations can be found in the literature [1–3]. Simulations concern either single processes or a group of processes engaged to achieve a certain result (a product). Different simulation software is applied to create models of processes and perform simulations. In this paper attention is focused on manufacturing processes, especially on manufacturing lines composed of several production processes. However, the goal of the paper is not to solve any specific industrial problems, but to present how modeling and simulations can be applied in order to obtain specified Intended Learning Outcomes (ILOs).

In the paper the authors first presented the benefits of the use of modeling and simulation in engineering education. Then, in Sect. 2, they focused on the simulation

A. Hamrol et al. (Eds.): *Advances in Manufacturing II* - Volume 3, LNME, pp. 169–184, 2019.
https://doi.org/10.1007/978-3-030-17269-5_12

application in educational programs. The authors believe that it is important to teach students (future employees of companies) how to create models in different environments to be ready to perform further simulations. In Sect. 3, the ILOs expected to be obtained in a didactic process are proposed. The ILOs were developed based on Conali ontology and Bloom's Taxonomy, which are first explained in this section. Then, two sets of the proposed ILOs are presented. The first set concerns a didactic process in which students learn, on the basis of data, how a computer model can be developed. The other set concerns the modeling and simulation application in problem analyses.

The fourth section of the paper presents how the planned ILOs can be achieved with the use of the presented educational programs. The first example concerns classes conducted in order to teach students a computer model development. Two other examples concern complex simulations realized in the diploma thesis development for the needs of industry. Additionally, a structure of possible tasks to be performed in the practical implementation of simulation is proposed.

The work ends with a summary where, among others, additional needs concerning student's knowledge are indicated, based on modern technologies which should be involved in the educational process.

2 Modeling and Simulation in Engineers Education

For many years, engineers' curricula have contained issues related to modeling, and in recent times one of the most popular types of modeling is computer or computational modeling and simulation. Essentially, computer modeling and simulation is used in education in three different ways [4]. The first way puts emphasis on students learning the art of modeling. In this case, students learn a modeling language and principles, often referred to as systems thinking [5]. For example, System Dynamics method allows students to analyze the cause and effect relationships and to perceive the system as a whole. This use of modeling occurs most often in engineering practice and is used, among others, to simulate and analyze existing or developed systems or products. The second way of introducing modeling into education is that students gain knowledge about the modeled area. Computer modeling is, in this case, used as a "vehicle for understanding" [6] in domains such as algebra, calculus, finite math, physics, etc. The third method of using computer models in education consists in discovering knowledge by working with ready-made models, e.g. simulation of experiments in virtual laboratories [7]. In this type of models, after entering the values of input variables and registering the values of the output variables, the task of students is to identify the mechanisms constituting the foundation of the analyzed model. West and Graham [8] give five factors that positively influence the learning process: visualization, interactions, reflection, authenticity & engagement and practice. These factors of the technology impact on the learning process can be directly related to modeling and simulation in education of engineers:

Visualization - Commonly used programs for Computer Aided Manufacturing (CAM) allow for simulation and visualization of machining processes, what gives the opportunity to correct each operation.

Interactions - Simulations carried out in the form of games teach students how to interact in a team and perform different roles. In [9] a simulation game in which students learn the cooperation in the areas of supply, production, planning and control, quality control, sales and marketing was presented. It is worth noting that simulation games can be implemented with the support of IT technology (network environment, web applications) as well as in a traditional form (e.g. board games).

Reflection - Simulations used in education teach critical thinking and expanding knowledge. Teaching simulations requires learners to be active participants in order to determine parameter values, try to predict results and formulate new questions. Simulations allow to extend knowledge to a new problem and build understanding through analogical reasoning [10].

Authenticity and Engagement - Simulations are powerful tools that enable students to perform authentic, real tasks in a safe environment. The example presented in the work [11] is the use of simulations in training in an atomic power plant operation. The training is important as an operator's error in such a plant may cause disastrous consequences.

Practice - Computer simulations allow the analysis of existing or newly developed devices or systems in a wide range of input variables determining their behavior. A simulation analysis allows to determine the values of input parameters providing the optimization of the system operation according to the selected criterion (e.g. cost, efficiency, reliability). The examples developed in [12] allow to test the manufacturing system operation parameter values in order to ensure the expected system performance and identification of threats to the production plan implementation.

The learning process should be constructed in such a way that it allows to obtain certain learning outcomes by students, to be sure that the achieved knowledge can be applied in real problem analyses. The research problem analyzed in this paper concerns the development of Intended Learning Outcomes connected with modeling and simulations of manufacturing processes.

3 Intended Learning Outcomes (ILOs) Development

3.1 Assumptions for ILOs Development

An educational process is undertaken to achieve certain Intended Learning Outcomes (ILOs). In the area of learning, the so-called "constructivist" approach [13] is currently emergent. In this approach, a learner is perceived as the main factor in the process of acquiring knowledge. The "constructivist" approach to the learning process is contrasted with the "objectivist" approach, in which the main emphasis is put on the teacher passing knowledge to a student. In the constructivist approach, the teacher's role is only auxiliary, supporting students in the learning process. Therefore, the construction of ILOs presented in this work is based on Conali (Constructive Alignment) ontology [14] which structure is presented in (Fig. 1). The ILO is composed of a verb, has a content and a context. The verb is an educational verb related to the level of understanding a content. The content presents an object of the learning process. It is applied in a certain context.

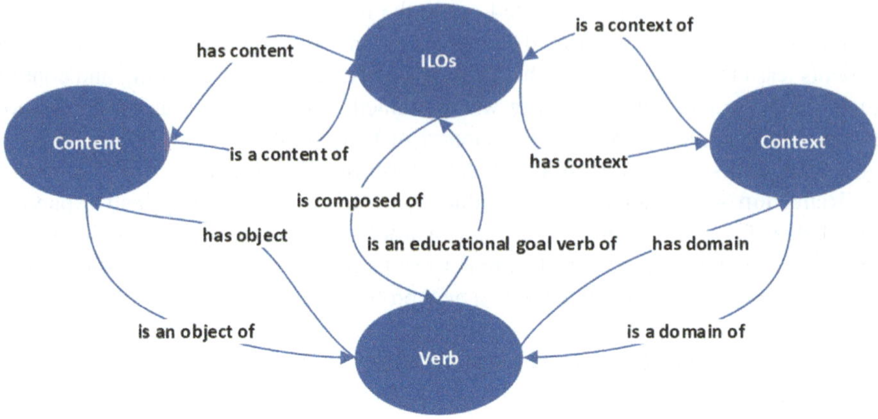

Fig. 1. Construction of constructive alignment according to Conali ontology. Based on [14, 15].

In ILOs construction Bloom's Taxonomy was applied [16–18]. According to the taxonomy, there are six levels of understanding which are connected to certain skills such as: knowledge, comprehension, application, analysis, synthesis and evaluation. The verbs used in ILOs creation are connected with certain levels of understanding. The verbs used in this work in the development of ILOs for learning of modeling and simulation are underlined in (Table 1).

Table 1. Verbs used in ILOs development.

Level	Skills	Verb
Level 1	Knowledge	Select, list name, **define**, **describe**, memorize, label, **identify**, locate, **recognize**, **measure**, **be aware**
Level 2	Comprehension	Match, rewrite, express, illustrate, **explain**, distinguish, **summarize**, **interpret**, extend, **build**, **test**
Level 3	Application	Organize, generalize, prepare, produce, choose, **apply**, solve, draw, show, **implement**
Level 4	Analysis	Compare, **analyze**, classify, **differentiate**, select, **prioritize**, point out, categorize, **validate**
Level 5	Synthesis	Compose, **hypothesize**, develop, combine, **construct**, **produce**, **create**, organize
Level 6	Evaluation	**Judge**, **predict**, criticize, support, **evaluate**, summarize, compare, consider, **recommend**, **discriminate**

In the learning process two types of knowledge can be achieved: declarative and functional knowledge. Declarative knowledge is called open knowledge [19]. A person can be aware of it and can express it clearly [20]. Functional knowledge is the knowledge that can be adapted to different circumstances.

The developed ILOs present the type of knowledge to which the ILO is connected.

The presented assumptions were taken into consideration in the development of:

- a set of ILOs for the course in which students learn how to create a model of a system and perform simulations;
- a set of ILOs for the preparation of a diploma thesis in which students use simulations to perform system analyses.

3.2 Proposal of ILOs Set for Modeling and Simulation

In this study, we deal with the role of computer simulations in the learning process. The motivation to take up such a topic is the fact that learning with the use of computer simulations contributes well to the constructivist approach to the learning process. In addition, simulations are a very popular form of computer-aided learning.

A simulation can be defined as a representation of reality, some known processes or phenomena. It is a mathematical or algorithmic model, usually implemented in the form of a computer program with an appropriate set of constraints, which allows the analysis of the system behavior. Simulations allow for "what-if" analysis in different conditions that are difficult to achieve in reality. Therefore, it applies a "practice in safety" to make optimal decisions in the given restrictions.

Simulations can significantly contribute to learning outcomes, provided that they are strongly related to the content of the course and to the learning objectives. Simulations should support learning objectives in a clear and direct way, otherwise they will be a flat and uninteresting experience for students and teachers [21].

In [22], the analysis of educational goals was conducted in a group of academic teachers of engineering and science using simulation methods in their classes. The study was attended by 14 instructors educating students in Electrical and Computer Engineering, Materials Science Engineering, Chemistry and Physics. The participants of the study had been selected based on the intensity of the use of simulation tools available on the nanoHUB.org portal [22]. As a result of a careful analysis of transcripts in the group of instructors under study, eight main categories of the learning objectives related to the use of simulation methods in the training of engineers were identified.

The categories selected in the study led to the formulation of two main learning goals:

- Building simulations to understand the possibilities of modeling and computational techniques for solving engineering tasks - learning by modeling (LBM);
- The use of simulations for identifying and describing the fundamental physical rules that govern the behavior of systems, devices or materials - learning from models (LFM).

The following categories of the learning objectives given in [22] are organized according to the growing level of understanding and skills. Moreover, a set of learning objective categories was supplemented by the authors of this article with additional verbs used in Bloom's taxonomy:

1. To recognize (identify, define) and to be aware (interpret) of the potential role of computational simulations in a particular field of science or engineering (LFM, LBM).
2. To measure (describe) materials or devices by collecting data as in a laboratory experiment (mostly LBM).
3. To explain (summarize) the cause-effect relationship of the underlying model governing a simulation (mostly LFM).
4. To test (interpret) the accuracy of a given model and/or its computational implementation (LBM).
5. To validate (analyze, differentiate) the results or performance of the product of a design task (LFM).
6. To implement (apply) computational techniques in a modeling task (LBM).
7. To predict (evaluate) the results of an experiment in a design task (LFM).
8. To discriminate (judge, recommend) between different models to represent a given physical phenomenon (LFM, LBM).

It is worth noting that the above-mentioned categories of educational objectives were developed on the basis of transcripts prepared by academic teachers who had no formal preparation in the use of simulations as teaching tools in the training of science and engineering.

Nevertheless, based on the presented categories of learning objectives, it is possible to develop the ILOs set, compliant with the Bloom's taxonomy, for the simulation models creation and analysis, summarized in (Table 2).

Table 2. Structure of knowledge for simulation models creation and analysis.

No	Kind of knowledge	Level	Verb	Content	Context
1	Declarative	1	Identify	Role of simulations (pros and cons)	Field of science or engineering
2	Functional	2	Explain	Cause and effect relationships	Simulation model
3	Functional	5	Create	Simulation model	Cause and effect relationships
4	Functional	3	Apply	Computational techniques	Modeling task
5	Functional	4	Analyze	Simulation results	Design task
6	Functional	5	Hypothesize	Results of a simulation	Design task
7	Functional	6	Recommend	Suitable model	Physical phenomenon (System, Process)

The presented ILOs achievement is the basis for further practical analyses. It ensures the well-established knowledge that can be next applied in any process analyses with the use of simulations.

3.3 Proposal of ILOs Set for Simulation Practical Application

The knowledge about modeling and simulation can be used in a diploma thesis preparation where simulations can be applied in practice to achieve certain goals of an educational process and, more specifically, to prove that a student is able to use the knowledge achieved in an educational process in a practical case study. The student will prove that he or she knows how to apply modern technology – simulations, being an element of Industry 4.0 concept, in the system analyses.

Input knowledge for the students who are going to perform computer simulations in a diploma thesis includes ILOs obtained in the course of simulations and other courses concerning manufacturing processes and production engineering. Through a diploma thesis preparation, a student obtains further competences which can be described by a set of ILOs. The ILOs were developed with the use of the mentioned Conali ontology and Bloom's Taxonomy. The structure of knowledge is presented in (Table 3). Then, eight intended learning outcomes (ILOs) which can be the results of an educational process were identified.

Table 3. Structure of the knowledge obtained during a diploma thesis preparation involved computer simulations.

No	Kind of knowledge	Level	Verb	Content	Context
1	Declarative	1	Identify	Key elements of a system	Specified process
2	Functional	2	Build	A model	Specified process
3	Functional	3	Apply	A chosen computer program	A computer model of a specified process creation
4	Functional	1	Identify	Possible problems	Specified process
5	Functional	4	Analyze	A system behavior	Specified disturbances
6	Functional	6	Recommend	Improvements	Specified process
7	Functional	4	Prioritize	Improvements	Specified process
8	Functional	5	Create	A computer model of improved system	Recommendations

After the diploma thesis defense a student should be able to:

ILO 1: Identify key elements of a system for a specified process.
ILO 2: Build a model of the specified process.
ILO 3: Apply a chosen computer program to model the specified process.
ILO 4: Identify possible problems in the specified process.
ILO 5: Analyze the system behavior under specified disturbances.
ILO 6: Recommend improvements for the analyzed process.
ILO 7: Prioritize improvements for the analyzed process.
ILO 8: Create a computer model of the improved system based on recommendations.

In the next section, the examples of teaching programs realized by students are presented.

4 Application of Modeling and Simulations in the Learning Program

4.1 An Example of Modeling and Simulation Application in a Didactic Process

The example of the application of modeling and simulation in the education of engineers presented below is carried out as a part of the subject of "Risk management in IT enterprises", at master's studies in the field of "Management and production engineering". The main objective of education was specified in the course sheet: The module aims to familiarize students with issues related to risk management in an accessible way, based on practical examples in which the emphasis is placed on risk management in IT projects related to software manufacturing and implementation. As the outcome of the course the student: knows, understands and applies correctly the concepts related to risk management; identifies the sources of risk; models and simulates risk; assesses the impact of risk on the project; selects remedial actions and assesses the effects of the risk management system.

Risk management is a very important aspect of running a business. In practice, risk management is identified with risk assessment and control that are intended to provide proper conditions for the organization development. The risk assessment process usually involves assessing the probability of potential threats and losses it may bring. Modeling and simulation of threats that may occur in project activities is particularly useful in situations where it is not possible to estimate the losses of a specific threat due to the lack of experience (the threat has not occurred in the past).

An example from [23] was used in the classes. The students' task was to assess the impact of an employee fluctuation on the project workflow. In the considered example, a series of simplifications was adopted: an employee leaving the project causes an immediate adoption of a trainee, full integration of the trainee lasts two months, the productivity of the trainee is 50% of an employee's productivity throughout the internship period, staff turnover is 4% in the first half of the project and 8% in the second period. (Figure 2) shows the workers' fluctuation model.

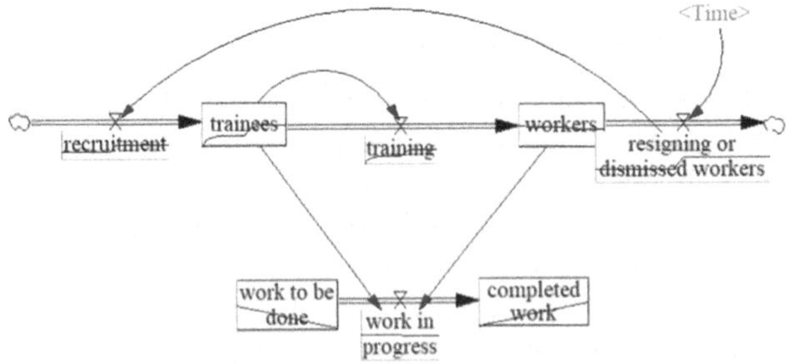

Fig. 2. A model of workers' fluctuation prepared by a student in Vensim program.

The analysis of the model was carried out in order to estimate the feasibility of project work realization at the assumed time. The obtained results are illustrated in Figs. 3, 4 and 5. The presented graphs show that in a situation where 4 employees leave the project every month, 8 trainees are trained (Fig. 3). This is due to the fact that the training time is 2 months. The result is a permanent shortage of employees in the project team (Fig. 4) and the inability to implement all the planned work (Fig. 5).

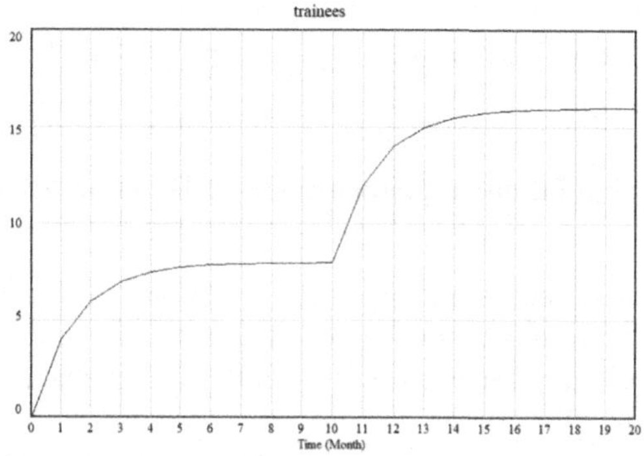

Fig. 3. A number of trainees in the project.

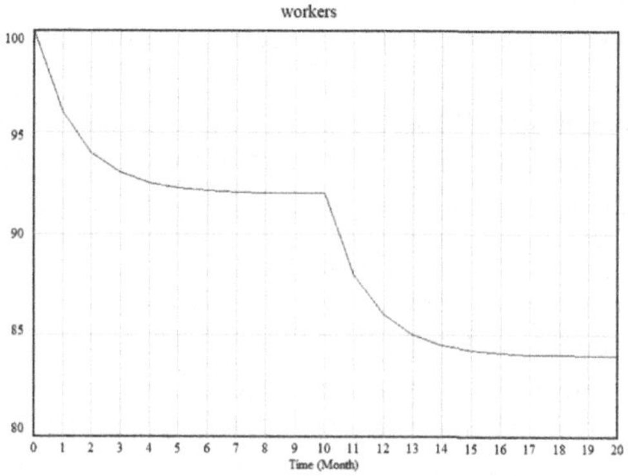

Fig. 4. A number of workers in the project.

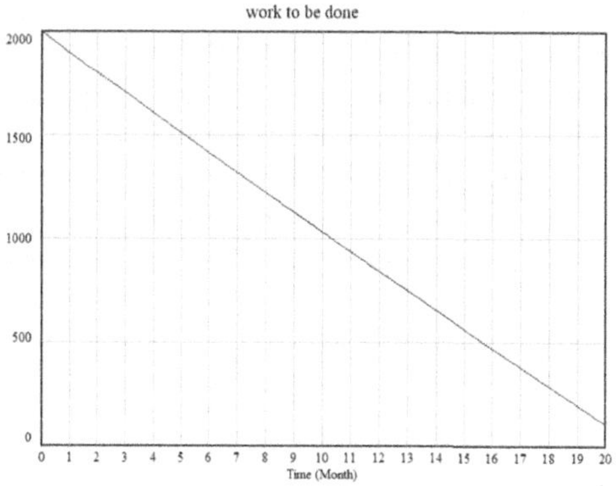

Fig. 5. The flow of work in the project.

Modeling and simulation of the risk is an important element of the Risk Management program. It allows for the implementation of several educational goals:

1. Understanding the System Dynamics method.
2. Application of modeling of the cause and effect relationships governing a given phenomenon.
3. Awareness of the importance of simulation in the study of complex objects, processes or phenomena.
4. Analysis of the possibilities of the improvements of the phenomenon or process studied.

4.2 Examples of Simulation Implementation in Diploma Thesis Realization

Simulations can be successfully implemented in a diploma thesis realization in relation to industrial problems. In this section two examples in which the ILOs presented in Sect. 3.3 were achieved are discussed. In two different diploma theses the simulations were carried out by no IT engineers.

The first topic was: "The use of IT tools for modeling and simulation of production line operation" [24].

The topic was realized according to the following educational program:

1. Review of computer programs used for modeling and simulation of a manufacturing line.
2. Selection of a simulation software.
3. Development of a manufacturing line model based on the data provided.
4. Simulating the operation of the manufacturing line in various circumstances.
5. Summary and conclusions.

The work was done on the basis of the data obtained from a manufacturing company in which certain problems exist. On the basis of the computer programs review, the student chose Vensim to create a model and to perform simulations. Then, based on the input data provided, a computer model was created (Fig. 6). The analyzed process concerns the production of metal-rubber sleeves. It includes 11 consecutive operations such as: degreasing, sandblasting, phosphatizing, glue coating, cutting, vulcanization, adhesion test, visual control and packaging of product elements and/or a ready product.

The model presents processes and a material flow. The model was used in the simulations of the manufacturing line operations under specific circumstances. The research question was: Is it possible to manufacture 420 pcs of products in the current work organization. If not, what should be done to achieve the target. After the simulations, it was discovered that it is impossible to achieve 420 pcs of products in available working time (4,200 min per a shift). The student proposed and then tested two possible solutions:

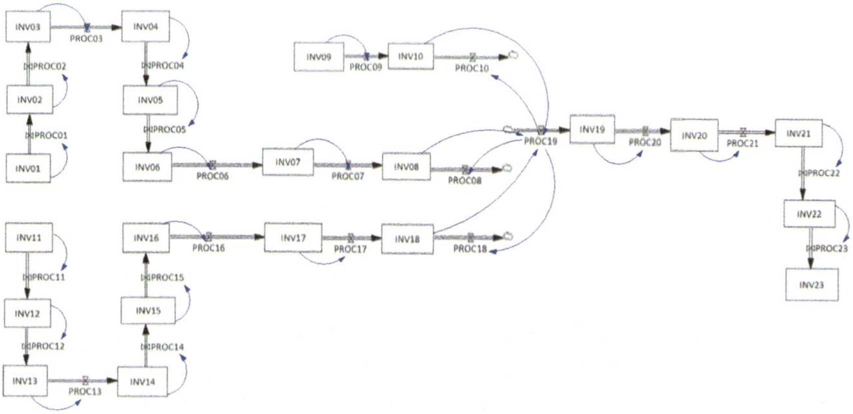

Fig. 6. A Vensim model of a metal-rubber sleeves manufacturing line.

1. Reduction of a production batch in a vulcanization operation to 20 pieces.
2. Introduction of a new machine for the glue coating process of internal sleeves with the efficiency of 70 pcs (a batch) to be done in 34 min.

The simulations with the use of the improved model confirmed that the second solution can give better results.

The other topic was as follows: "Using Any Logic to simulate the production process" [25].

The topic was realized under the following educational program:

1. Analysis of the input data delivered in the form of a value stream map.
2. Development of a manufacturing line model based on the data provided.
3. Development of graphics to represent simulation parameters and simulation results.
4. Simulating the operation of a manufacturing line in various circumstances.
5. Discussion on results.
6. Summary and conclusions.

On the grounds of the delivered value stream map, a computer model (Fig. 7) in AnyLogic software was developed together with the necessary graphics (Fig. 8). AnyLogic was indicated to be applied from the beginning.

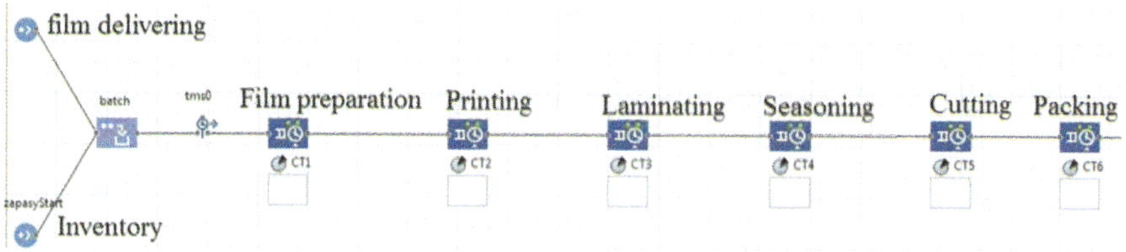

Fig. 7. A model of the manufacturing system developed in Any Logic.

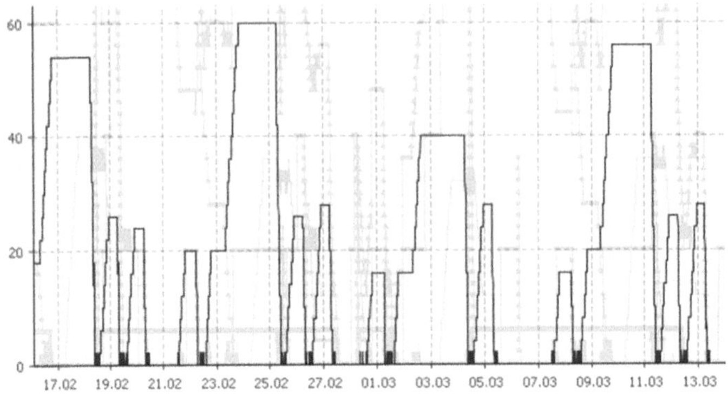

Fig. 8. A time plot depicting an emerging inventory in the production process before the packaging process.

The analyzed process concerns the production of printed laminated films and it includes 6 consecutive operations such as: film preparation, printing, laminating, seasoning, cutting and packing.

In this case, a student performed several experiments implementing various changes and assessing the system performance. On the basis of the obtained results the student proposed changes that would bring higher efficiency along with a lower amount of work in process (WIP).

4.3 Steps of Simulation Implementation

On the basis of the analyzed case studies we discovered that, depending on the context, the simulation implementation can be realized in different ways in the diploma thesis realization. In (Table 4) potential situations and tasks to be performed by a student are presented.

As the presented structure shows, adequate tasks should be chosen based on the existing circumstances connected with the analyzed problem and processes. The project realized in the indicated task performance will allow the student to answer the research question which should be put at the beginning of the work. It will also prove that the student is prepared to undertake important and complete analyses for the needs of companies.

Table 4. The structure of possible tasks to be performed by a student in the diploma thesis in which simulations are applied.

Step	Task	Description of circumstances	Possible options
Step 1	Review of computer programs possible to be applied in simulations and selection of the simulation software	Only one program available for a student	Option 1: Review is not performed
		A company indicates a computer program	Option 1: Review is not performed
		The student has to choose a program depending on the possessed data and complexity of the system	Option 2: Review is required
Step 2	Data collection and analysis	A set of data is available and complete	Option 1: The student uses existing data
		A set of data is available but not complete	Option 2: The student identifies the missing data and collects them
		A set of data is not available	Option 3: The student has to define the kind of data necessary and collect the data
Step 3	Development of a manufacturing line model	A general structure of the model is available (e.g. a value stream map)	Option 1: The student builds a computer model on the basis of a general structure or adapts the existing model
		No structure available	Option 2: The student builds the model from scratch
Step 4	Development of graphics	Standard graphics are available in the software	Option 1: The student uses the available graphics or chooses from the available options
		Graphics has to be created in the software	Option 2: The student creates the graphics
Step 5	Development of a simulation plan and simulations	There are identified problems in the manufacturing line	Option 1: The student proposes solutions and tests them
		Manufacturing line efficiency is not sufficient	Option 2: The student makes a simulation of the current state to identify the source of problems, plans and realizes an experiment to see how the efficiency changes along with the changes introduced
		The changes are planned to be implemented	Option 3: The student implements changes and assesses the results

(*continued*)

Table 4. (*continued*)

Step	Task	Description of circumstances	Possible options
Step 6	Discussion and conclusions	Different problems solutions were tested	Option 1: The student indicates appropriate solutions
		Weak points of the manufacturing line were identified	Option 2: The student indicates the source of problems and their consequences
		Planned changes were tested	Option 3: The student confirms a possible improvement of efficiency

5 Summary

Intended Learning Outcomes (ILOs) should be defined in order to ensure a high quality of an educational process. In the paper, the authors analyzed the ILOs which can be defined for a didactic process in which simulations are involved. In the ILOs development Conali otology and Bloom's taxonomy were used.

The authors proposed two sets of ILOs that can be applied for no IT engineers' education. In the learning process the students apply existing IT tools. The first set of ILOs concerns a didactic process in which students learn how modeling and simulation can be realized with the use of available software. The other set of ILOs concerns a diploma thesis realization in which students apply simulations in order to solve certain problems.

The proposed sets of ILOs were achieved in a didactic process. The presented examples show how simulations can be applied in education to ensure the highest level of knowledge acquisition by engineering students.

Finally, the steps of the simulation implementation in a diploma thesis realization are proposed. The list of steps can be a basis for choosing a set of tasks for a student working on a certain problem to solve it with the use of simulations.

Nowadays, there are also more advanced technologies that could be used in a didactic process to illustrate better the reality of the simulated work, that is virtual reality (VR). Therefore, it has to be emphasized that this is a proper direction for the didactic process improvements in terms of process simulations. Unfortunately, the infrastructure for applying VR in a didactic process to be available for all the interested students is still too expensive. However, to some extent, VR is already applied in students' education. At Rzeszow University of Technology, students have already built VR in the frame of a diploma thesis realization. Therefore, this will be the area for further research.

Acknowledgments. The research work reported here has been partially supported by the "TIPHYS - Social Network based doctoral Education on Industry 4.0" project No 2017-1-SE01-KA203-03452 co-funded by ERASMUS + of the European Commission.

References

1. Gola, A.: Reliability analysis of reconfigurable manufacturing system structures using computer simulation methods. Maint. Reliab. **21**(1), 90–102 (2019)
2. Kłos, S., Trebuna, P.: Using computer simulation method to improve throughput of production systems by buffers and workers allocation. Manag. Prod. Eng. Rev. **6**(4), 60–69 (2015)
3. Stadnicka, D., Litwin, P.: Value stream mapping and system dynamics integration for manufacturing line modelling and analysis. Int. J. Prod. Econ. **208**, 400–411 (2019)
4. Magana, A.J., de Jong, T.: Modeling and simulation practices in engineering education. Comput. Appl. Eng. Educ. **26**(4), 731–738 (2018)
5. Booth, S.L., Sterman, J.D.: Bathtub dynamics: initial results of a systems thinking inventory. Syst. Dyn. Rev. **16**, 249–286 (2000)
6. Yaşar O., et al.: Computational math, science, and technology (CMST): a strategy to improve STEM workforce and pedagogy to improve math and science education. In: International Conference on Computational Science, pp. 169–176 (2006)
7. de Jong, T., Linn, M.C., Zacharia, Z.C.: Physical and virtual laboratories in science and engineering education. Science **340**, 305–308 (2013)
8. West, R., Graham, C.: Five powerful ways technology can enhance teaching and learning in higher Education. Educ. Technol. **45**(3), 20–27 (2005)
9. Bringelson, L.S., Lyth, D.M., Reck, R.L., Landeros, R.: Training industrial engineers with an interfunctional computer simulation game. Comput. Ind. Eng. **29**(1–4), 89–92 (1995). ISSN 0360-8352
10. Podolefsky, N.S., Perkins, K.K., Adams, W.K.: Factors promoting engaged exploration with computer simulations. Phys. Rev. Spec. Top.-Phys. Educ. Res. **6**(2), 20–117 (2010)
11. Badulescu, A., Lyon, R.: Classroom simulators. User friendly education with nuclear reactor simulators. IAEA Bull. **43**(1), 25–28 (2001)
12. Stadnicka, D., Litwin, P.: VSM based system dynamics analysis to determine manufacturing processes performance indicators. DEStech Trans. Eng. Technol. Res. (2017). (icpr)
13. Cooperstein, S.E., Kocevar-Weidinger, E.: Beyond active learning: a constructivist approach to learning. Ref. Serv. Rev. **32**(2), 141–148 (2004)
14. Maffei, A., Daghinia, L., Archentia, A., Lohseb, N.: CONALI ontology. a framework for design and evaluation of constructively aligned courses in higher education: putting in focus the educational goal verbs. In: 26th CIRP Design Conference, Procedia CIRP, vol. 50, pp. 765–772 (2016)
15. Stadnicka, D.: A multi-aspects approach to increase the efficiency of enterprises. Oficyna Wydawnicza Politechniki Rzeszowskiej, Rzeszow (2018). (in Polish)
16. Bloom, B.S., College, C., Examiners, U.: Taxonomy of Educational Objectives. David McKay, New York (1956)
17. Munzenmaier, C., Rubin, N.: Bloom's taxonomy: what's old is new again. the learning guild research. the elearning guild (2013). http://onlineteachered.mit.edu/edc-pakistan/files/best-practices/session-2/Pre-Session-Munzenmaier-Rubin-2013.pdf. Accessed 28 Nov 2017

18. Anderson, L.W., Krathwohl, D.R., Airasian, P.W., Cruikshank, K.A., Mayer, R.E., Pintrich, P.R., Raths, J., Wittrock, M.C.: A taxonomy for learning, teaching, and assessing: a revision of bloom's taxonomy of educational objectives, abridged edition. White Plains, Longman, New York (2001)
19. Anderson, J.R.: Cognitive Psychology and its Implications, 4th edn, p. 234. W. H. Freeman and Company, New York (1995)
20. Baumard, P.: Tacit Knowledge in Organizations, pp. 62–98. Sage publication, London (1999)
21. Wedig, T.: Getting the most from classroom simulations: strategies for maximizing learning outcomes. Polit. Sci. Polit. **43**(3), 547–555 (2010)
22. Magana, A.J., Brophy, S.P., Bodner, G.M.: Instructors' intended learning outcomes for using computational simulations as learning tools. J. Eng. Educ. **101**(2), 220–243 (2012)
23. de Marco, T.: The Deadline: A Novel About Project Management. EMKA, Warsaw (2002). (in Polish)
24. Wolanska, M.: The use of IT tools for modeling and simulation of production line operation. Diploma thesis under supervision of Dorota Stadnicka, Rzeszow (2018). (in Polish)
25. Gromadzki, F.: Using any logic to simulate the production process. Diploma thesis under supervision of Dorota Stadnicka, Rzeszow (2018). (in Polish)

An Ontological Framework for the Analysis of Constructively Aligned Educational Units

Antonio Maffei[1(✉)], Eleonora Boffa[1], and Cali Nuur[2]

[1] Department of Production Engineering, KTH Royal Institute of Technology, Stockholm, Sweden
maffei@kth.se

[2] Department of Industrial Economics, KTH Royal Institute of Technology, Stockholm, Sweden

Abstract. Higher education institutions are increasingly abandoning traditional teaching methods based on transmissive pedagogy and embracing a model based on the constructivist theory of learning. This is a necessary shift to face current challenges in their domain such as shrinking resources and need to reach students with different level of motivation and learning styles. Constructive Alignment (CA) combines the new pedagogical approach with aligned design for outcome-based teaching education and it is the leading approach underpinning current requirements for program specification and single course description. Another important trend emerging from digitalization of higher education is the collection and use of educational data to improve all the aspects of the learning process. This research community has produced a large body of valuable tools and methods, but there is no specific contribution targeting the analysis of constructively aligned education. The CONALI ontology is an ongoing modelling effort that currently provides a framework to describe the CA body of knowledge including all the relevant semantic relationships. This paper presents a new version of such ontology that supports collection and analysis of educational data from constructively aligned educational units as a first step to fill the gap described above. The specific requirements for this iterative upgrade have been inferred from a literature review of the typical analysis conducted in the domain of educational big data.

Keywords: Constructive alignment · Learning analytics · Ontology

1 Introduction

Modern pedagogy has recognized the value of constructivist approach to learning in contrast with traditional approaches based on passive transmission of knowledge between teacher and learner. Constructivist work on the assumption that knowledge is build (or construct) by the learner through personal experiences and hypotheses on the environment [1, 2]. School should then allow exposure to suitable activities with a certain degree of variation and able to enact the desired learning outcome through meaningful feedback to the learner, rather than assuming that the learner is a blank slate (*tabula rasa*) to be "filled" with knowledge. In other words, every learner brings past experiences and cultural factors to the learning arena. In this context Constructive

A. Hamrol et al. (Eds.): *Advances in Manufacturing II* - Volume 3, LNME, pp. 185–193, 2019.
https://doi.org/10.1007/978-3-030-17269-5_13

Alignment (hence CA) has been proposed by Biggs [3, 4] as practical approach to design outcome-based educational units that support construction of knowledge by the learners. In detail, CA is based on two principle:

1. the constructivist understanding of the learning process as described above, and
2. the practical need for aligned and outcome-based curricula designing: learners need to know up front what are the Intended Learning Outcomes (ILO) of the educational unit they will be attending. Those ILO must then be enacted by suitable Teaching and Learning Activities (TLA) and tested through a fair Assessment Task (AT). In brief, CA this is achieved by using the same action verb for ILO, TLA and AT.

The CONALI ontology, proposed by [5], is an ongoing modelling effort aimed at representing the semantic behind the Constructive Alignment body of knowledge. This model has recently been adopted as reference for the modularization of courses in the EC funded project TIPHYS (www.tiphys.eu). The first version of the CONALI ontology came as an answer to the growing interest toward CA and it was aimed mainly at unifying the terminology and making explicit the assumptions of the domain. This was a necessary reference to avoid an unorganized and diverging growth of the concept amid a larger and larger community of academics interested in the topic. In view of this, the first iteration in the development of the CONALI ontology was mainly concerned with covering CA's key elements and their semantic relation. However, ontological models have a potentially much broader domain of application. Ontology can be used to collect and maintain knowledge as well as separate the operational one from the theoretical one. This, in turn, allows sophisticated analysis on the represented domain. In other words, the process of populating an ontology with real instances can yield a knowledge base: i.e. a great deal of structured data that, once interpreted correctly, can be of great importance to improve and optimize the learning process.

In recent years, the digitalization in the domain of education has produced a large amount of unstructured and semi-structured educational data accumulating in the databases of schools. In addition to that, the academic community is producing sophisticated contributions that allow correlating student's discourses, social network interactions and online activities with the outcome of the learning process. A new research field, focusing on analysis of big educational data was born [6]. In order to work proficiently with educational big data there is a need of model that add semantical meaning to the otherwise simple numerical data collected. In this sense CA is potentially an important enabler of meaningful research endeavors related to educational big data. However, it is evident from literature that educational big data analysis and constructive alignment have so far grown separately. There are currently no contribution or tools specifically targeting courses designed following the CA approach. Considering that constructivism is a leading concept in modern designing of educational units the development of specific tools to analyze them is a necessary step to obtain the best from both words. In this sense, the CONALI ontology can be potentially used to fill this gap. However, in order to use the CONALI ontology as a tool for collecting and analyzing educational big data it is necessary to enhance the model including all the information and semantical relationships necessary to study the learning process. The aim of this work is to produce a second version of the CONALI ontology to fill this gap. Addressing this challenge requires two separate research activities:

1. Identification of the current purposes for analyzing educational big data in terms of what kind of questions are being investigated and which would be the data and information required to answer them.
2. Expand the current version of the CONALI ontology to accommodate the requirement identified above in terms of collection and fusion of specific learning data from the constructively aligned educational units.

The first point is addressed in the background section through a literature review of the state of the art in learning analytics. The result of this survey are used as basis for the modelling process behind the new version of the CONALI ontology presented in the result section.

2 Research Problem

This section presents an overview of the literature in the domain of educational Big data analysis aimed at identifying the key concepts and questions that this research topic is seeking to investigate. The identified issues will then be translated into the information required and the CONALI ontology will be expanded to include such information. Literature is however not the only input to this re-modeling process: the work aimed at building the specific knowledge base for the aforementioned TIPHYS project in early 2018 has elicited discussion and benchmark that have contributed to formulate the requirements for the new COANLI ontology.

The analysis of big data for educational purposes is currently performed using two set of tools developed within two distinct research communities having different objectives and perspectives: Educational Data Mining (EDM) and Learning Analytics (LA). These fields are somehow complementary: the following Table 1 shows the differences between the two areas:

Table 1. A brief comparison of LA and EDM (adapted from [7])

	Learning analytics	Educational Data Mining
Discovery	Leveraging human judgment is key; automated discovery is a tool to accomplish this goal	Automated discovery is key; leveraging human judgment is a tool to accomplish this goal
Reduction & Holism	Stronger emphasis on understanding systems as wholes, in their full complexity	Stronger emphasis on reducing to components and analysing individual components and relationships between them

(*continued*)

Table 1. (*continued*)

	Learning analytics	Educational Data Mining
Origins	Has stronger origins in semantic web, "intelligent curriculum," outcome prediction, and systemic interventions	Has strong origins in educational software and student modelling, with a significant community in predicting course outcomes
Adaptation & Personalisation	Greater focus on informing and empowering instructors and learners	Greater focus on automated adaptation (e.g. by the computer with no human in the loop)
Techniques & Methods	Social network analysis, sentiment analysis, influence analytics, discourse analysis, learner success prediction, concept analysis, sense-making models	Classification, clustering, Bayesian modelling, relationship mining, discovery with models, visualisation

The main areas where LA and EDM have produced contribution are [8]:

- Pedagogy-oriented issues: e.g., student modeling, prediction of performance, assessment and feedback, reflection and awareness.
- Contextualization of learning: e.g., multimodality, mobility, etc.
- Networked learning: e.g., MOOCs, social learning platforms, etc.
- Educational resources handling: organizing and recommending educational resources from data pools, and selecting the most appropriate algorithmic method for making suggestions.

Formulation and test of hypothesis in both LA and EDM domain requires qualitative and quantitative data about all the activities related with the education process. As the main objective of this techniques is to improve learning it is necessary to formulate metrics to measure the effectiveness of the learning process. As direct measures are not available it is necessary to work with proxies such as improved performance of the students or higher efficiency of the process in using resources [9]. CA is, in theory, able to support LA/EDM researcher in finding such proxies and by offering:

1. A detailed breakdown of the single learning outcomes in each educational unit
2. A connection between learning outcomes and related teaching and learning activities based on the classification of the level of understanding required for each learning outcome (as seen in the Bloom's taxonomy [10–12])
3. A connection between learning outcomes and assessment task based on the desired structure of the knowledge to be transferred (as described in the SOLO taxonomy [13])
4. Possibility to relate learning outcome ordering them according to precedence constraints.

The limited scope of this contribution allows only to focus only on a reduced part of the identified requirements. In general CA can contribute in defining and implement personalized learning strategies through a model of the learner and the learning environment. Leveraging on that, the new version of the CONALI ontology will focus on providing the necessary data to understanding the practical aspects of the learning process. The current version already supports the connection between ILOs and related TLAs and ATs. It needs to be expanded to answer the following questions:

- How can the educational goal verbs used in the ontology be fully characterized in term of level of understanding and expected outcome from the learners?
- How are different ILOs related with each others?
- Which are the characterizing features of the teaching and learning activities for both functional and declarative knowledge?
- Which one are the characterizing features of the assessment tasks for both functional and declarative knowledge?

Further improvement of the expressivity of the ontology are to be addressed in future studies by the authors of this paper or by other academicians that will consider partaking this endeavor.

3 Results

This section presents the most important updates in the new version of the CONALI ontology. The basic taxonomy has not changed as the initial set of classes and *is_a* relationships have been found suitable to represent the new set of requirements. None of the original classes, semantic relationships and attributes have been eliminated. The work has been exclusively expanding the current set of attributes for some of the classes and add some semantic relationships connected with the question proposed at the end of the background section. Figure 1 represent two semantical relationship that have been added to the ILO class: i.e. *isPrerequisiteOf* and *hasPrerequisite*. These two relationships allow a temporal ordering of the ILO according to their position in an optimal process of knowledge building. In other words, they allow constraining different pieces of knowledge in a pedagogically sound and technically feasible chain. It is important to remark that ILO and related TL and AT are like educational unit at the lowest independent level of granularity. They are, thus, allowed three possible states in relation to each other: precedent, indifferent, following.

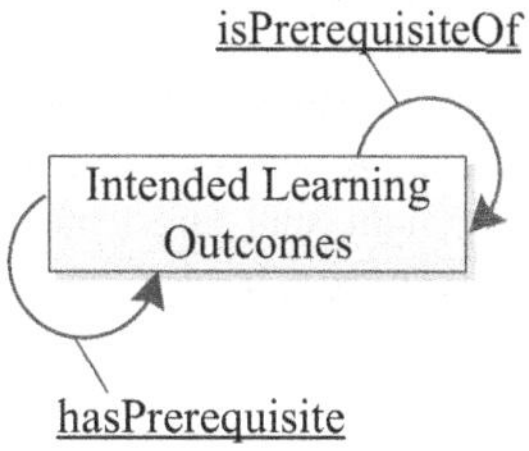

Fig. 1. Summary of the semantic relationships for ILO: isPrerequisiteOf and hasPrerequisite.

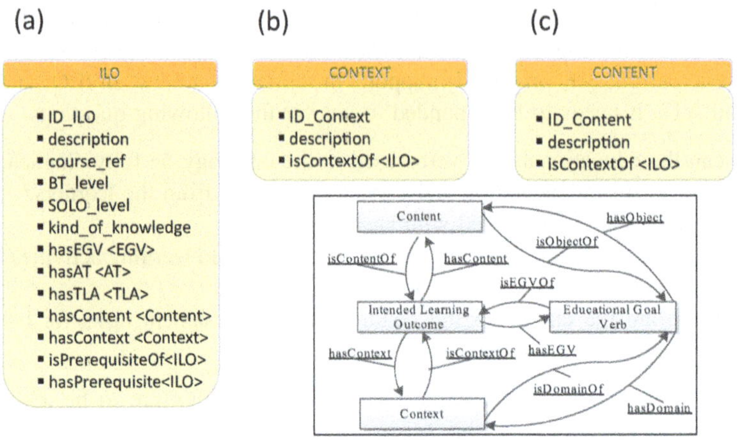

Fig. 2. Specification and visualization of the classes ILO (a), Context (b) and Content (c)

The following Fig. 2 presents the update of: (a) ILO, (b) Context and (c) Content as well as the graphical representation of their link. The attributes that have been added aims at identifying univocally the ILO and single context and content. The Educational Goal Verb (EGV) has also been updated accordingly to be univocally identified through the following attributes: ID_EGV, description, kind_of knowledge, BT_level (Level of understanding in the Blooms Taxonomy).

The following Table 2 represent the update of the TLA class for both Functional and Declarative kind of knowledge.

Table 2. TLA attributes in the new CONALI ontology.

Teaching and learning activity				
Declarative knowledge		Functioning knowledge		
Lecture	Seminar	Peer teaching	Tutorial	Group activity
ID_lecture	ID_Seminar	ID_peer_teac	ID_tutorial	ID_GrAct
type	topic	name	name	type
date	date	date	date	date
time	time	time	time	topic
topic	frequency	topic	topic	language
lecturer_name	location	language	language	num_stud_gr
num_lecturer	num_attendants	num_stud_gr	lecturer_name	num_stud_target
num_stud_target		num_stud_target	num_lecturer	support_used
support_used		support_used	num_stud_gr	
language			num_stud_target	
online_source			support_used	
			online_source	

As the process only involved the attributes, the table only represent the related static slot of the ontology. The characterization of the TLA is completed in this ontology, as in the previous version, by the semantic connection with *Course*, *ILO* and *EGV*. The following Table 3 represent the update of the AT class.

Table 3. AT attributes in the new CONALI ontology.

Assessment task				
Individual assignment	Homework	Presentation	Assignment in class	Exam
ID_IndAssD	ID_AssHome	ID_Pres	ID_AssClass	ID_Exam
type	date	name	date	date
date	exam_session	date	topic	exam_session
exam_session	time_given	exam_session	type	time
time_given	topic	topic	lecturer_name	type
topic	type	compulsory	compulsory	lecturer_name
lecturer_name	kind_of_delivery	time_given	time_given	max_points
compulsory	lecturer_name	lecturer_name	num_stud_gr	max_grade
num_registered	compulsory	num_stud_gr	feedback	min_grade
max_points_bonus	num_stud_gr	num_registered	support_used	average_grade
max_grade	num_registered	max_points_bonus		num_registered
min_grade	max_points_bonus	min_grade		num_passed
average_grade	max_grade	max_grade		num_failed
kind_of_delivery	min_grade	average_grade		support_used
	average_grade	support_used		open_book

AT characterization in its full version includes also reference to *course*, *ILO* and *EGV*.

4 Conclusion and Future Work

The work presented a new version of the CONALI ontology including attributes and new semantical relationships that will enable the use of this model as a framework to collect educational data from activities designed using CA as reference. The current embodiment of the model allows to target the questions identified as a priority in the background section however this contribution must be further developed into a fully-fledged tool to collect and analyze big data from constructively aligned educational unit and provide all the possible support to LA and EDM effort in the domain. A few areas of improvement have emerged during the study. In detail from the point of view of fully representing the CA framework the CONALI ontology should be expanded to include:

- A distinction between Formative and Summative Assessment where the former should be framed as TLA
- Grading and grading criteria for each ILO through the related AT (summative)

- Definition of the process stakeholders: student, teachers, institution etc.,. This would for example allow to integrate this approach with mainstream tools for educational management.

In addition to that the model should be developed technically to allow:

- Web based support for populating the ontology to reach a wider audience
- Definition of an expert system based on historical data
- Support to interrogate the knowledge database: this in turn requires definition of specific metrics.

Finally, given its nature, this contribution can be used to model any constructively aligned educational unit: its application is not limited to the domain of advanced manufacturing. However, this ontology has ties with the advanced manufacturing domain because both the development and first instantiation of this contribution have been conceived and supported by production related education activities. The first target of this work is thus to improve education in the domain of advanced manufacturing.

Acknowledgements. This work has been partially supported by the "TIPHYS 4.0 – Social Network based doctoral Education on Industry 4.0", Project No 2017-1-SE01-KA203-03452 funded by ERASMUS+ of the European Commission.

References

1. Cooper, P.A.: Paradigm shifts in designed instruction: from behaviorism to cognitivism to constructivism. Educ. Technol. **33**(5), 12–19 (1993)
2. Ertmer, P.A., Newby, T.J.: Behaviorism, cognitivism, constructivism: comparing critical features from an instructional design perspective. Perform. Improv. Q. **6**(4), 50–72 (1993)
3. Biggs, J.: Enhancing teaching through constructive alignment. High. Educ. **32**(3), 347–364 (1996)
4. Biggs, J.: What the student does: teaching for enhanced learning. High. Educ. Res. Dev. **18** (1), 57–75 (1999)
5. Maffei, A., et al.: CONALI ontology. A framework for design and evaluation of constructively aligned courses in higher education: putting in focus the educational goal verbs (2016)
6. Siemens, G.: Learning analytics: envisioning a research discipline and a domain of practice. In: Proceedings of the 2nd International Conference on Learning Analytics and Knowledge. ACM (2012)
7. Siemens, G., d Baker, R.S.: Learning analytics and educational data mining: towards communication and collaboration. In: Proceedings of the 2nd International Conference on Learning Analytics and Knowledge. ACM (2012)
8. Papamitsiou, Z., Economides, A.A.: Learning analytics and educational data mining in practice: a systematic literature review of empirical evidence. J. Educ. Technol. Soc. **17**(4), 49–64 (2014)
9. Romero, C., Ventura, S.: Data mining in education. Wiley Interdisc. Rev.: Data Min. Knowl. Discovery **3**(1), 12–27 (2013)

10. Anderson, L.W., et al.: A Taxonomy for Learning, Teaching, and Assessing: A Revision of Bloom's Taxonomy of Educational Objectives, Abridged Edition. Longman, White Plains (2001)
11. Bloom, B.S., et al.: Taxonomy of Educational Objectives: Handbook I: Cognitive Domain, vol. 19, p. 56. David McKay, New York (1956)
12. Krathwohl, D.R.: A revision of Bloom's taxonomy: an overview. Theory Pract. **41**(4), 212–218 (2002)
13. Biggs, J.B., Collis, K.F.: Evaluating the Quality of Learning: the SOLO Taxonomy (Structure of the Observed Learning Outcome). Academic Press, Cambridge (2014)

Cognitive Methods of Manager Behavior Formation in the Conditions of International Enterprise Activities

Alla Polyanska and Roman Psiuk(✉)

Ivano-Frankivsk National Technical University of Oil and Gas, Ivano-Frankivsk, Ukraine
psyuk@ukr.net

Abstract. The article discusses the cognitive approach to management; in particular, the cognitive characteristics of the manager in the field of international activity of the enterprise are highlighted. Attention is drawn to the necessity to study the cognitive style of the manager, his individual style and style of leadership, which eventually forms his behavior in solving the assigned functional tasks and the results of the activities of the headed division or team. The cognitive methods of shaping the manager's behavior are determined; the economic evaluation of their effectiveness is carried out. In the article it is proposed to apply the manager's work complexity factor to study the prerequisites for the formation of his behavior. Reducing the work complexity is achieved by improving cognitive characteristics, namely the implementation of such methods of shaping the manager's behavior as: creativity stimulating method, professional development method, cognitive learning method, intellectual mobility method. It has been determined that on the complexity of task solving affect both professional competencies and personal characteristics of the manager, which can be improved by cognitive methods that develop the ability and skills of the manager to perform tasks with the highest efficiency. It is summed up that in the sphere of professional activity knowledge plays an important role in the development of cognitive methods of manager's behavior shaping. The potential opportunities to improve the performance of the enterprise, in particular, taking into account the reduction of the complexity of such work, are substantiated.

Keywords: Cognitive methods · Behavior · Enterprise · International activity · Manager

1 Introduction

In the formation of a modern management model that is able to provide the enterprise with long-term success, the dominant role is given to human. The human is the main subject of the organization (employee) and at the same time the subject of management (manager). Their activity is closely related to the concept of comprehensive personality development, which allows forming a type of employee who is able to determine long-term development goals, solve adequate tasks, make effective decisions and perform

A. Hamrol et al. (Eds.): *Advances in Manufacturing II* - Volume 3, LNME, pp. 194–206, 2019.
https://doi.org/10.1007/978-3-030-17269-5_14

other work, without which the company cannot operate relying only on technical, financial and material resources. Thus, a new view on people was formed as on the most important enterprise resource. Its dissemination takes place in the context of the implementation of European integration policy and the growing interest of international organizations in the potential of industrial enterprises at the international markets. This puts before the managers of domestic enterprises the task of establishing mutually beneficial partnerships at the international level and organizing proper business processes within the enterprises. Achievement of professional competence determines the success of the manager which is displayed on the results of the work of the headed division. However, achieving of professionalism requires not only formal knowledge (hard skills), professional skills (soft skills) of working in difficult situations, but also the ability to produce new ideas that are relevant to international development trends. The use of cognitive methods of manager's behavior shaping is aimed at identifying the characteristics necessary for his work and determining the possible ways of their formation, if necessary.

2 Research Methods

Representatives of the School of Behavioral Sciences researched that the motives for the productive work of people are not only external material but also internal awards, and they themselves emphasized the role of human behavior in organizational development. In general, human behavior is considered as a form of human response to external and internal stimuli. Scientists define behavior as a combination of rational and irrational actions and deeds of people under the influence of personal and situational factors and are aimed at adapting to the environment [1–3]. If rational actions are determined by the knowledge and competencies of a person to perform their functional duties, then irrational actions are mainly determined by the knowledge obtained intuitively as derived from professional knowledge and experience, using intelligence and additional information [4]. J. Schumpeter noted that "the success of an enterprise depends on intuition, the ability to read the situation correctly, even if the initiative result is not backed up, from the ability to realize the essence of things and reject the secondary without realizing the principles of selection" [5]. The questions of the cognitive approach research in management remain relevant and are considered through new opportunities to form an effective manager for the development of the organization, on the one hand, and the favorable conditions creation for the development of personality creativity – on the other. Experts in the field of human intelligence believe that to ensure the creative process, a combination of convergent (logical, sequential, linear) and divergent (intuitive, relative) thinking is necessary [6–8]. The ability to combine such characteristics of thinking is determined by psychological, social and professional skills, knowledge and experience of the individual, and the results of such a combination of characteristics of thinking are manifested in the corresponding behavior of the manager, which is characterized by mental flexibility and originality [9]. And their formation and use should be carried out taking into account the following approaches to knowledge management. The first approach aims to put into practice the knowledge that already exists and its further use requires the use of knowledge coding technologies and its distribution to the destination. The second approach involves the creation of new knowledge in the processing of existing knowledge, when its use in activities allows identifying new problems and generating demand for new knowledge and finding ways to satisfy it [10–12].

Both approaches require creativity implementation, which means the creative abilities of the individual, characterized by a willingness to create fundamentally new ideas deviating from traditional or accepted patterns of thinking. Today a special group of people is singled out – a creative class, whose driving force of life activity is the realization of professional knowledge based on internal motives [13]. Richard Florida notes that the right approach to creative personnel management helps to increase the performance of companies by 30–40% [14]. The main methods of mastering creativity include bisociation, intellectual mobility, tolerance for uncertainty, the use of various types of thinking.

P. Drucker, considering the new paradigm of the management of the 21st century, connects the work of people with knowledge, calling them knowledge workers. He identifies six main factors that determine the productivity of labor of modern workers: awareness of the tasks set; assigning responsibility for their own productivity to the knowledge workers themselves, giving them autonomy; innovations should be part of the work, tasks and responsibilities of knowledge workers; continuing education of the knowledge worker; priority is not only the quantity but also the quality of work performed; considering the knowledge worker as an "asset" rather than "expense" [15].

Knowledge of the situation according to the decision of the manager determines the appropriate control technology. In particular, we attribute management style to knowledge management technologies. In our opinion, leadership style is such formalized way for a manager to perform his work, which allows not only to take into account technical and professional tasks, but also to make decisions in the area of using the psychological and social context about the influence, interaction, motivation of employees to perform the set tasks. Today, the economic literature deals with a wide range of leadership styles that determine their behavior. The PAEI-concept of Ichak Adizes [16] deserves attention, according to which each person has his own personal style, which is determined by his ability to realize the following functions: production of results for which the company exists and which determine its effectiveness; administration; entrepreneurship that is responsible for managing change; integration like combining elements of the organization to ensure its viability in the long term. The method developed by the scientist for assessing a person's strengths and weaknesses makes it possible to understand how a person's personality characteristics, approaches to work, and way of thinking influence the interaction with a team of people whose behavior in different from the manager's styles.

Another model of leadership was proposed by Koestenbaum [17], highlighting such personality traits: caring for people, adherence to moral principles; application of analytical thinking skills; the choice of new ways of motivation to work; manifestations of initiative, responsibility, freedom of decision; purposefulness and focus on performance results.

R. Tannenbaum and V. Schmidt have investigated the factors influencing the choice of leadership style. In their opinion, the choice depends on the situation, and the success of its implementation depends on the manager's ability to assess the problem based on an assessment of the factors influencing his behavior. We are talking about personal factors of both the leader and the subordinates, interaction with the working group, organization, understanding of the problem and means of managing it. The identification and assessment of these factors influence on the manager and determines the tactics of their behavior towards subordinates [18, 32–34].

Consideration of these methods of studying the behavior of people on the basis of identifying and taking into account their desire and ability to solve assigned tasks is closely related to the development of advanced training for managers of enterprises working in the field of international activities. After all, a modern manager is a manager who, in addition to basic competences and skills in performing tasks and experience in solving business tasks, must constantly expand his knowledge in various areas of life, with which his functions are related. The result of this not only depends on his work, but on the whole department, he heads. Choosing ways to change the behavior of a manager in the field of international activities, it is necessary to realize how changes in behavior can potentially affect the results of the work of department headed by the manager. We offer to use the indicator of the complexity of the tasks performed [19, 20]. The content of this indicator is that the behavior of a manager, formed both on the basis of professional knowledge and on the basis of cognitive and individual personality styles, affecting his ability to perform the tasks assigned and achieve improved results. Under complexity, we consider the characteristics of tasks that differ from those typical for a particular department and require additional time and additional efforts to solve them. Estimating the complexity of the work, we characterize in particular the level of uncertainty of the tasks being solved, which may also arise as a result of the behavior of the manager [21].

So, we believe that the level of complexity of the tasks will determine the behavior of the manager and influence the choice of methods for its formation in order to achieve the best results. The more complex the task, the more knowledge of management technology must be used, which take into account not only the care for work or people in accordance with traditional leadership styles, but also the desire to learn, change, adapt to new working conditions [22–25].

Thus, the study of the people's behavior forming methods allows us to conclude that the cognitive approach in management activity is constantly in the purview of scientists. They are interested in the cognitive methods selection problem for personal development to balance personal and organizational interests. Despite the fact that these issues remain unchanged, globalization processes bring new challenges and demands to them. Therefore, today is the task of finding ways to harmonize personal development based on the interests of all interested in the issue of effective management of the parties, taking into account the connection: man-team-organization-international environment.

The purpose of this study is to summarize the theoretical foundations of substantiation of the choice of shaping manager behavior cognitive methods based on identifying ways to develop cognitive characteristics necessary to solve problems in the sphere of international activities of an enterprise and grouping them using the work complexity factor.

3 Research Results

The development of management science based on the study of human behavior in the organization, has intensified the interaction of sciences, which focuses on the personality. The study of factors that determine the actions of people in the environment in which they are located is the subject of study of many sciences, in particular: psychology which develops knowledge about the mental state of a person, their behavior in accordance with physiological characteristics; sociology aims to explain the motives of interaction in a team, taking into account the interaction and communication between people at the level of formal and informal associations; management undertakes the task to form such models of human behavior in the organization, as a result, will allow to solve effectively the goals and objectives set before the organization. The generalization of knowledge in the afore-mentioned spheres allows us to highlight cognitive-behavioral characteristics of the leader in a modern organization and consider possible ways to improve it or even form it at all (Fig. 1).

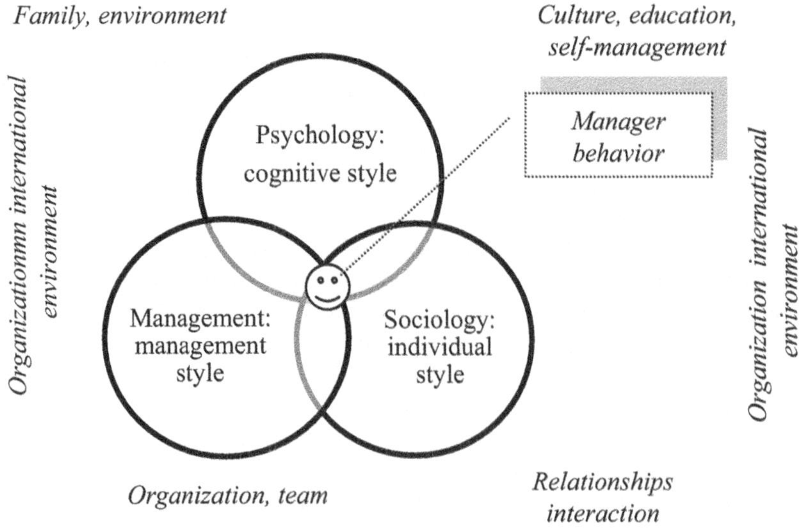

Fig. 1. An interdisciplinary approach to the choice of cognitive methods of shaping the behavior of a manager in the field of international activity, source: own work.

In general, the cognitive style of human behavior is formed on the basis of taking into account its physiological and somatic characteristics. Cognitive style is a prerequisite for the formation of an individual style [26], which a person manifests in the appropriate environment. These styles are decisive in shaping the style of behavior, taking into account its application in interaction with the environment.

Globalization, which today turns out to be in new trends of economic development, contributes to the spread of international interaction, and therefore new ways of partnership and teamwork. International cooperation is becoming a phenomenon of everyday life and requires the establishment and strengthening of international relations, the expansion of cultural boundaries and the interweaving of traditions, customs and behavior of the partner countries. Differences in the perception of the surrounding world by carriers of different cultures, entail misunderstanding, conflict and, as a result the possible loss of customers and potential partners. In order to effectively overcome intercultural contradictions and misunderstandings, the ability to build constructive relationships with clients, partners and colleagues, and representatives of other cultures acquires extraordinary weight [27]. Cross-cultural in business is based on the ability of partners to recognize alternative values and patterns of thinking and behavior of another participant in international relations. Decision making by managers in the face of uncertainty, in addition to personal experience and knowledge, takes into account tolerance for uncertainty and risk [20].

Taking into account the foregoing, we define that the cognitive methods of shaping the behavior of a manager in the field of international activities are aimed at improving the process of the manager's perception of the situation in which international activities take place based on awareness of their role in it and the role of the other party or parties involved in the formation of various situations. The development of the cognitive style of the manager is based on the developed intelligence and formed productive ways and methods of working with information, which ultimately determines the success of his professional activities. It can be assumed that the effectiveness of a manager in the field of international activity can be achieved by developing their ability to process information, which is manifested in the ability to perceive and evaluate the situation, analyze it, make decisions; ability to self-development, self-organization; the ability to identify situations that arise in the subject area of activity, as well as to determine methods of influence on it. Hence, the growth of the productivity of the manager's work with information at an enterprise is possible through the use of IT systems with various types of knowledge structures, which may include: document structures (forms, templates, reports, charts, diagrams) images (photographs and graphic files), videos (presentations and video files), sounds and signals, data, cases (case studies, best practices, lessons learned), processes (resources, specifications), models. Knowledge based on management information systems creates opportunities for the creation, storage, exchange and use of documentary knowledge of the organization.

The use of cognitive methods in shaping the manager's behavior is associated with the exchange of knowledge through interpersonal interaction. This allows you to manage knowledge through their exchange between people, in social networks, through professional groups and teams. To make a decision, a manager can use a wide range of knowledge management information technologies, in particular: e-mail; Data Base & Data Warehouse; Groupware; Browsers; Intranet & Internet; Expert Systems & Knowledge Base; Intelligence Systems; artificial intelligence systems.

It is also advisable to consider the cognitive methods of shaping the manager's behavior through training. In particular, the socio-cognitive theory of learning allows forming a new human behavior based on the study of the behavior of other people or groups. The following factors contribute to social cognitive learning: observational learning, self-control, and efficiency. In the field of international activities of familiarization with the work of international companies, observation of operations, complicity, cooperation and partnership are effective ways of learning through observation.

Self-control forces workers to work better, relying on the new experience gained not on the basis of external control and coercion, but at the expense of taking responsibility for learning and behavior. Self-efficiency is based on an individual assessment of one's own ability to perform certain tasks. The manager with high self-efficiency seeks to expand the boundaries of international cooperation, makes maximum efforts to solve new problems and shows perseverance and confidence in achieving the objectives set.

Decision making in difficult situations is sometimes accompanied by ambiguity or difference of opinion regarding the goals and objectives of the unit headed by the manager, as well as the lack of agreement on the method of action required. In such conditions, it is advisable to practice the methods of stimulating the creativity of both individual workers and the team or group. In particular, the method of horizontal thinking is based on the conscious process of generating new ideas by changing the typical logical pattern that a manager or team uses to process and store information. The combination of this method with the method of vertical thinking aimed at the logical step-by-step process of developing an idea by consistently moving from one event to another by analyzing of the relevant information, allows developing the generated ideas and approaches and find their necessary application [28].

Cognitive methods of shaping the manager's behavior are accompanied by the development of lateral (non-standard) thinking necessary for creative work associated with the development of new ideas and the liberation from old ones. In particular, in international activities, the creative process of thinking is primarily manifested in overcoming typical thinking with the help of creative deviations, that can be traced in leadership styles, teamwork, culture and ethics, responsibility, performance evaluation parameters, cooperation image evaluation methods, qualitative indicators of activities.

The notion of intellectual mobility [29–32] associated with the issue being investigated, under which we consider the integration characteristic of the manager, which forms their behavior in training, development, communication, interaction, which allows to move from one type to another type of tasks within his professional activities, or the ability to make such transitions in different areas of activity. Such mobility of the manager allows to consider the situation from different points of view, sometimes rather unusual or far from the initial ones. In the field of international activity, intellectual mobility allows you to take into account cognitive diversity [33, 34] and use them to achieve the desired results, particularly taking into account age, gender, race, ethnicity, physical capabilities, as well as such categories of cognitive diversity as education, work experience, religious beliefs.

Summarizing the above, we note that the effectiveness of the international activity of the enterprise depends on the behavior of the manager, formed on the basis of the development of cognitive perception of the environment, the development of an individual style of behavior, the choice of leadership style. The insufficient level of skills and abilities, the inability to solve problems taking into account the international environment increases the level of complexity of the tasks and, consequently, reduces the actual results of international activities. The potential opportunities to achieve the results of the international activities can be represented using formula (1):

$$C_{pr} = C_{ar} \times (1 + C); \tag{1}$$

where C_{pr} – a potential result of international activity, adjusted in view of the complexity of the work performed; C_{ar} is the actual result of such activity; C – the coefficient of difficulty, is determined in the range (0; 1).

The substantiation of the content of the ability indicator we have made in the source [21].

In the context of the international activities implementation involving different types of work, the complexity factor will take into account the complexity level of each type, with an emphasis on the share of work in total volume. Hence, the weighted average complexity of the work execution for the international activities of the enterprise, taking into account the various types of work can be calculated by the formula (2):

$$C_{corr} = C_1 \times w_i + C_{i+1} \times w_{i+1} + \cdots + C_n \times w_n = \sum\nolimits_{i=1}^{n} C_n \times w_n; \tag{2}$$

where C_i is the complexity factor for the *i*-th type of work, w_i is the share of the *i*-the type of work in the total volume of performed work in the field of international activity.

An example of taking into account the impact of the complexity of tasks on the results of international activity of the enterprise is presented in Table 1.

Table 1. Actual and potential export volumes of the enterprise's products, thousand US dollars.

Period	The actual sales volume	Potential sales volume, calculated on the base of formula 2 (take into account the complexity factor)
04.2018	44 060	66 618,72
05.2018	49 440	74 753,28
06.2018	33 544	50 718,528
07.2018	53 970	81 602,64
08.2018	34 810	52 632,72
09.2018	43 332	65 517,984
10.2018	42 600	64 411,2
11.2018	49 990	75 584,88
12.2018	46 850	70 837,2
Total	398 596	446 428

Source: Formed by the data of industrial (petrochemical) enterprise's export activity

The gap may occur between the actual results of operations, which are represented by the enterprise's export volumes and the potential result, which is achieved by reducing the complexity, is shown in (Fig. 2). It should be noted that the increase in the actual volume is possible by reducing the uncertainty of the environment, the risks arising in international activities due to such behavior of the manager, allows you to make decisions taking into account these restrictions of activity.

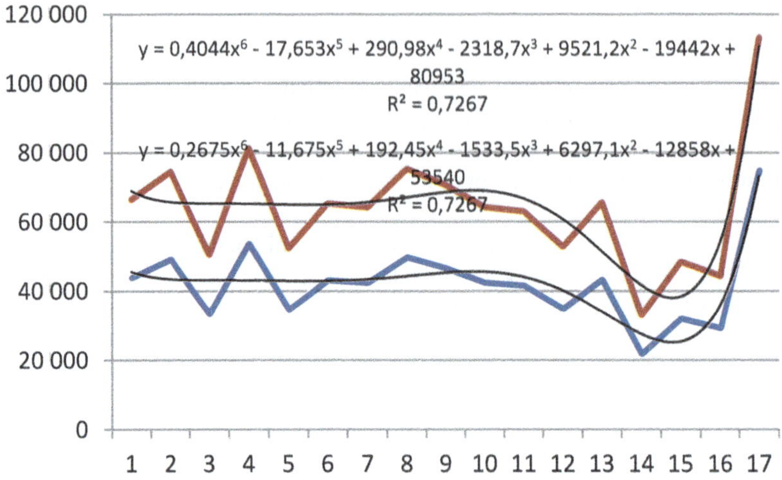

Fig. 2. Dynamics of actual and potential export volumes. Source: Formed on the basis of data from the (Table 1).

The use of mathematical tools with the use of a power function to approximate the results obtained allows us to quantify the gap in the results of international activities, which is determined taking into account the complexity of the work.

To determine the minimum and maximum value of the function that describes the dynamic of factual and potential volume of export on segments from 1 to 17 (each segment is equal to month's activity), we find its derivative by the formula:

$$f'(x) = \lim_{\Delta x \to 0} \frac{f(x+\Delta x) - f(x)}{\Delta x}, \tag{3}$$

As we see from Fig. 2, an increase in sales volume potentially forms the export potential of an enterprise, the calculation of which we propose to implement taking into account the maximum and minimum results of international activity (export) of enterprise by formula:

$$E_p = \sqrt{(\max_{forc} - \min_{forc}) \times (\max_{act} - \min_{act})}, \tag{4}$$

where max_{forc}, min_{forc} – the maximum and minimum value of the forecast export volumes of products respectively; max_{act}, min_{act} – the maximum and minimum value of the actual export volumes of products respectively.

According to formula (4) we define the volume of export activity potential on the investigated enterprise and present the obtained values in the Table 2.

$$E_p = \sqrt{(31340,56 \times 19989,123)} = 26199,96\,(t).$$

Table 2. Calculation of export potential of the investigated enterprise.

Object	Period	Critical value	Potential of export activity, t
Enterprise	According to the actual volume	f_{max} = 45723.299	Ep (3) = 26199,96
		f_{min} = 25734.176	
	According to the forecast volume	f_{max} = 68959.126	
		f_{min} = 37618.566	

Source: Critical value is calculated on the basis of formula 3; Potential of export activity is calculated on the basis of formula 4.

Consequently, after studying the results of enterprise's international activity on the basis of accounting of such work complexity level, we are able to evaluate the content, quality and volume of work performed in the investigated field. This, in turn, will

Table 3. The influence of complexity on the choice of cognitive methods of manager's behavior formation in the field of international activity of the petrochemical enterprise.

Tasks complexity level	Characteristics of the tasks of international activities	Cognitive methods of manager's behavior formation
Elementary tasks that require basic social skills (communication with colleagues, manager)	Knowledge of basic competences, knowledge in the field of foreign trade activities; work experience. Fulfillment of official duties in accordance with the instructions; ability to work with information technology; knowledge of legislation, regulations, internal instructions for the performance of assigned functions; knowledge of the language if it is necessary to perform direct functions	Associated with the expansion of practical experience, the desire for learning, the improvement of professional skills, the expansion of competencies, the stimulation of the internal motives of self-organization and self-development
Ordinary tasks requiring access to the information base at the level of the structural subdivision and communication to coordinate activities and conduct negotiations	Understanding the tasks that are assigned to the employee; ability to communicate; knowledge of the organizational structure and the ability to communicate in it within the functional responsibilities; coordination of international strategies and tasks	

(*continued*)

Table 3. (*continued*)

Tasks complexity level	Characteristics of the tasks of international activities	Cognitive methods of manager's behavior formation
The task with a high level of uncertainty that arises as a result of atypical (unusual) situations	Preparatory period for signing a contract: searching of a Buyer, collecting information about him, organizing a meeting; approval of price, range, features and conditions of delivery; agreement of terms of cooperation	Associated with the development of creative work, creativity, lateral thinking, intellectual mobility

Source: Formed by authors

determine the peculiarities of work planning in such a way that the performance of export operations will be distributed optimally in relation to the urgency, priority and professionalism of the functional tasks fulfillment. In addition, the manager will be able to assess the ability of the structural unit to perform the tasks and decide on the use of outsourcing and outstaffing, as well as to develop measures for the development of cognitive methods for forming the behavior of the manager and staff which are associated with the development of their creativity, lateral thinking and intellectual mobility.

Thus, the work of a manager in the field of international activity is connected with the complexity, the overcoming or reduction of which will allow the enterprise to potentially increase the volume of activity. Summarizing the material presented, as well as taking into account the experience in the field of international activities at the petrochemical company the Table 3 presents and characterizes the complexity of the implementation of tasks in the field of international activity (in the example of export) and proposes cognitive methods of forming manager's behavior aimed at work complexity reducing and at improving of the results of such activities.

4 Conclusions

Thus, in conditions the enterprise working environment complexity, the interest to using the cognitive approach in management is increasing. The considered three-level structural and functional organization of the manager's behavior integrates the knowledge gained in the course of his life activity and described in studies about the well-known today cognitive styles, factors of individual personality and factors of professional readiness of the manager to perform functional tasks and combines them with the environmental factor of professional activity. Taking into account these conditions, features of leadership styles opens opportunities for professional training of specialists. And the cognitive methods of forming a manager's behavior are hypothetical tools for acquiring and using by a specialist the methods of perception, thinking, and actions in certain situations to achieve the desired result.

Working in the sphere of international activity, an enterprise is faced with the need to involve specialists in its work who are capable of working in conditions of cross-cultural relations. Work in the field of international activity (and not only) is accompanied by certain difficulties defined by cognitive diversity. Cognitive methods of personality formation are cross-cutting in nature, which is reflected in the combination of all levels of manager's behavior, ranging from the cognitive style, influencing their individual style and affirming in the style of leadership.

It is substantiated that the complexity of the tasks is mainly due to the ability of managers to ensure targeted action on the employees of the subdivision, as well as the opportunities that exist to reduce this complexity and increase the effectiveness of the unit, and thus of the whole enterprise. The complexity of the work performed potentially affects the results of such work, in particular the volume of international activity. We can conclude that reducing the complexity of the work, the manager forms the potential of international activities by increasing the possible volume of sales in the conditions of reduction of cooperation uncertainty. Generalized cognitive methods of forming a manager's behavior in a given area of activity are aimed at reducing such complexity and are realized through training, development, self-improvement, intellectual mobility, non-standard thinking, and creative approach to solving complex issues.

References

1. Haselton, M.G., Nettle, D., Andrews, P.W.: The evolution of cognitive bias. In: Buss, D.M. (ed.) The Handbook of Evolutionary Psychology, pp. 724–746. WileyInc., Hoboken (2005)
2. Gavetti, G., Greve, H.R., Levinthal, D.A., Ocasio, W.: The behavioral theory of the firm: assessment and prospects. The Academy of Management Annals, pp. 1–40 (2012)
3. Tom, C.W., Lin, A.: Behavioral Framework for Securities Risk (2011). http://scholarship.law.ufl.edu/facultypub/114
4. Kahneman, D.: Thinking, fast and slow, 480 p. (2017). (in Ukrainian)
5. Schumpeter, J.A.: Economic Development Theory: Profit, Capital, Credit, Percentage, and Economic Cycle, 242 p. (2011). (in Ukrainian)
6. Petrenko, V.: Intellectualization Management of Intellectual Resources - Urgent Need of Modern Society (2016). www.worldscientificnews.com
7. Petrenko, V.: Problems of intellectualization of management: an intellectual model of the person of manager. Sci. J. Herald Khmelnitsky National University Econ. Sci. **6**(1), 43–46 (2007). (in Ukrainian)
8. Sytnyk, Y.: Intellectualization of Enterprise Management Systems: Concept, System Monitoring and Modeling: Monograph, 380 p. Lviv Polytechnic Publishing House (2017). (in Ukrainian)
9. Wiig, K.: Supporting knowledge management: a selection of methods and techniques. Expert Syst. Appl. **13**(1), 15–27 (1997). http://www.hcs.science.uva.nl
10. Regan, E.: The evolving concept and practice of knowledge management: seeking a common understanding and definition, Managing Worldwide Operations & Communications with Information Technology (2007). http://www.irma-international.org/viewtitle/33390/
11. Bohdan, O.I., Aristarov, Y.M.: Problems and development of cognitive methods in strategic management. Econ. Bull. Donbas **2**, 96–99 (2007). (in Ukrainian)
12. Nonaka, I., Takeuchi, H.: A company – the creator of knowledge. In: The Origin and Development of Innovations in Japanese Firms, 384 p. (2011). (in Russian)

13. Rindermann, H., Sailer, M., Thompson, J.: The impact of smart fractions, cognitive ability of politicians and average competence of peoples on social development. Talent Dev. Excellence **1**(1), 3–25 (2009)
14. Florida, R.: Bohemia and economic geography. J. Econ. Geogr. **2**, 55–71 (2002)
15. Drucker, P.: Management Challenges for the 21st Century, 210 p. HarperCollins Publishers Inc. (2001)
16. Adizes, I.: The ideal executive. Why you cannot be one and what to be about it? 266 p. (2006). (in Ukrainian)
17. Koestenbaum, P.: Leadership: The Inner Side of Greatness, A Philosophy for Leaders, New and Revised Hardcover, 369 p. Jossey-Bass Inc. Publishers (1991)
18. Tannenbaum, R., Shmidt, V.: How to choose a leadership style, p. 72 (2017). (in Ukrainian)
19. Maitland, E., Sammartino, A.: Managerial cognition and internationalization. J. Int. Bus. Stud. **46**(7), 733–760 (2015). https://www.researchgate.net/publication/276280274_Managerial_Cognition_and_Internationalization
20. Aharoni, Y., Tihanyi, L., Connelly, B.L.: Managerial decision-making in international business: a forty-five-year retrospective. J. World Bus. **46**(2), 135–142 (2011)
21. Psiuk, R.: Asymmetries overcoming in the sphere of foreign economic activity of enterprises. SEPIKE JUNIOR **1** (2018). http://www.sepikecloud.com/
22. Liker, J.K.: The Way of Toyota. 14 Management Principles From the World's Greatest Manufactures, 424 p. (2017). (in Ukrainian)
23. Bok, L.: Work Rules! Insights from inside Google that will Transform How You Live and Lead, 424 p. (2017). (in Ukrainian)
24. Thompson, A.A., Strickland III Jr., A.: Strategic Management: Concepts and Cases, 12 edn., 928 p. (2005). (in Russion)
25. Sierikov, D.O.: Development of managerial competence of business leaders, 20 p. (2018). (in Ukrainian)
26. Tihomirova, I.V.: Style and productive characteristics of abilities: typological approach. Psychol. Issues **3**, 106–114 (1988). (in Ukrainian)
27. Gallagher, T.: The value orientations method: a tool to help understand cultural differences. J. Extension **39**(6) (2001). http://www.joe.org/joe/2001december/tt1.php
28. Helrihel, D., Slokum, D.V., Vudmen, R.V., Breninh, N.S.: Organizational Behavior, 726 p. (2001). (in Ukrainian)
29. Hutmacher, W.: Key Competencies for Europe. Report of the Symposium, Berne, Switzerland, 27–30 March 1996. Council for Cultural Co-operation (CDCC) a Secondary Education for Europe, Strasburg (1997). https://files.eric.ed.gov/fulltext/ED407717.pdf
30. A Memorandum on Lifelong Learning. Commission of the European communities. http://arhiv.acs.si
31. The Intellectual and Policy Foundations of the 21st Century Skills Framework. http://www.p21.org
32. Tuning Educational structures in Europe: Final report pilot project–phase 2. http://www.tuning.unideusto.org
33. Karlgaard, R.: The soft edge. Where great companies find lasting success, 336 p. (2017). (in Ukrainian)
34. Goleman, D.: Emotional intelligence, 512 p. (2018). (in Ukrainian)

Management of Personnel Development in Conditions of Change

Lesya Verbovska(✉)

Ivano-Frankivsk National Technical University of Oil and Gas,
Ivano-Frankivsk, Ukraine
verlesya@gmail.com

Abstract. The article deals with issues of personnel management in the conditions of a changing environment. It has been ascertained that the key factor in management of professional development is the formation of an employee's personality with competitive abilities, which include vocational training, education, counselling, preparation, adaptation, competence and psycho-physiological factors. The application of the graphic and analytical model in management of personnel development has been proposed. It has been proven that the use of appropriate tools and technologies has an impact on the development of personnel, and is the basis for improving the efficiency functioning of economic agents in the process of change.

Keywords: Development · Personnel · Change · Management · Organization

1 Introduction

It is generally known that the core of any organization is the man working in it, which must be skillfully managed and from whom the effectiveness of the activity of business entities directly depends on. Today there is an actual question of which level of personnel is needed by modern business entities. Openness of boundaries, informative society, creation of modern working conditions, considerable opportunities for realization of their knowledge and skills give an opportunity to think about an important problem of modern Ukrainian enterprises, as it is necessary to properly organize the process of managing the development of personnel in conditions of change. The development of personnel is an important condition for the success of any business entities in the context of rapid changes and competency requirements. Clarification of its essence involves determining the content of the basic concepts, personality, harmonious and professional development. Among many qualities which personnel should possess under conditions of changes of environment, and which play a key role: knowledge of the spheres of activity of an organization; professional knowledge and skills in personnel management; leadership and management of changes; ability to learn and develop.

The purpose of the article is to determine the essence of management development of personnel and the need of applying of the model of change management of B. J. Fogg for business entities.

A. Hamrol et al. (Eds.): *Advances in Manufacturing II* - Volume 3, LNME, pp. 207–217, 2019.
https://doi.org/10.1007/978-3-030-17269-5_15

2 Research Methods

Over the state [1] of the modern environment, on its variability, inconsistency and contradiction of issues related to personnel, management and changes are highlighted in the works of many domestic and foreign scientists, in particular: according to A. Yegoryshyna, "the staff" unites all the components of the labor collective of an organization and includes all employees who perform production or management functions, which are engaged in processing of labor objects using means of labor [2]. Guseva states that "At the end of the 20th century the main problem of management became an essentially new, systematic nature of changes in the conditions of economic activity. Now these changes have begun to occur continuously and even to superimpose on one another. They are closely interconnected and each of them in a chain reaction may entail a series of new components [3] that unfold in an outside environment with increasing speed". The Authors [4] note that human resources management is a system of interdependent, organizational, economic and social measures for creation of conditions for normal functioning, development and effective use of labor potential at the organization level.

It is worth noting that the concept "development" in philosophical dictionaries is defined as irreversible, directed, natural change of material and ideal objects [5, 6]. As Lukyanykhin states Personnel development is a systemically organized process of continuous professional training of employees for preparing them to carry out new production functions, professional and qualification promotion, formation of reserve managers and improvement of a social structure of staff [7, 8].

Scientific researchers Marra and Schmidt consider development of personnel as training and raising of qualification [9], and Schlender explains the development of human resources as a system of measures aimed at supporting workers capable of training, expansion of knowledge and up-to-date experience, training of young qualified employees [10].

According to Khmil personnel development is a repeated process leading to achievement of such indicators of the level of skills of employees that are correspond the organization's development strategy [11]. Savchenko asserts that [12] personnel development is a many-sided and complex concept which covers a wide range of interrelated psychological, pedagogical, social and economic issues. All the above mentioned gives an opportunity to assert that development acts as a qualitative change in stable states, and the process of development management is carried out through the change of states of internal and external environment.

In Drucker opinion it is impossible to manage changes, but they can be to left behind [13]. As Clark fnotes "a frequent mistake of many organizations is underestimation of the impact of changes on people" [14].

Regarding the issue of management, in the opinion of the author [15] - the transformation of information about the state of the object into command information from the subject. It is purposeful programmable or arbitrary impact on objects in order to achieve the ultimate goal with the help of processors, phenomena, processes, when there is interaction with them in the mode of a deterministic or arbitrary program [16]. Having analyzed the concepts of "personnel", "personnel management", "personnel development" and "change", in our opinion, management of personnel development in a changing environment is a combination of tools and methods including innovation and investment which are aimed at identifying and disclosing of possibilities and competency of an employee through improving of his qualitative characteristics.

Speaking of human-factor changes [17], Zamyshlyaev, which distinguishes four types of situations that determine the current perception of introduction of changes by workers [18]:

1. Changes that suppress.
2. Toy changes.
3. Fancy changes.
4. Real changes and specificity of plans - at a high level.

Taking into account new realities, the system of personnel development at enterprises should be flexible, able to change the content, methods and organizational forms according to the needs of production and the situation, which is formed on the labor market. In this regard, personnel management should concentrate its efforts on solving such problems [17, 19]:

- development of a strategy for the formation of skilled personnel;
- definition of training needs in terms of specialties and profession;
- selection of forms and methods of professional development of personnel;
- selection of software, methodological and logistical support of the learning process as an important condition of its quality;
- financial support of all types of training in the right amount.

A key point in the management of professional development is the organization of interrelated processes for the formation of an employee's personality with competitive abilities, which includes professional training, education, counseling, preparation and adaptation. Professional development contributes to the general, intellectual and professional growth of a person, expands her erudition and strengthens self-confidence [13, 20].

The most difficult moment in the learning process is to master the methods of self-knowledge. Using different development methods, you can manage the staff and increase the efficiency of its work Table 1.

Table 1. Modern methods of personnel development of an enterprise [21].

Method	Possibility of use
Training	Minimum information and maximum exercise skills. It is used to develop skills for a small group of people
Case-study	Interactive technology for short-term management training based on business situations
Coaching	A new form of advisory support, a means of assistance, assistance in finding solutions or overcoming any difficult situation
E-learning	A massive method of distance learning in the form of e-books, video tutorials, computer exercises
Self-study	Individually under the condition of stimulating learning using a variety of materials: books, documentation, audio, video and multimedia programs

3 Research Results

Many researchers have proven that a motivated and highly skilled worker is characterized by greater mobility in the labour market, more competitive, easily adaptable to changes, skilfully applying his professional abilities.

From 2014, Ukraine has begun the latest development processes that have affected all spheres of life, provide new opportunities for the development of economic entities and personnel, including Only in the years 2015–2017 in Ukraine were opened 27 Industrial parks; 194 operating factories (including Ivano-Frankivsk region 12); 52 factories being built (including in IF region 3); 206 power plants (including IF region 5); 70 elevators/terminals (including IF region 1); 6 gas fields; 10 livestock buildings (including IF region 1) and 5 other areas of activity [22].

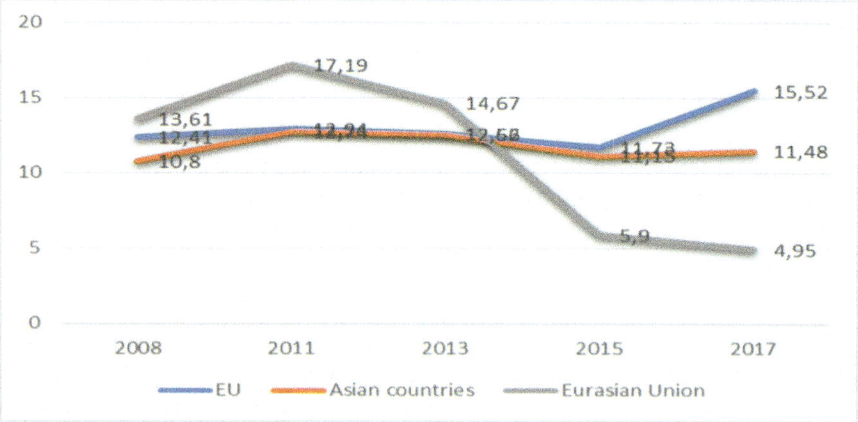

Fig. 1. Reorientation of export of goods from Ukraine in the period from 2008–2017, € billion. [23].

Significant number of companies have begun a process of reorientation to European markets (Fig. 1) [23]. According to statistical data [22], the number of enterprises is growing, and with it and the demand for relevant qualified specialists. Business entities faced personnel famine regarding highly skilled personnel ranging from managers to low-skilled workers. There was also a demand for new highly specialized specialists.

Based on statistics, the number of higher and vocational schools in the Ivano-Frankivsk region and their trained employees is at a rather high level [24], but they often release specialists who are not in demand on the labour market.

According to the data provided Tables 1, 2 and 3, the market is full of trained personnel, however, technical and technological changes require new approaches from both employees and managers. Employees require appropriate working conditions, motivational environment, creating comfortable recreation areas, opportunities for development, continuous improvement and career development, “participation in profits.” Employers require employees to be committed, relevant qualifications,

Table 2. Higher educational establishments [24].

	Number of HEIs, per		Number of persons graduated from universities, thsd. Persons	
	I-II accreditation of levels	III-IV accreditation of levels	I-II accreditation of levels	III-IV accreditation of levels
2015/16	11	5	2,4	8,0
2016/17	11	5	2,11	6,32
2017/18	11	5	2,11	8,92

Source: own work.

Table 3. Vocational educational institutions [24].

	Number of establishments, unit	Skilled workers prepared/ issued, thousand persons
2015	20	5,6
2016	20	5,0
2017	20	4,4

Source: Formed by the author

participation in the organization's life, qualitative performance of their work, readiness to apply creative/creative approaches to increase the level of economic activity of the enterprise.

In the global world, the most important factor in the efficient operation of the enterprise is the continuous retraining of personnel, which requires enterprises to make permanent investments in the staff. The constant need of the enterprise in increasing the productivity of labour and theoretical knowledge makes it necessary to keep records and analysis of training, to systematically improve their qualifications, which directly affects the development of personnel (Table 4).

Table 4. Number of employees involved in the implementation of research and development, by category of staff [24].

Year	Number of employees - total	Including (persons)				
		of them have a degree		Researchers	Technology	Auxiliary staff
		Doctor of sciences	Doctor of Philosophy/Ph.D.			
2015	674	60	128	574	28	72
2016	524	46	120	440	31	53
2017	580	46	125	483	29	68

Source: Formed by the author

Training, retraining, advanced training and training of management personnel should be carried out at least once in 5 years, such principals should be followed by directors of many modern enterprises, in particular, in the last 3 years in higher bulk establishments of the region they have increased their professional level and received diplomas near 620 persons, including 273 persons at the Institute of Postgraduate Education of IFNTUNG (2016 - 23; 2017 - 119; 2018 - 131).

Thus, the reason for changes for Ukrainian business entities in today's environment is the significant impact of the internal and external environment. Knowledge of models, tools and technologies of modern leaders' influence on staff, in the process of changes, is the basis of the maintenance of the subjects of highly qualified employees, and their future acquisition of new qualifications.

For efficient personnel management, senior managers need to be able to model the process of personnel development and take into account internal and external factors. The author is recommended to use the graphoanalytic model below to simulate the situation. The model makes it possible to simulate in the space the basic characteristics which staff should meet and which could be managed under conditions of environmental changes.

3.1 Graph-Analytical Models of Changes in the Management of Personnel Development

The knowledge of the known models of management of changes in organizations and people's behavior in the processes of these changes is necessary for managers of all levels to make reasonable and weighed decisions concerning the formation of appropriate influences on personnel of an organization which is the executor of the plan of changes envisaged by this plan of tasks and achieving goals by them.

Among the most well-known models used as a tool for managing people's behavior in the process of changes, one can distinguish, for example, the equation of change by C. Carnall, who believes that the change occurs only when people recognize that positive results from their implementation will exceed the cost of their implementation, and the conditions of this implementation are described by the Eq. (1):

$$EC = A - B - C \tag{1}$$

where EC is the energy of change, A is the level of dissatisfaction with the current state, B is a well-defined desirable state, C is practical action for achieving the desired state. Almost similar in content is the formula of changes introduced by D. Glatcher, and which was modified and presented by a number of other authors in the following form (2) [25, 26]:

$$C = D \cdot V \cdot F > R \tag{2}$$

where C is the success of any change; D – dissatisfaction of people with the current situation; V – the vision of the future offered to people; F – concrete steps which have been taken to realize this vision with a simultaneous demonstration of at least partial success that demonstrates that the product of factors D, V and F should be greater than the anticipated resistance to changes R. In this case, the planned changes are feasible and will be successfully implemented. In the case of the formation of a deficit in any component D, V or F, there is a risk of occurrence of such resistance R, which will not allow implementing the changes. Therefore, for the success of changes it is important: to convince people of their necessity, to explain how the proposed changes will improve the situation, explain the goals and demonstrate the first positive results. The formula of changes by Backhard and Harris (3) is as follows:

$$C = [A \times B \times D] > X \tag{3}$$

where C is the changes; A is the level of dissatisfaction with the existing state; B is the desire to implement the proposed changes or to reach the final state; D is the practicality of changes (minimal risk and devastating consequences); X is the "cost" of changes. In this case, factors A, B, and D should outweigh the estimated costs for making changes. If one of the factors in parentheses is equal to zero, the value of the change tends to zero.

However, in the following formula 3 is described as the conditions under which changes can occur and provides guidance on how to influence people through beliefs, explanations and demonstrations of success, and does not make it possible to assess the state and readiness for changes in human resources of enterprises or organizations, as well as the impact of these factors on the development process.

From this aspect, the model of change suggested by B. J. Fogg, the professor of Stanford University (USA), whose formula (4) has the following form, looks more informative:

$$B = M \cdot A \cdot T, \tag{4}$$

where B is the behavior of people; M is the motivation of people (low–high); A is the ability (or inability) to change something; T is the trigger – the incentive condition, influence, circumstances, causes or occasional events which push a person to radically change his position, attitude, behavior concerning the subject of change, etc.

The graphic interpretation of the B. J. Fogg model in the form of a surface model with axes of coordinates M and A (Fig. 2) has gained popularity in the processes of managing changes first of all due to the fact that it shows the possible directions and means of influencing people in order to obtain the desired changes in their behavior [18, 27].

However, in order to make good management decisions about processes and procedures for implementing changes in organizations of this model, there is insufficient information about the state of awareness of staff in the necessity, validity and loyalty of decisions made on changes and, as a consequence, support of the latter or counteraction and resistance to their implementation.

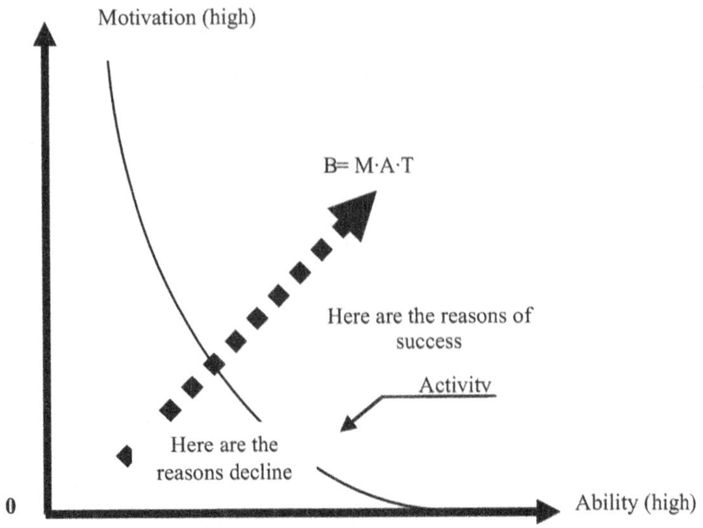

Fig. 2. B. J. Fogg's change management model [25].

Therefore, on the basis of our experience of practical work with human resources of many organizations of various spheres of life of the Ukrainian society, we consider it expedient to additionally offer for use during decision-making processes concerning the impact on human resources, which plays a more or less important role in the processes and procedures for implementing planned organizational changes than the state of motivation and the ability of people to implement the proposed changes.

Thus, on condition of modification of the well-known surface model by B. J. Fogg by adding to it the third coordinate that reflects the level of integration (low–high) of organization's personnel around the idea of the necessity, goals, and tasks of necessary changes, the improved spatial model of management of behavior of people during the process of change has been proposed (Fig. 3) [25, 28, 29].

From our own experience with personnel, I think it is advisable to propose, in addition to motivating also a few important characteristics, which during the processes of making changes play an important role and can influence the process of change. These are characteristics which the worker must meet, namely personal goals, stress resistance, level of experience and competence, reliability, intelligence, ability to influence, knowledge of technologies in relation to the goals of their joint actions with effective and successful achievements during the time of changes.

The following analytical model corresponds to the proposed graphical model:

$$G_{org} = A \cdot M \cdot L \cdot I \tag{5}$$

where G_{org} are the goals of creation and activity of an organization, A is the ability of human resources of all the constituents of an organization to make necessary changes, M is motivation of the personnel to implement changes, L is the level of integration of

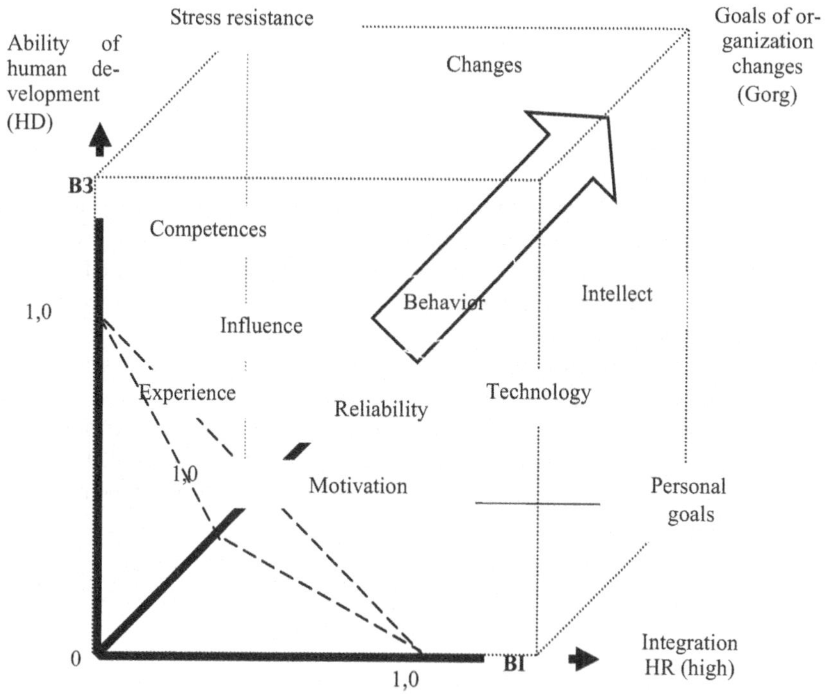

Fig. 3. Spatial modification of the change management model B. J. Fogg, changes by L. Verbovska on the basis of Panasyuk and Petrenko [28].

human resources concerning objectives of an organization, I is the incentive conditions, the source of which must act the initiators and leaders of the process of change.

Thus, active activity of initiators, managers and supporters of the process of changes in an organization with report to its staff about the necessary, reliable and convincing information about their goals, benefits and expected results is intended to contribute to the achievement of high integration (HI), high ability (HA) and high motivation (HM) of all human resources (HR) to changes [25, 28].

In today's conditions, the very management of personnel development acts as one of the most important factors of stability of a company in the market, therefore, it must be based on harmonious combination of personal, professional, social, moral and other human qualities.

Improvement which ones takes place under the influence of the continuous mastering of new knowledge and professional experience acquired by a person in the process of his life and the development of managerial skills and abilities.

4 Conclusions

In summary, it can be concluded that modern business entities in the majority of cases do not pay due attention to the skills of the personnel involved in the organization, and the global conditions lay demand for highly skilled personnel of the appropriate level and profile that is capable of development and creativity in the work. Considered the problems of using instruments by business entities, they proved that they are not used in the development of personnel. Working in a challenging environment, the leaders of Ukrainian enterprises faced an acute problem - the lack of qualified staff. The graphoanalytic model proposed by us allows us to design the basic competency requirements for the personnel.

We can conclude that in order to be competitive in the market, it is necessary to constantly improve their professional skills and competencies, be able to adapt to work on new equipment, be creative. The successful application of classical development management tools should be accompanied by appropriate methods that will allow enterprises to improve the quality of the staff, and the use of advanced methods will be the focus of further research.

References

1. Muromets, N., Melnychenko, O.: Modern Problems of Management of Changes in Enterprises of the Region, Monograph. Limited Liability Company, Kharkiv (2016)
2. Egorishin, A.: Personnel Management: Textbook for High Schools. NIMB, Nizhny Novgorod (2005)
3. Guseva, O.: Management of strategic changes: theory and applied aspects, p. 395, Donetsk (2014)
4. Vinogradsky, M., Vinogradskaya, A., Shkanova, O.: Personnel Management, Teaching Manual. ZUL, Kyiv (2009)
5. Verbovska, L.: Competence as the main aspect of human resources development, monograph making the organization more interesting, pp. 111–120 (2013)
6. Podoprigora, S., Podoprigora, A.: Philosophical Dictionary. Phoenix, Rostov n/D (2012)
7. Lukyanykhin, V., Smetanenko, Yu.: Improvement of the personnel development system in the sphere of hotel business. http://www.essuir.sumdu.edu.ua/retrieve/25280/510045.pdf
8. Savchenko, V.: Management of Personnel Development, Manual. KNEU, Kujv (2002)
9. Marra, R., Schmidt, G.: Personnel Management in a Social Market Economy. Moscow State University, Moscow (1997)
10. Shlender, P.: Personnel Management [and others]. UNITIDANA (2005)
11. Khmil, F.: Personnel Management: Textbook for Students at Higher Educational Institutions. Akademydav, Kujv (2006)
12. Savchenko, V.: Personnel Development Management: Teaching, Manual. KNEU, Kujv (2002)
13. Lukyanihin, V.: Personnel Management: Teaching manual. University Book, Sumy (2004)
14. Clarke, L.: Changes Management. Gebethner & Ska, Warsaw (2007)
15. Latham, J.: Leading change formula revisited. http://johnlatham.me/change-formula-revisted/
16. Guseva, O.: Management of Strategic Changes: Theory and Applied Aspects: Monograph, Donetsk (2014)

17. Daniuk, V., Petyukh, S., Tsimbalyuk, O.: Personnel Management: Teaching Manual. KNEU, Kujv (2004)
18. Fogg, B.J.: A behavior model for persuasive design. http://www.mebook.se/images/page_file/38/Fogg%20Behavior%20Model.pdf
19. Webinar: History, statistics and forecasts. http://www.trainings.ru/library/articles/?id=15267
20. Jasinska, J.: Changes in Organizations: Efficient Management, Crisis Situations and Conditions for Success. FREL, Warsaw (2015)
21. Lizunova, O., Ishchenko, Ya., Kondrashov, G.: Using innovative methods of personnel management of an enterprise. http://economyandsociety.in.ua/index.php/journal-14
22. Companies opened in Ukraine since the beginning of 2015. https://www.google.com/maps/d/viewer?mid=1kI7M2ngkfqQ78HuA0sCKGEaENzg&shorturl=1&ll=48.60824904063167%2C26.73258966114065&z=7
23. The Official Website of the Weekly Electronic Magazine. Trade with the EU: Reorientation Succeeded. https://tyzhden.ua/Economics/219969?fbclid=IwAR2Xib1NIdwOYn1tj_cr-_osiifFq0BXH1ia-EYsI1HgdZux6vO8dMawglk
24. Official site of the Main Directorate of Statistics in Ivano-Frankivsk Oblast. http://www.ifstat.gov.ua/
25. Verbovska, L., Dzvinchuk, D., Petrenko, V., Yasinska, J.: Nine Essays on Management of Changes, Teaching Manual. Misto PH, Ivano-Frankivsk (2018)
26. Kotter, J.P.: Leading Change, 208 p. Harvard Business Review Press, Brighton (2012)
27. Fogg, B.J.: What causes behavior change? http://www.behaviormodel.org/index.html
28. Panasyuk, V., Petrenko, V.: On modification of the B.J. Fogg model for the management of the processes of creation of united territorial communities in the regions of Ukraine, Theory and practice of strategic management of the development of branch and regional social systems, Ivano-Frankivsk, pp. 113–116 (2017)
29. Pichugina, T., Tkacheva, S., Tkachenko, O.: Change Management. KhDUHT, Kharkiv (2017)

A Model of Production Process Stability Measurement and Control with Use of Shewhart Control Charts

Łukasz Łampika[1], Anna Burduk[1](✉), and Tomasz Chlebus[2]

[1] Faculty of Mechanical Engineering,
Wroclaw University of Science and Technology, Wrocław, Poland
anna.burduk@pwr.edu.pl
[2] Faculty of Computers Science and Management,
Wroclaw University of Science and Technology, Wrocław, Poland

Abstract. The paper presents the idea, way of building and implementation results of a model of stability measurements and control of a production process. The presented methodology makes it possible to detect disturbances in the process and facilitates detection of root causes of these disturbances. In the model, Shewhart control charts were used for stability analysis and control. The concept of stability of production systems and way of building the model subdivided to stages are presented. Finally, the model was practically verified in a production company.

Keywords: Control model of a production system · Shewhart charts · Stability of a production system · Parameterization of a manufacturing process

1 Introduction

Efficient functioning of a production system is possible, if stability of its processes is guaranteed by proper control. It is necessary that decisions are taken quickly and are charged with possibly smallest risk. Present-day production systems are, to a large extent, metered and monitored. Thanks to using IT systems, models can be supplied with suitable data coming from a real system. The biggest problem to ensure proper control is acquiring the data, their processing and analysis [1]. Innovative products or processes and changes occurring in the company environment cause that new methods of controlling production systems are searched for [2]. Figure 1 schematically shows the way of managing a production system with use of the model.

Companies are characterized by high complexity and integration, both inside and in their environment. It is necessary to apply a systematic approach. However, there is no possibility to consider, in a short time, all the elements required for taking a decision, without building models of production systems. The models are simplified representations of the reality. On one side, they make it possible to reduce the degrees of complexity, complication and interactions with the other elements of the real system and, on the other side, they include all elements and features significant from the viewpoint of the research goal [1, 3, 4]. Ensuring stability to production systems

A. Hamrol et al. (Eds.): *Advances in Manufacturing II* - Volume 3, LNME, pp. 218–230, 2019.
https://doi.org/10.1007/978-3-030-17269-5_16

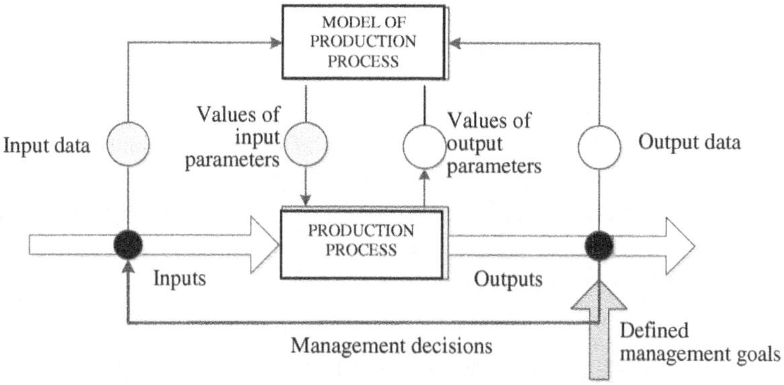

Fig. 1. Model of stability measurement and control of a production process.

becomes the problem particularly important nowadays, at the time of the global market requiring variable, innovative products with short life cycles.

2 Stability of Production Systems

Stability of a production system, in the assumed variability margin, can be defined as maintaining the state of equilibrium for a certain assumed time. The production system is stable if values of its describing parameters are within the ranges defined in the planning function and written in the pattern being the production plan.

Each system is situated in a certain state and, since production systems are of dynamic nature, the transformation (*TA*) resulting in their transition from one state to another proceeds continuously. Considering any time interval (a unit accepted conventionally), it can be said about a sequence of states of inputs or outputs [5]. The more effective the transformation rule, the more effective will be its result, i.e. sizes of components and total size of the vector Y. Size of the vector Y is connected with the vector X by the following functional relationship (expressed by the transformation) (1):

$$\mathrm{Y} = \mathrm{TA}\,(\mathrm{X}, \mathrm{Z}) \tag{1}$$

For course of a process, the concept of a pattern is introduced, describing sequence of stages in the system. If a pattern is available, control of the process is reduced to approximation of its functioning to the pattern by means of a network of feedbacks. The feedbacks make it possible to compare states of the real system with those desired (patterns) and, next, to correct the detected deviations.

In the case of production processes, the pattern is usually a production plan of final products with specified parameters or criteria and in proper time scale [1, 6]. However, in production practice, the system is always accompanied by disturbances that throw it out of balance [6].

If the parameter value $P(t_i)$ at the moment t_i is within the fixed range $P_1 \leq P(t_i) \leq P_2$, this evidences correct course of the process. Otherwise, corrective actions should be taken-up. Usually, these actions consist in such changes of controlling variables (inputs to the system X) that the parameters characterising the controlled values (outputs from the system Y) return to the patterns fixed in the planning function [5, 7, 8].

It can be also said that stability of a production system is its ability to realise the targets assumed in the plans. This makes it possible to refer stability of the system to the level of strategic and tactical planning. Apart from their technical nature, production systems have economic nature and one of their existing purposes is to generate a profit by continuously increasing market share.

3 Concept of a Model for Stability Measurement and Control of a Production Process

One of the basic methods that can be used for monitoring, inspecting and then controlling a production process are process control charts. The most popular are Shewhart control charts used in the Six Sigma systems [5, 9].

The idea of a multilevel model of stability control, suggested in the paper, describes defined production processes whose representative values can be shown in function of time. The model is aimed at demonstrating relationships between sub-processes of a manufacturing process and their parameters.

The created network of relationships permits, at the analysis stage, faster and more effective detection of the disturbance source that caused results at the level n = 1. Each

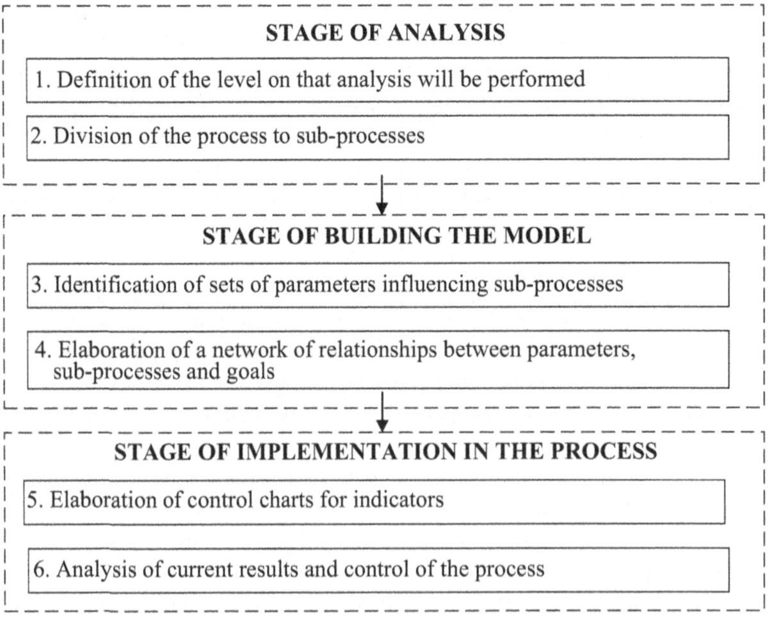

Fig. 2. Model of stability measurement and control of a production process.

disturbance has its root cause, located in one of the higher-level parameters. The more exact model with more interdependent parameters, the larger area of the manufacturing process is recognised and the more probable is finding the root cause resulting from disturbance of stability of a parameter or a sub-process [10].

The suggested idea of building the model for stability measurement and control of a process includes three stages, as shown in Fig. 2.

The following assumptions were accepted in the model:

1. The processes, in that values can not be determined but only demonstrated quantitatively, will not be considered.
2. The model does not include the processes that are not repeatable.
3. The model can not include the processes about that the company has no knowledge.

3.1 Stage of Analysis Before Building the Model of Stability Measurement and Control

Definition of the level for that the model will be built, will affect also further decomposition of goals and scopes of the given parameters of production processes. The sub-processes (P) resulting from division of the main process will belong (2) to the level n = 1 of the process tree. Each subsequent division of a lower level will accept the number n + 1. Defined sub-processes $P_n^{(k)}$ will create the set P of disjoint elements.

$$P = \left\{ P_n^{(k)} \right\}, \quad k = 1, 2, \ldots, l; \quad n = 1, 2, \ldots, m, \tag{2}$$

where:

k – number of sub-process,
n – division level.

At decomposition of the main process, a technical, technological or organisational criterion can be accepted. Depending on the production system level, goals can have various time horizons. It was assumed that, for the needs of the described model, only the goals expressed numerically and related to the production process will be subjected to decomposition. Decomposition of goals will create a network of goals with n levels, where n designates the division level. Each subsequent division of lower level will accept the number n + 1. Defined goals $C_n^{(k)}$ will create the set C of disjoint elements (3):

$$C = \left\{ C_n^{(k)} \right\}, \quad k = 1, 2, \ldots, l; \quad n = 1, 2, \ldots, m, \tag{3}$$

At this stage, to each goal $C_n^{(k)}$, where $C_n^{(k)}$ k = 1,2,. . ., l; k-number of goals; n – division level, at least one sub-process $P_n^{(k)}$ with the same n value will be assigned.

3.2 Stage of Building the Model of Stability Measurement and Control

For each sub-process $P_n^{(k)}$, the parameters $x_i^{(k)}$ should be determined, where i = 1, 2, . . ., o; o = o-th parameter of the k-th sub-process (4):

$$X^{(k)} = \left\{x_i^{(k)}\right\}, \tag{4}$$

where:

i – number of parameter (i = 1, 2, . . ., o)
o – number of parameters of k-th sub-process.

In other words, a parameter of a sub-process is a control variable or, in the model expression, input to the process [10]. The decision about selection of parameters depends on technological specificity, available measurements, level of experts' knowledge and level of the analysed production system. In most cases, selection of the parameters is based on experience and knowledge of external experts and specificity of the production process. Selected sets of parameters of the sub-processes X(k) will create a set of parameters X of the entire examined process (5):

$$X = \left\{X^{(k)}\right\}, \tag{5}$$

where:

k – number of sub-process (k = 1, 2, . . ., l),
l – total number of sub-processes.

At subsequent stages, a diagram is worked-out, on that relations between individual parameters, sub-processes and goals is illustrated. Exact and individual time intervals t_j (j = 1, 2, . . ., m) of measurements should be determined. It is suggested to aim at unification of time intervals for all elements of the model (parameters, sub-processes, goals), which would univocally connect a change of the given parameter value in function of time with another parameter.

3.3 Stage of Implementation of the Model in a Production Process

For each parameter, a control chart must be plotted. To establish the central line and control limits (LCL and UCL), proper number of data and proper sampling is required [9]. Once determined limits and central line for a given parameter do not change and are not calculated again, unless a signal appears, indicating a permanent change in the process, and root cause of this change is found. If it is a desired change, it will be permanently entered to the process, with no prospects to be changed. At this stage, control charts are added to graphs of the elements and parameters tree. These charts will measure stability of the manufacturing system and its individual elements.

If the process is thrown out of balance, the signals defined in literature appear [9]. The basic signals of the described methodology are: exceeding the upper or lower control limit or 9 points in sequence below or above the central line. The first type of the signal informs about appearance of a momentary special cause and the second type informs about appearance of a permanent special cause changing the observed process. Observation consists in detecting signals at the sub-process level and defined goals, i.e. at the possibly lowest level n. Such an approach makes it possible to focus on a few or a dozen most important control charts and, from the viewpoint of the management process, to monitor stability of the entire observed manufacturing process in a safe way.

Analysis of the model consists in searching for the signals that appeared at the same time t_j on control charts of interconnected parameters. Appearance of a signal on the n = 1 level chart at the time t_l implicates searching for a signal on interconnected charts of the n = 2 level at t_l. If the signal appeared also on the second level, searching for a subsequent signal can be moved to the level n = 3 and this way the analysis can be continued. The last found signal is probably the root cause of the signal on the first-level chart. Depending on complexity of the model and numbers of levels and links, the analysis path can be longer or shorter.

After finding the signal on the chart of the highest level at the time t_l, technological analysis of its origin can be started. In order to improve the model, it should be checked, whether a subsequent level can be added to particularize source of the signal. In order to improve the manufacturing process, it is necessary to define cause of the highest-level signal, to implement trial corrective and preventive actions, to verify effectiveness of the solution and to standardize it finally.

4 Case Study

The suggested model of stability measurement and control of a production process, described in the previous chapter, was implemented in a company manufacturing aluminium wheel rims for passenger cars of numerous world-wide makes. The basic production process in the company consists of 3 sub-processes:

- casting,
- machining,
- painting.

The below-presented example will concern only the sub-process of casting, which consists of two sub-processes: melting the material and pouring the mould. Melting consists in converting the initial solid material together with alloying additives to liquid state in melting furnaces. Pouring the mould consists in pumping liquid alloy to the mould together with the solidification process.

4.1 Stage of Analysis Before Building the Model for the Casting Sub-process

For analysis, the level of the entire company was accepted and, more precisely, of the manufacturing process. However, the described example concerns only the sub-process of casting. According to the methodology described in Sect. 3, the designation $P_n^{(k)}$ was assigned to each sub-process. The mould-pouring sub-process was subdivided to 48 sub-processes, each of that corresponding to pouring at individual casting machines. The melting process was subdivided to 3 sub-processes, each of that dedicated to another workstation. The considered divisions are schematically shown in Fig. 3.

Production goals were determined by the Management Board for the entire company (n = 1) in three categories: product quality, capacity of the production process

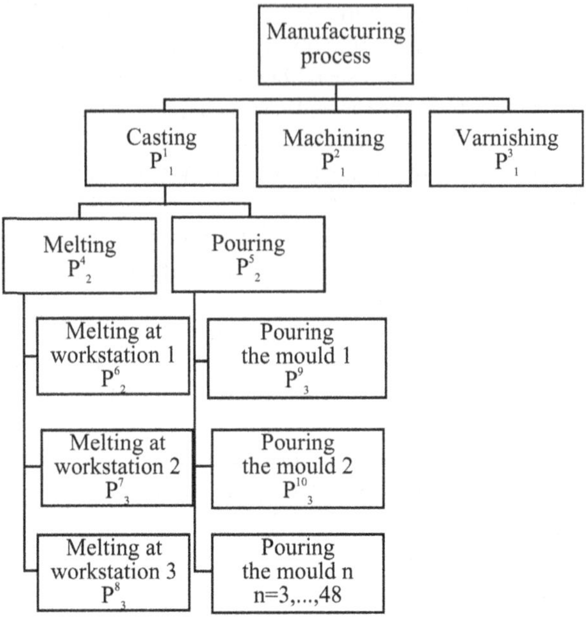

Fig. 3. Decomposition of the manufacturing process to sub-processes, accepted in the model.

(productivity) and availability of production resources. The presented example is limited only to realisation of quality goals, which are:

- C_1^1 – tightness of the castings,
- C_1^2 – positive result of X-ray inspection,
- C_1^3 – no defects at visual inspection.

The above goals are determined as maximum number of defects of the given type acceptable by the Management, guaranteeing satisfactory business results.

4.2 Stage of Building the Model for the Casting Sub-process

The sub-process for that parameters will be defined, is pouring the mould $\left(P_2^5\right)$. Together with specialists from the casting field, 13 parameters influencing the casting process were determined:

x_1 – temperature of the mould,
x_2 – material density in the machine,
x_3 – percentage of Mg in the material,
x_4 – percentage of Ti in the material,
x_5 – percentage of Si in the material,
x_6 – run of the pouring curve (pressure),
x_7 – humidity of mould cooling air,
x_8 – casting cycle time,

x_9 – temperature of mould cooling air,
x_{10} – temperature of material poured to the machine,
x_{11} – density of material in the ladle,
x_{12} – air pressure in the system,
x_{13} – refining time.

Next, the parameters were given numerical values and were covered with measurements in the manufacturing process. The defined sub-processes, parameters and goals were placed in form of a graph in Fig. 4. Pouring the mould is a sub-process divisible to 48 sub-processes of the higher level, which was marked in the graph with one branch and means that the influencing parameters are the same for all the casting machines.

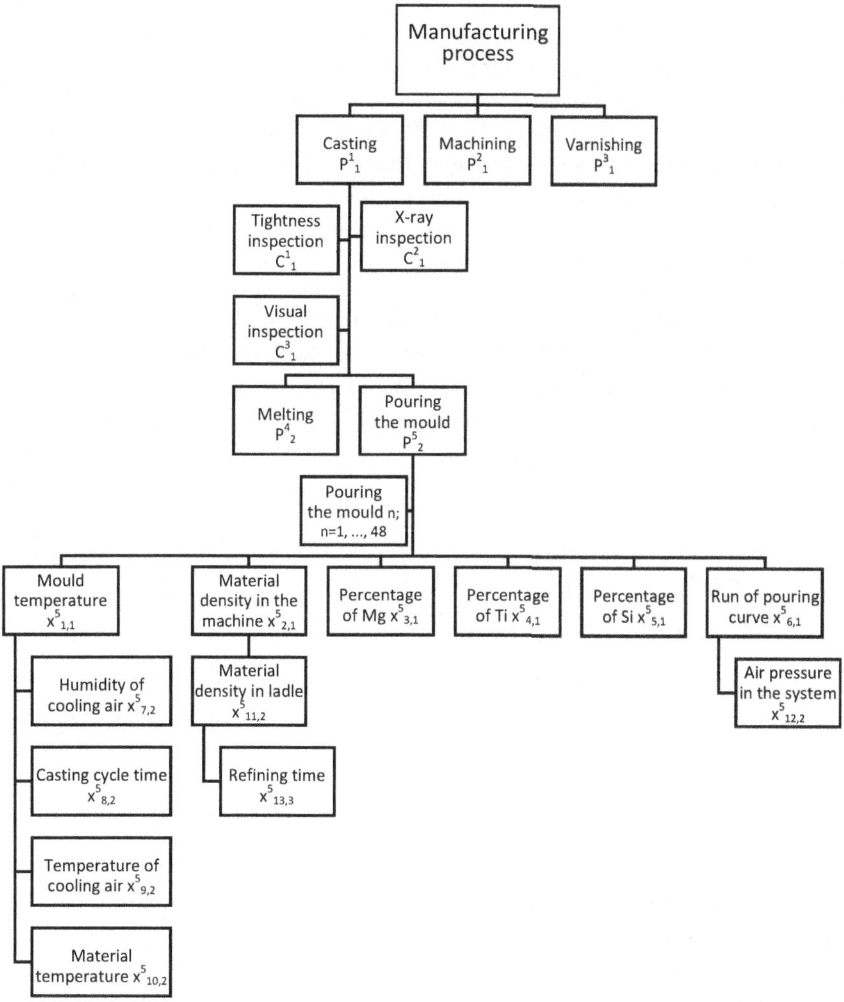

Fig. 4. Network of relationships between parameters, sub-processes and goals.

Definitions of the parameters distinguished also between those variable and constant in time. In this case, the constant parameters are the mould (its shape and dimensions), casting program and way of inspection.

4.3 Stage of Implementation of the Model

For each parameter defined at the previous stage, a control chart was prepared. It was assumed that the minimum number of data from that the limit can be drawn, is 25 measurements, and that each parameter can have another time interval of measurement. For example, time of the casting cycle $x_{8,2}^{5}$ will be measured every one production cycle, cooling air temperature $x_{9,2}^{5}$ will be measured every 5 s and material density in the machine $x_{2,1}^{5}$ will be sampled and examined once a day.

The most important element of the presented model idea is analysis of current results. When the model is ready and data for all elements of the model are regularly acquired, course of the data from sub-processes of the level n = 1 can be observed on control charts. Observation of the level 1 charts makes it possible to manage the entire manufacturing process with a relatively small number of data. The process observed on the control charts can be stable or unstable. In the first case, the process runs between the control limits and numbers of the points positioned above or below the central line must be approximately equal. In the second case, one of the two defined signals indicating occurrence of incorrectness appears.

On 14th of October 2018, a signal of exceeding the upper control limit was identified. Exceeding tightness of castings, at the point 62 for the control chart, is shown in Fig. 5. Value of the excess was 0.59% with the upper control limit (UCL) being 0.55%. The target chart C_1^1 is kept for the ratio of the number of incompatible products detected on the given day to the number of compatible products manufactured that day.

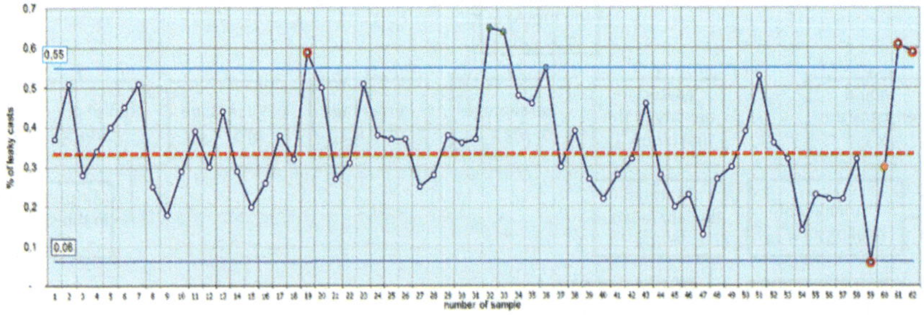

Fig. 5. Control chart for the goal C_1^1 – tightness of castings.

After detecting the signal at the first level, according to the developed network of relationships between the parameters, sub-processes and goals, searching for the signal in one of the related level 2 charts was started. According to Fig. 3, it is necessary to review 48 control charts measuring tightness of the castings manufactured in individual workstations. In one of such charts for the sub-process P_3^{29} the same type of signal was found for the same day. In Fig. 6 for the sample 22, the signal of 5.5% can be seen at the upper control limit at the level 4.98%. This means that, probably, the signal in the chart of the goal C_1^1 was caused by faulty products from the machine 21, where the production process was unstable. The next step is reviewing the parameters affecting P_3^{29} among which the special cause could occur.

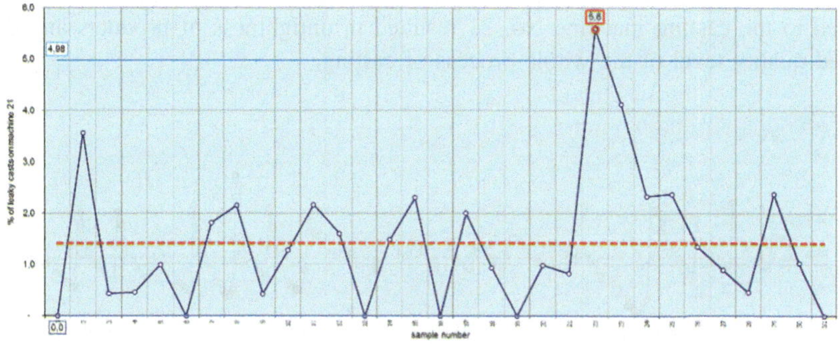

Fig. 6. Control chart for P_3^{29} – tightness of castings for the machine 21.

After analysis of all the 6 parameters of the level 1 connected with P_3^{29} for the date 14th October 2018, a signal was found in the chart of the parameter $x_{1,1}^{29}$. The chart in Fig. 7 shows temperature course of the mould from 13th to 14th October. Temperature was sampled every casting cycle, i.e. every ca. 6 min. It can be seen that the process in unstable. Signals of excessing both the upper and the lower control limit appear. It was found that, for untight castings, the dominant negatively affecting factor is too low temperature. It was decided to check first the strongest signal, i.e. the points from 46 to 56. Approximate times of manufacture of faulty products, known for each 8-h shift, were compared. The points from 46 to 56 were located within the production shift at that 5 of 7 faulty pieces were produced. No other similarly strong signals were found within that shift, lasting from point 1 to point 77.

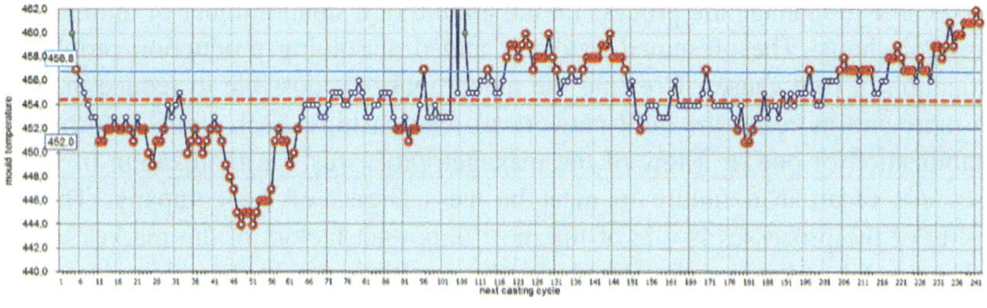

Fig. 7. Control chart for $x_{1,1}^{29}$ – temperature of the casting mould.

One more level of parameters influencing temperature of the mould was written in the model. To know the reason why the mould cooled down to 444 °C, control charts of the level 2 parameters had to be inspected. Signals were found in the chart of the parameter $x_{10,2}^{29}$ – temperature of liquid material. Figure 8 shows the signal between the points 48 and 52 corresponding to 672 and 674 °C. Signals for both charts occurred at the same time, which, in connection with their technological relation, demonstrates the root cause of the rejects qualified as untight castings. Too low temperature of metal poured to the casting machine No. 21 resulted in untightness of products and destabilized quality level of the whole process of casting.

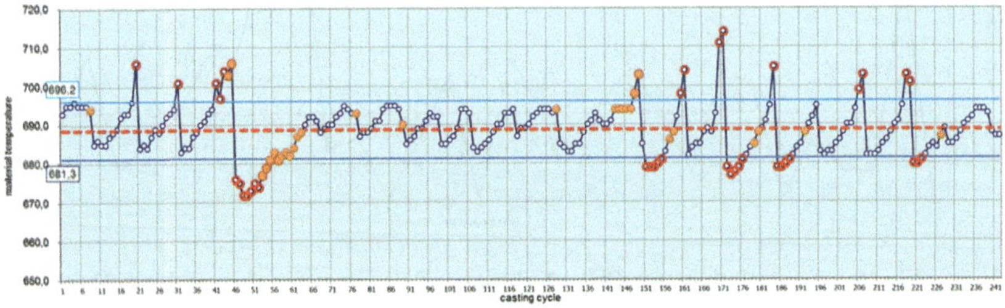

Fig. 8. Control chart for $x_{10,2}^{29}$ – material temperature.

After situating the root cause, solutions aimed at ensuring, for that product, material temperature at a level above 680 °C were introduced. The main changes were increasing material temperature in the transfer ladle and reducing time of holding the material in the ladle between the melting furnace and the casting furnace.

5 Conclusions

The model of stability measurement and control makes it possible to connect defined parameters of sub-processes with goals of the company in such a way that the manufacturing process can be managed with use of a few to a dozen control charts. Each subsequent level can be managed by another group of managers. Each of them, at his level, is able to manage the process on the ground of a small number of indices shown in control charts. The presented model is aimed at solving production problems by elimination of root causes. Creating the model makes it possible to reduce the search area more and more by the built network of relationships. Each signal informs about destabilization of one or a few of the managed processes. Destabilization of the processes can result in a failure to meet business targets of the company. Therefore, stability of the processes is one of the most important factors of the market success.

The model can be used not only in the production field, but also in service or financial areas. For each process, for that measurable and time-variable parameters can be determined, it is possible to plot a control chart and develop a model of stability measurement, analogous to that presented here.

A significant restriction is the fact that defined sub-processes and parameters only are observed in the model. The methodology assumes that one of the defined parameters influences the given sub-process. Unfortunately, in practice, the given sub-process is affected by countless potential disturbances. It is impossible to predict extraordinary situations and the model does not consider them. On the other side, most of extraordinary situations are quickly and easily identifiable because of their unusualness, and standard parameters and processes are always the most important factor of a success or defeat of the company. A subsequent restriction is knowledge of the people creating the company and their technological know-how. The model will be as effective and complete, as much it is allowed by knowledge of the experts employed in the company. Nevertheless, with each analysed signal, the model can be developed by subsequent parameters or new relationships. Such a situation appears, when no higher level exists in the model and, in the given case, process engineers are convinced that the influencing parameter was not measured so far.

In the future, further research works are planned on relationships between signals from control charts of connected parameters, determination and calculation of control limits, as well as on the method of determining weights of parameters and their influence on interpretation of the signals.

References

1. Kalinowski, K., Krenczyk, D., Paprocka, I., Kempa, W., Grabowik, C.: Multi-criteria evaluation methods in the production scheduling. In: IOP Conference Series: Materials Science and Engineering, vol. 145 (2016)
2. Saniuk, A., Jasiulewicz-Kaczmarek, M., Samolejova, A., Saniuk, S., Lenort, R.: Environmental favorable foundries through maintenance activities. Metallurgy **54**(4), 725–728 (2015)
3. Kłos, S., Patalas-Maliszewska, J.: An approach to buffer allocation, in parallel-serial manufacturing systems using the simulation method. In: Rocha, Á., Adeli, H., Reis, L., Costanzo, S. (eds.) Trends and Advances in Information Systems and Technologies, WorldCIST 2018. Advances in Intelligent Systems and Computing, vol. 747. Springer, Cham (2018)
4. Sobaszek, Ł., Gola, A., Świć, A.: Predictive scheduling as a part of intelligent job scheduling system. In: Burduk, A., Mazurkiewicz, D. (eds.) Intelligent Systems in Production Engineering and Maintenance – ISPEM 2017. Advances in Intelligent Systems and Computing, vol. 637. Springer, Cham (2018)
5. Burduk, A., Grzybowska, K., Kovács, G.: Hybrid model of an expert system for assessing the stability of a production system. LogForum **14**(4), 507–518 (2018)
6. Foit, K, Banaś, W., Gwiazda, A., Hryniewicz, P.: The comparison of the use of holonic and agent-based methods in modelling of manufacturing systems. In: 5th International Conference on Modern Technologies in Industrial Engineering (ModTech). V Book Series: IOP Conference Series-Materials Science and Engineering, vol. 227 (2017)
7. Burduk, A.: Artificial neural networks as tools for controlling production systems and ensuring their stability. In: Saeed, K., Chaki, R., Cortesi, A., Wierzchoń, S. (eds.) Computer Information Systems and Industrial Management, CISIM 2013. Lecture Notes in Computer Science, vol. 8104. Springer, Heidelberg (2013)

8. Sankar, N., Prabhu, B.: Modified approach for prioritization of failures in a system failure mode and effects analysis. Int. J. Qual. Reliab. Manag. **18**(3), 324–336 (2001)
9. Hamrol, A.: Strategies and practices of efficient operation. Lean Six Sigma and Other. PWN, Warsaw (2015)
10. Rojek, I., Dostatni, E., Hamrol, A.: Ecodesign of technological processes with the use of decision trees method. In: Pérez García, H., Alfonso-Cendón, J., Sánchez González, L., Quintián, H., Corchado, E. (eds.) Proceeding of the International Joint Conference SOCO 2017-CISIS 2017-ICEUTE 2017, León, Spain, 6–8 September 2017. Advances in Intelligent Systems and Computing, vol. 649. Springer, Cham (2018)

Evaluating and Improving the Effectiveness of Visual Inspection of Products from the Automotive Industry

Krzysztof Knop[1(✉)], Ewa Olejarz[2], and Robert Ulewicz[1]

[1] Department of Production Engineering and Safety, Faculty of Management, Czestochowa University of Technology, Armii Krajowej 19B, 42-200 Czestochowa, Poland
krzysztof.knop@wz.pcz.pl

[2] Faculty of Management, Czestochowa University of Technology, Czestochowa, Poland

Abstract. The article presents the results in the field of assessment and improvement of the effectiveness of visual inspection of products - airbag modules, manufactured by a company from the automotive industry (Tier 1 supplier). The research was carried out using the Kappa index and auxiliary indicators proposed by the automotive industry for the assessment of the inspection system consisting of 3 controllers involved in the inspection of the above-mentioned products. The results showed numerous interesting interdependencies in the inspectors' classifying the specially selected sample of products into two qualitative categories (OK, NOK). The findings included, among other things, poor or average compliance of the controllers' decisions with the expert's decision, an unacceptably high share of the type II errors consisting in the acceptance of a non-conforming product - this applied to all controllers, and a poor result of the entire control system (3 controllers verified in total). Furthermore, the conducted research allowed to develop a ranking of controllers. The obtained research output led to actions intended to improve the alternative product control system including the trainings for all controllers, the modification of the visual assessment standard and implementation of the Andon system.

Keywords: Visual inspection · Effectiveness · Cohen's Kappa coefficient · Auxiliary indicators · Improvement

1 Introduction

Operations of each production enterprise are closely related to the continuous improvement in the quality of the manufactured products and reduction of costs of own production, which is reflected in high profitability [1]. Production costs can be limited by eliminating or restricting those activities (in their quantity and/or time) which do not create any added value and reducing the losses in those processes which do not create

A. Hamrol et al. (Eds.): *Advances in Manufacturing II* - Volume 3, LNME, pp. 231–243, 2019.
https://doi.org/10.1007/978-3-030-17269-5_17

added value, but are inevitable [2]. A process that does not provide for any added value but is necessary, is the quality control process [3]. This process allows to obtain some valuable information on whether the quality of the process/product is appropriate, thus it cannot be eliminated [4]. Control points adopted in the production process allow to detect and eliminate any non-conformities, and to improve manufacturing by identifying and eliminating the causes of such non-conformities [5]. Therefore, the quality control can make some significant contribution to development of the added value by suggesting new improvements [3].

Quality control takes on various forms and is implemented with diverse methods [6]. Despite the development of measurement solutions based on increasingly more reliable instruments, still the dominant method of quality control in numerous enterprises is the inspection adopting only the senses of a man (organoleptic control) [7]. Human presence during the control activities is desirable due to the diversity of the manufactured product and flexibility [8]. This is the human with extensive knowledge must adequately respond in a manner compliant with the occurring conditions. The variety of features that must be checked and that change quickly is not conducive to automatic control. This is the visual inspection what provides for higher quality in product assessment than the machine [9].

Detailed inspection, especial in high value manufacturing is a workload-intensive task, dominated by controller's mental demand and effort [10, 11]. At alternative inspection, manufactured part is often visually evaluated for defects and a decision is made to pass or fail the part [12]. The quality controller's work is affected by plenty of factors which determine the effectiveness of the control. These are factors which depend on, among others, the controller's physical and mental state, workplace organization, environmental conditions, types of non-conformities [13–16]. These factors can degrade performance. They functioned in the past and functioning now, unless care is taken to design the inspection process appropriately [17]. Errors pose the undesirable result of the influence of these factors. The errors in inspection can be detected through special tests, studies, experiments and results can be used to improve the inspection process [18–21]. Visual inspectors, like NDI inspectors, make errors of both missing a defect and calling a non-defect (misses and false alarms respectively) [22]. The errors in visual inspection are referred to as type I and type II errors [23]. The consequences of these errors are different, while the type II errors, consisting in the acceptance of a non-conforming product, have much more dangerous effects. The errors in quality control generate unnecessary costs and should therefore be kept to a minimum [3]. For this purpose, it is first of all necessary to assess their severity by adopting some appropriate indicators [24]. It is important that the controllers agree on the qualification of a product, both with each other and with the expert. The Kappa indicator may be employed to determine this [25] - a common compliance measure adopted by the automotive industry and recommended for use in the MSA manual [26, 27]. Feedback of information is really important in human (visual) inspection [28]. It should be remembered that each indicator is only a number, and it should be treated as a starting point to initiate the improvement activities [24]. Effectiveness at the level of 100%, in particular in the case of "human" quality control (visual inspection) is not possible (the digital brain scores higher in repeatability [29, 30]), but it is worth undertaking actions that will aim at this kind of excellence. Never cease in the efforts to improve the quality of processes and their results, because the benefits of these activities often outweigh the costs [31]. The effectiveness and efficiency of quality control processes is also reflected in the effectiveness and efficiency of the production process [32]. Therefore, it should be taken actions to improve the performance of visual control, which may be focused on training, inspection procedures, and final apparatus [33, 34].

2 Research Problem

Providing high quality of products not covered by defective products is an important part of the supply chain between the manufacturer of the automotive parts and the final assembly company [35]. The aim of the research is to evaluate the effectiveness of visual inspection of airbag modules and to improve this process. The evaluation and improvement of the effectiveness of visual inspection was carried out in an enterprise manufacturing products for the automotive industry, located in the Silesian Voivodeship, and the research was carried out on three controllers involved in the assessment process performed in the mentioned enterprise. The Kappa index and auxiliary indicators of the effectiveness of alternative inspection were employed in the course of the research.

The visual inspection in the analysed enterprise allows to specify a number and type of defects per given inspected unit. During the visual inspection, the "error catalogue" is used, which was developed and approved by engineers is. Additionally, based on the error catalogue, finished product templates are available at the control stands on each production line (lid and logo emblem), divided into permissible errors and NOK errors which should be rejected (i.e. put in a designated place on the production line). Visual inspection of the products is carried out by appropriately trained Inspectors of the Stationary Quality Control against the applicable templates available on the production line. Each inspection is carried out in accordance with a quality control plan.

Visual inspection of the airbag module (lid and logo emblem) consists in a proper interpretation and comparison of the product with an available boundary template by the inspector who qualifies it as an OK or NOK product. The use of individual control points in a visual inspection allows for quick detection and elimination of a conforming product that does not conform to the customer's requirement. The costs of production depend on the effectiveness of the inspection performed by the Quality Control Inspector.

In the event of any non-conformities on the production lines, a "QOS Chart - Monitoring the level of defects on the production line" is developed every 10 weeks.

Among the nonconformities created or rejected in the first stage of production, a significant share are the defects associated with the aesthetics of the lid, as well as the emblem of the customer's logo. The most greatest number of non-conformities arise due to damage to flexible and plastic surfaces in the lid of the driver's airbag module, therefore the effectiveness of detecting this type of non-conformities by the controllers is very important.

3 Results and Discussion

In the enterprise, there were examinations carried out in order to assess the effectiveness of visual inspection of the airbag modules elements. The Opel driver's airbag modules were adopted in the examination. Based on the available documentation: job instructions, a catalogue of errors and existing defects, as well as available technological documents, a list of defects that occur during the manufacturing of the product and which should be noticed by controllers at a given production site was developed. The non-conformities which should be detected include: improper TOX connection, damaged perforations, damage to the surface of aesthetic elements, damage to the protective layer.

In order to evaluate the effectiveness, a sample of 30 pieces of airbag module covers was prepared, with the logo emblem already fixed. The samples were prepared by the enterprise's quality engineers in advance, and there were some products with the OK (conforming) status among them, constituting 1/3 of the total, and with the NOK (non-conforming) status, also 1/3 of the whole, and 1/3 of the products were between conforming or non-conforming. The examination was carried out by three persons assessing the level of compliance in the conditions of the plant; they were the controllers working at a given workstation, with appropriate training and qualifications. In addition, the products were evaluated by a qualified expert holding high qualifications in this field (Quality Engineer). During the examinations, a double series of evaluations of a sample of 30 products was carried out. The developed results sheet is presented in Table 1.

Table 1. Results of the conformity assessment of the fixed airbag module cap among the 3 controllers and an expert (ref.).

Part	*Controller A*		*Controller B*		*Controller C*		*Ref.*
	A - 1	*A - 2*	*B - 1*	*B - 2*	*C - 1*	*C - 2*	
1	P	P	P	P	P	P	P
2	P	P	P	P	P	P	P
3	P	P	P	P	P	P	P
4	P	P	P	P	P	P	P
5	P	P	P	P	P	P	P
6	P	P	P	P	P	P	P
7	P	P	P	P	P	P	P
8	P	P	P	P	P	P	P
9	NEG	NEG	P	P	P	P	NEG
10	NEG	NEG	P	P	NEG	NEG	NEG
11	P	P	NEG	NEG	P	NEG	NEG
12	P	P	P	P	P	P	P
13	P	P	NEG	NEG	NEG	NEG	NEG
14	P	P	NEG	NEG	NEG	NEG	NEG
15	NEG	P	P	P	P	P	P
16	P	P	NEG	P	P	P	P
17	P	P	P	P	P	P	P
18	P	NEG	P	P	P	P	NEG
19	NEG	NEG	P	P	NEG	P	NEG
20	P	P	P	P	P	NEG	P
21	NEG	NEG	NEG	NEG	NEG	NEG	NEG
22	NEG	P	P	P	NEG	P	NEG
23	P	NEG	P	NEG	NEG	NEG	NEG
24	P	NEG	NEG	NEG	NEG	NEG	NEG
25	NEG	NEG	NEG	NEG	NEG	P	NEG
26	P	P	P	P	P	P	P
27	P	P	NEG	NEG	P	P	P
28	NEG	NEG	P	P	NEG	P	NEG
29	P	P	NEG	P	P	P	NEG
30	P	P	P	P	P	P	NEG

The level of compliance of the controllers' decisions among themselves and with the expert was verified using the Kappa Cohen coefficient. This indicator, although generally accepted and recommended for use in the automotive industry, has its drawbacks and limitations [36–38]. The Gwet's AC1 coefficient [39, 40] seems a better alternative, which gets rid of numerous limitations that condition the correct use and conclusions based on the Kappa index.

The results of assessing the conformity of decisions among controllers by means of the Cohen's Kappa indicator are presented in (Table 2).

Table 2. Cohen's Kappa coefficients for pairs of controllers.

A with B cross-tab

			B		
			NEG	P	Sum
A	NEG	OBS	6	11	17
		OBL	3.2	8.1	11.3
	P	OBS	11	32	43
		OBL	8.1	20.5	28.7
	Sum	OBS	17	43	60
		OBL	11.3	28.7	40.0

Kappa A - B

Po	0.63		
Pe	0.59		
	Po-Pe	1-Pe	Kappa
	0.04	0.41	0.10

B with C cross-tab

			C		
			NEG	P	Sum
B	NEG	OBS	11	6	17
		OBL	3.4	7.9	11.3
	P	OBS	7	36	43
		OBL	8.6	20.1	28.7
	Sum	OBS	18	42	60
		OBL	12.0	28.0	40.0

Kappa B - C

Po	0.78		
Pe	0.59		
	Po-Pe	1-Pe	Kappa
	0.20	0.41	0.48

A with C cross-tab

			C		
			NEG	P	Sum
A	NEG	OBS	10	7	17
		OBL	3.4	7.9	11.3
	P	OBS	8	35	43
		OBL	8.6	20.1	28.7
	Sum	OBS	18	42	60
		OBL	12.0	28.0	40.0

Kappa A - C

Po	0.75		
Pe	0.59		
	Po-Pe	1-Pe	Kappa
	0.16	0.41	0.40

Kappa

<0.40	denotes poor level of compliance between controllers
0.40-0.75	denotes average level of compliance between controllers
>0.75	denotes good or very good level of compliance between controllers

The results showed a poor or average level of compliance among controllers. The poorest agreement was recorded for the A and B pairs of controllers, and the average for the B and C pairs of controllers.

In the next stage, the level of compliance of the controllers' decisions with the decision of the expert was determined. The obtained results are presented in Table 3.

Table 3. Cohen's Kappa coefficients for controller A/B/C & expert.

A with expert (REF) cross - tab

			Expert		
			NEG	P	Sum
A	NEG	OBS	16	1	17
		OBL	5.7	5.7	11.3
	P	OBS	14	29	43
		OBL	14.3	14.3	28.7
	Sum	OBS	30	30	60
		OBL	20.0	20.0	40.0

Kappa A - Expert

Po	0.75	
Pe	0.50	
Po-Pe	1-Pe	Kappa
0.25	0.50	0.50

B with expert (REF) cross - tab

			Expert		
			NEG	P	Sum
B	NEG	OBS	14	3	17
		OBL	5.7	5.7	11.3
	P	OBS	16	27	43
		OBL	14.3	14.3	28.7
	Sum	OBS	30	30	60
		OBL	20.0	20.0	40.0

Kappa B - Expert

Po	0.68	
Pe	0.50	
Po-Pe	1-Pe	Kappa
0.18	0.50	0.37

C with expert (REF) cross - tab

			Expert		
			NEG	P	Sum
C	NEG	OBS	17	1	18
		OBL	6.0	6.0	12.0
	P	OBS	13	29	42
		OBL	14.0	14.0	28.0
	Sum	OBS	30	30	60
		OBL	20.0	20.0	40.0

Kappa A - Expert

Po	0.77	
Pe	0.50	
Po-Pe	1-Pe	Kappa
0.27	0.50	0.53

Kappa

<0.40	denotes poor level of compliance between controllers and expert
0.40-0.75	denotes average level of compliance between controllers and expert
>0.75	denotes good or very good level of compliance between controllers and expert

The results showed an average and poor level of compliance of the controllers' decisions with the expert's decision. The C controller was turned out the best, followed by the A controller, while the worst was the B controller. For all the controllers it is necessary to take improvement actions, as the results turned out to be far from satisfactory.

In the next stage of the analysis, auxiliary indicators were calculated in order to obtain additional information on the level of effectiveness of the alternative control performed by individual controllers. The results of calculations are presented in Table 4.

Table 4. Auxiliary indicators to assess the effectiveness of the alternative control system.

Auxiliary indicators	*Controller A*	*Controller B*	*Controller C*
% Appraiser	83.33%	90.00%	80.00%
% Score vs Attribute	63.33%	60.00%	66.67%
System % Effective Score		43.33%	
System % Effective Score vs Reference		40.00%	
False negative	0.00%	6.67%	0.00%
False positive	33.33%	46.67%	26.67%
False alarm rate (acc. to products)	6.67%	13.33%	6.67%
Miss rate (acc. to products)	60.00%	60.00%	60.00%
False alarm rate (acc. to decisions)	3.33%	10.00%	3.33%
Miss rate (acc. to decisions)	46.67%	53.33%	43.33%
Bias (acc. to products)	0.111	0.222	0.111
Bias (acc. to decisions)	0.071	0.188	0.077

The results showed that the best repeatability i.e. "internal" compatibility of decision for the same products ("% Appraiser") was demonstrated by the controller B, then by the controller A, and the controller C was at the end. In turn, the highest correctness of the results, i.e. the internal compliance between the decision made by the controller and the model decision (made by the expert) for the same products ("% Score vs Attribute") was proven for controller C, then for the controller A and B (who was the worst). The correctness results ("% Score vs Attribute") proved to be much worse and different from the repeatability results ("% Appraiser") for individual controllers. The effectiveness of the control system consisting of 3 inspectors measured as the repeatability of their results, i.e. the internal compliance of their decisions for the same products ("System % Effective Score") was 43.33%, and when measured as the correctness i.e. the internal compliance of their decisions with the model decision (expert) for the same products ("System % Effective Score vs Reference") was worse and amounted to 40.00%. The tendency to rejection indicator, i.e. the indicator of the controller's tendency towards type I errors (rejection of the conforming product) ("False negative"), showed that only the B controller made such errors. On the other hand, the tendency to acceptance indicator, i.e. the indicator of the controller's tendency towards type II errors (acceptance of the non-conforming product) ("False positive"), proved that such errors were most often made by the B controller, then A and C. The frequency indicator of the conforming rejected products acc. to products and decisions ("Miss rate") showed that the majority of the products were rejected (in other words: the greatest number of such decisions were made) by the controller B; both the controller A and C rejected the same number of good products (they made such decisions). In turn, the frequency indicator for the type II errors acc. to products ("False alarm rate acc. to products") showed the same share of type II errors (60%) in relation to all controllers, while the frequency indicator for the type II errors acc. to decisions ("False alarm rate acc. to the decision"), specifying the number of wrong acceptance decisions in relation to all possible decisions, showed that these errors were made mostly by the B controller, then by A and C controllers. The Bias index, i.e. the controller's tendency indicator in the classification to one of the qualitative categories, calculated for products and decisions as the share of type I errors in relation to share of type II errors, showed a definite dominance of type II errors (more dangerous) over type I errors (values well below 1) for all controllers, the largest for B, then A and C.

While comparing the obtained results of the selected indicators (Table 5) with the criteria of the alternative control system assessment according to the MSA 4th Edition 2010 guidelines (Table 6), it should be stated that since the results are acceptable or at the boarder of acceptance in the case of effectiveness and the type I errors, all results were totally unacceptable for type II errors. It was decided to undertake improvement actions intended to reduce the number such errors - these actions are presented in the part containing the conclusions from the article.

Table 5. Effectiveness indicators according to MSA 4th Edition 2010.

Appraiser	*Effectiveness = % Appraiser*	*False alarm rate (acc. to decisions) ($P_{false\ alarm}$)*	*Miss rate (acc. to decisions) (P_{missed})*
A	83.33	3.33	46.67
B	90.00	10.00	53.33
C	80.00	3.33	43.33

Table 6. Acceptance criteria for assessing the effectiveness of the alternative control system according to MSA 4th Edition 2010.

Decision / *Measurement system*	*Effectiveness*	*False alarm rate*	*Miss rate*
Acceptable for the appraiser	≥ 90%	≤ 5%	≤ 2%
Marginally acceptable for the appraiser – may need improvement	≥ 80 - < 90%	> 5 - ≤ 10%	> 2 - ≤ 5%
Unacceptable for the appraiser – needs improvement	< 80%	> 10%	> 5%

To sum up the results obtained in the field of the effectiveness of visual inspection, a ranking of controllers was developed, which is as follows:

- 1st place - controller C - good internal compliance, the largest compliance with the expert, the lowest tendency to accept a non-conforming product, the lowest share of type I and II errors.
- 2nd place - controller A - good internal compliance, worse (but relatively good) compliance with the expert, greater inclination to accept a non-conforming product, the lowest share of type I errors, a relatively higher share of type II errors.
- 3rd place - controller B - the best internal compatibility, but the worst compliance with the expert, the relatively highest inclination to accept a non-conforming product, relatively highest share of type I and type II errors.

4 Conclusions

Detailed assessment of the effectiveness of the visual inspection system of airbag modules (consisting of 3 controllers) in the analysed enterprise proved that plenty of results remain outside the permitted ranges. These results indicate the necessity to introduce improvement actions. These activities were related to the re-training of all controllers, and a thorough analysis of the developed catalogue of errors, including all available templates on a given production line.

After a thorough analysis of the calculated indicators (Kappa, auxiliary indicators), it was found out that the additional impact on the results was exerted by the assessment made by the controllers of the studied samples, where in some cases the controller classified a product as a NOK incorrectly during a visual inspection of a small incision on the lid, which in reality the controller found hard to assess as OK and preferred to evaluate such a product as non-conforming. Evaluation of the visual inspection system against the adopted indicators led to implementation of a standard, which defined correct performance of the inspection, i.e. the controller must hold the module cover at a right angle (70°) and 50 cm from the eye during the visual inspection.

The Andon system was implemented in the analysed enterprise's process of product quality control, which significantly contributed to the increase in the detection of specific types of non-conformance of products (difficult or even impossible to detect by a human). The Andon system increased the speed of response to emerging irregularities during the production process. The implementation of this system provided for a great improvement in the speed of reporting problems encountered during the production, a faster response to existing problems and improved productivity with increased involvement of relevant services.

Benefits resulting from the improvement activities undertaken in the area of visual inspection processes confirm the validity of such analyses. Thanks to the introduced changes, there was an increase in effectiveness in detecting non-conformance of products, which contributed to improvement in the efficiency of the production process and an increase in the satisfaction of external customers (OEM) with the quality of the products delivered.

In conclusion, it should be emphasized that projects in the field of quality control (visual inspection) effectiveness improvement should not be of a one-off nature. It is important to maintain the effects of actions taken, immediately responding to emerging or reported problems and periodically checking the effectiveness [41] and the efficiency (which can be an indicator of the maturity of a quality management system (QMS)) [42] of the quality control system. As Plato said, "quality is a certain degree of perfection", hence the level of quality and effectiveness of the visual inspection system should only be the next step in the pursuit of perfection (i.e. 100% effectiveness of the inspection system).

References

1. Hamrol, A.: Strategies and practices of efficient operation. Lean Six Sigma and other (in Polish), (PWN), Warsaw (2015)
2. Czerska, J.: Improving the value stream (in Polish), (Difin), Warsaw (2009)
3. Borkowski, S., Knop, K.: Challenges faced in modern quality inspection. Manag. Prod. Eng. Rev. **7**(3), 11–22 (2016)
4. Webber, L., Wallace, M.: Quality Control for Dummies. Wiley, Hoboken (2007)
5. Ulewicz, R.: Practical application of quality tools in the cast iron foundry. Manuf. Technol. **14**(1), 104–111 (2014)
6. Czyżewski, B.: Internal quality control in a production company (in Polish), (Wielkopolski Quality Club FSNT NOT), Poznan (2007)
7. Rewilak, J.: Do you trust visual inspection? (in Polish), TQM Soft. http://szkolenia.tqmsoftcom/inspekcja. Accessed 20 Mar 2013
8. Kujawińska, A., Vogt, K.: Human factors in visual quality control. Manag. Prod. Eng. Rev. **6**(2), 25–31 (2015)
9. Schultz, D.: Psychology and the challenges of today's work (in Polish), (PWN), Warsaw (2012)
10. See, J.E.: Visual inspection reliability for precision manufactured parts. Hum. Factors **57**(8), 1427–1442 (2015)

11. Charles, R., Fletcher, S.R., Tailor, M.: Search patterns in human visual inspection. In: Contemporary Ergonomics and Human Factors, Proceedings of the International Conference on Ergonomics & Human Factors, 13–16 April 2015
12. Johnson, L.: Can You Improve Your Visual Inspection Process? https://www.minitab.com/uploadedFiles/Content/News/Published_Articles/inspection_process_improvement_breakthrough.pdf. Accessed 05 Dec 2018
13. Hamrol, A., Kowalik, D., Kujawińska, A.: Impact of selected work condition factors on quality of manual assembly process. Hum. Factors Ergon. Manuf. Serv. Ind. **21**(2), 156–163 (2011)
14. Vogt, K., Kujawińska, A.: Analysis of the effect of the type of nonconformity on the effectiveness of visual inspection (in Polish). In: Knosala, R. (ed.) Innovation in Management and Production Engineering, vol. 2, pp. 470–480 (2014)
15. Kujawinska, A., Vogt, K., Diering, M., Rogalewicz, M., Waigaonkar, S.D.: Organization of visual inspection and its impact on the effectiveness of inspection, In: Hamrol, A., Ciszak, O., Legutko, S., Jurczyk, M. (eds.) Advances in Manufacturing. LNME, pp. 899–909 (2018)
16. Purswell, J.L., Hoag, L.L.: Strategies for improving visual inspection performance. In: Proceedings of the Human Factors and Ergonomics Society Annual Meeting, vol. 18, no. 4, pp. 397–403 (1974)
17. See, J.E., Drury, C.G., Speed, A., Williams, A., Khalandi, N.: The role of visual inspection in the 21st century. In: Proceedings of the Human Factors and Ergonomics Society, pp. 262–266 (2017)
18. Schoonahd, J.W., Gould, J.D., Miller, L.A.: Studies of visual inspection. Ergonomics **16** (1973–4), 365–379 (2007)
19. Kavuri, S.H.: Size Effects in human visual inspection for micro/meso scale parts, mechanical (and materials) engineering. Dissertations, Theses, and Student Research, p. 81 (2015)
20. Thapa, V.B., Gramopadhye, A.K., Melloy, B., Grimes, L.: Evaluation of different training strategies to improve decision making performance in inspection. Int. J. Hum. Factors Manuf. **6**(3), 243–261 (1996)
21. Lee, J., Ko, K.W., Lee, S.: Visual inspection system of the defect of collets for wafer handling process. Int. J. Control Autom. **9**, 129–138 (2016)
22. Drury, C.G., Watson, J.: Good practices in visual inspection, Human factors in aviation maintenance-phase nine, progress report, FAA/Human Factors in Aviation Maintenance (2002). https://www.faa.gov/data_research/research/med_humanfacs/oamtechreports/1990s/media/AM91-16.pdf. Accessed 29 Oct 2018
23. Knop, K., Ingaldi, M., Śmiłek-Starczynowska, M.: Reduction of errors of the conformity Assessment during the visual inspection of electrical devices. In: Advances in Manufacturing, LNME, pp. 857–867. Springer, Heidelberg (2018)
24. Reinfus, R.: MBO - a simple and effective technique for managing your company (in Polish). Warsaw (2011). (Helion)
25. Knop, K., Borkowski, S.: The estimation of alternative control efficiency with the use of the Cohen's Kappa coefficient. Manag. Prod. Eng. Rev. **2**(3), 19–27 (2011)
26. Measurement System Analysis (MSA), Reference Manual, 4th Edn. AIAG Group (2010)
27. Diering, M., Kujawińska, A.: MSA-Measurement System Analysis. Guide to procedures (in Polish), Poznan (2012). (AR Comprint)
28. Gramopadhye, A.K., Drury, C.G., Sharit, J.: Feedback strategies for visual search in airframe structural inspection. Int. J. Ind. Ergon. **19**(5), 333–344 (1997)
29. Boucher-Genesse, A.: The Human Brain VS the Digital Brain - A Case for Visual Inspections (2016). https://blog.robotiq.com/human-brain-vs-digital-brain-what-is-so-different-case-study-for-a-visual-inspection. Accessed 30 Nov 2018

30. Smith, C.J., Adendorff, K.: Advantages and limitations of an automated visual inspection system. South Afr. J. Ind. Eng. **5**(1), 27–36 (1991)
31. Stoller, J.: The Lean CEO: Leading the Way to World-Class Excellence: Leading the Way to World-Class Excellence. McGraw-Hill Education, New York (2015)
32. Bożek, M., Rogalewicz, M.: The ineffectiveness of the final control the reason of the low efficiency of the production process (in Polish), Engineering of Machines (in Polish), R. 18, z. 1, pp. 84–96 (2013)
33. See, J.E.: Visual inspection: a review of the literature (Report SAND2012-8590). Sandia National Laboratories, Albuquerque (2012)
34. Clemons, A.J.: Training methods for visual inspection tasks. Iowa State University Capstones, Graduate Theses and Dissertations, 13005 (2013)
35. Ulewicz, R.: Outsourcing quality control in the automotive industry. In: MATEC Web of Conference, vol. 183, p. 03001 (2018)
36. Klaput, P., Vykydal, D.: Effect of the number of non-conforming samples on the Kappa indicator values. In: MATEC Web of Conference, vol. 183, p. 03014 (2018)
37. Diering, M., Dyczkowski, K., Hamrol, A.: Estimating the level of conformity of assessments in visual inspection - problems in determining the Kappa coefficients (in Polish), Innovations in management and production engineering (in Polish), pp. 257–268. Publishing House of the Polish Society for Production Management (2015)
38. Jarosz-Nowak, J.: Models of assessing the extent of agreement between raters using kappa coefficients (in Polish). Appl. Math. (in Polish) **8**, 126–154 (2007)
39. Wongpakaran, N., Wongpakaran, T., Wedding, D., Gwet, K.L.: A comparison of Cohen's Kappa and Gwet's AC1 when calculating inter-rater reliability coefficients: a study conducted with personality disorder samples. BMC Med. Res. Methodol. **13**(1), 61 (2013)
40. Starzyńska, B., Szajkowska, K., Diering, M., Rocha, A., Reis, L.P.: A study of raters agreement in quality inspection with the participation of hearing disabled employees. Advances in Manufacturing. LNME, pp. 881–888. Springer, Heidelberg (2018)
41. Anttila, J., Jussila, K.: Organizational learning in developing the integrated quality management. Prod. Eng. Arch. **18**, 3–13 (2018)
42. Grabowska, M., Takala, J.: Assessment of quality management system maturity. In: Advances in Manufacturing. LNME, pp. 889–898. Springer, Heidelberg (2018)

Statistical Process Control Using LMC/MMC Modifiers and Multidimensional Control Charts

Milena Markiewicz[1], Emilia Bachtiak-Radka[2], Sara Dudzińska[2], and Daniel Grochała[2(✉)]

[1] F.P. Wenglon Sp. z o.o. Sp.k., Dobra, Poland
[2] West Pomeranian University of Technology, Szczecin, Poland
daniel.grochala@zut.edu.pl

Abstract. The quality of manufactured products is often a combination of many different specifications. Tolerances of shape and position errors with additional modifiers are increasingly being used in coordinate metrology. The most commonly encountered modifiers are the least/maximum material conditions. The authors in this article present new and original tools for statistical control for specifications with associated tolerances.

Keywords: Statistical Process Control · Geometric dimensioning and tolerancing · Control charts · Coordinate metrology

1 Introduction

Control in manufacturing processes carried out on a mass scale is an extremely responsible and difficult task. Its aim is to guarantee the correctness of the dimensional and shape tolerances of an entire series of manufactured products. However, in the majority of supervised production processes, measurements of all manufactured products would be too expensive. This issue is addressed by Statistical Process Control (SPC) [1–10], used to forecast potentially dangerous situations and avoid conditions that could lead to the manufacture of products not meeting the required specification. The basic guidelines for planning activities aimed at maintaining the required quality level (and sometimes improving them gradually [2–5]) are included in Advanced Product Quality Planning (APQP), Production Part Approval Process (PPAP); Failure Mode and Effects Analysis (FMEA), Statistical Process Control (SPC), and Measurement System Analysis (MSA), all in accordance with ISO IATF 16949:2016.

Modern manufacturing is characterized by an increasing degree of complexity. At the same time, the aim is to obtain the functional characteristics of final products which are often interrelated [11–20]. This is especially visible in hybrid manufacturing [14–19], aimed at achieving a synergistic effect; here, the quality of the manufactured parts depends on the technological parameters of two types of machining combined in a single machine system [11, 13–22] - Fig. 1.

A. Hamrol et al. (Eds.): *Advances in Manufacturing II* - Volume 3, LNME, pp. 244–253, 2019.
https://doi.org/10.1007/978-3-030-17269-5_18

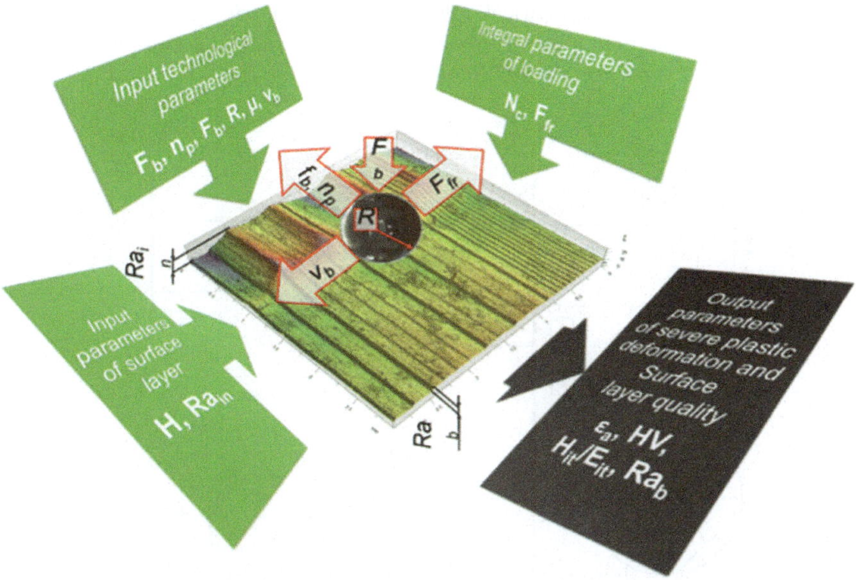

Fig. 1. Hybrid surface technology which combines shape machining with finish burnishing [13].

Its added value consists in improved surface layer and geometrical specification quality [13–20], shorter production time and lower price [11–22].

2 Associated Geometrical Tolerances

The development of modern measuring technologies has resulted in much lower measurement uncertainties than a few years ago [23–33]. More precise measurements leave more room to maneuver for technologists and quality engineers, allowing corrections limited by the tolerances defined by the constructor [24–28]. In addition, designs increasingly often use modifiers to combine length and angle tolerances with geometrical tolerances (most often concerning direction and position errors) [27–33] Fig. 2.

Fig. 2. Modifiers used in the design to combine length and angle tolerances with geometrical tolerances for direction and position errors.

Their use in control by means of coordinate measuring machines (CMM) allows to achieve an effect similar to that in control with Go/NoGo testing, used to check the functionality of products (most often their assemblability) without any distinction between the length and geometrical tolerances [28–30]. Here, length tolerance is associated with geometrical tolerance by using the least material condition (LMC) and maximum material condition (MMC) modifiers [27] - Fig. 3.

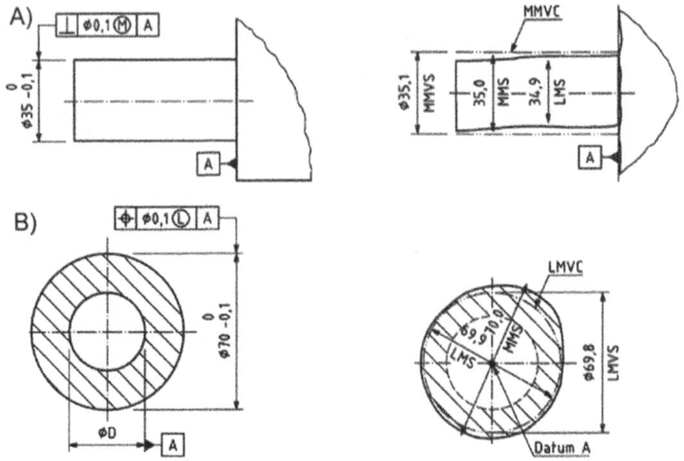

Fig. 3. Dependent tolerances for length and position using (A) maximum material condition and (B) least material condition, according to ISO 2692:2006(E) [27].

The maximum material condition (Fig. 3A) is used to determine the assemblability of the product. It is used when, for functional reasons, it is not important which part of the permissible range of dimensional and geometrical variation is used in the dimensional deviation and which part in the geometrical deviation [28–32]. MMC occurs when each point of the dimensioned element has a limit dimension corresponding to the largest amount of material, e.g. the largest shaft size and the smallest hole size. The maximum material virtual condition (MMVC) (1) is a dimension created from MMC of the dimensioned element and geometrical tolerance. For external features, MMVS is the sum of MMC and the geometrical tolerance, defined by the following equation [27]:

$$MMVS = MMC + \delta \tag{1}$$

and for internal features, the MMVS (2) is the difference between MMC and the geometrical tolerance, defined by the equation:

$$MMVS = MMC - \delta \tag{2}$$

where: δ - is the defined geometrical tolerance.

The least material condition (LMC) (Fig. 3B) is the combined requirement for the dimension of the dimensioned element, the geometrical deviation of this dimension and the position of the derived feature. LMC occurs when the dimensioned element in each place has a limit dimension corresponding to the smallest amount of material, e.g. the smallest shaft size and the largest hole size. The least material virtual state (LMVC) is a dimension made up of LMC and geometrical tolerance. For external features, LMVS is the difference between LMC and geometrical tolerance, defined by the Eq. (3) [27]:

$$LMVS = LMC - \delta \tag{3}$$

and for internal features, LMVS is the difference between LMC and the geometrical tolerance, defined by the Eq. (4):

$$LMVS = LMC + \delta \tag{4}$$

Tests from a production control area, where the control characteristics were described by the associated tolerances associated by modifiers (LMC/MMC), prove that slightly extending the value of LMC/MMC beyond the limits specified in the specification did not lead to a loss in product functionality (or assemblability of parts [29–32]), provided that the second independent component of the δ specification was close to the nominal value (as defined by the designer). Using tolerances associated by a modifier (LMC/MMC), part of the unused δ tolerance can be transferred to the LMC or MMC, reducing the amount of scrap without fear of escalating quality problems associated with the manufacture of products with compromised functionality (assemblability), i.e. exceeding MMVC or LMVC. Using the described specification tolerance methods, the results of the CMM control each time will be the same as the results of the Go/NoGo test control [1, 2].

3 Statistical Process Control of Associated Characteristics

SPC tools are designed to achieve the desired level of variability and/or position of a controlled characteristic in accordance with the adopted quality objective. In practice, the most common tool are the Shewhart control charts [1–8, 34–36]. One of their basic assumptions is that control tolerance limits remain unchanged over time [34–36]. On the other hand, control limits may change periodically; they are usually narrowed down. However, in the time horizon (on the control chart) adopted by the quality engineer, the control limits are not evaluated following changes in the standard deviation observed in the production process [1–8]. In rare cases where two specifications need to be controlled simultaneously, multidimensional T2 Hotelling charts are used [2, 37–40] - Fig. 4. This chart is a generalization of a classic one-dimensional control chart and should be understood as the distance from the centre of a multidimensional normal distribution [37–40]. In the case of measurements with a given LMC or MMC, we deal with two mutually associated variables. A one-dimensional chart is useless here, because even using more than one such chart would not present the relationship between variables. The T2 Hotelling chart can graphically represent two variables as non-associated with each other and combine these charts into a third, showing the relationship between the variables [2, 37–40]. However, also in this case, for the associated variable the time control limits assume constant values - Fig. 4.

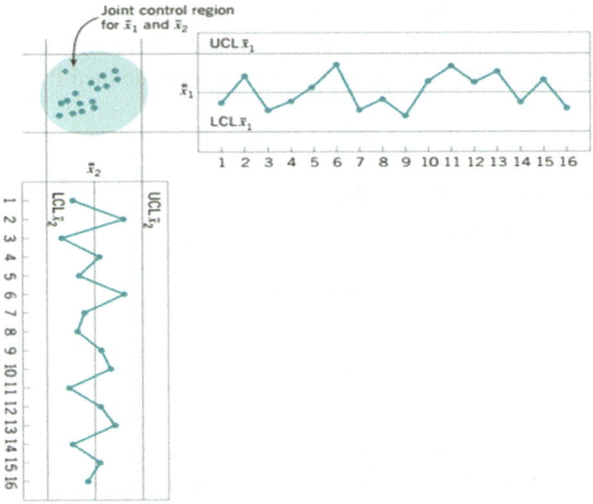

Fig. 4. Multidimensional T2 Hotelling control chart [2].

The use of associated geometrical tolerances causes the unused tolerance margin r of one component to change (widen) the tolerance for the other component. This makes it necessary to track two specification values at the same time, with one of the tolerance limits changing over time. The following classical formulae (5 and 6) used to determine the control limits in the specification's path of variation [1–8, 34–36],

$$UCL = \bar{x} + 3 \cdot s \tag{5}$$

$$LCL = \bar{x} - 3 \cdot s \tag{6}$$

where: $\bar{x}$- is the medium value and s- is the standard deviation
should be added the value of the remaining r - unused tolerance reserve of the non-associated part of the geometrical condition; then the control limits will assume the values:

$$UCL = \bar{x} + 3 \cdot s + r \tag{7}$$

where

$$r = UCL - \bar{x} \tag{8}$$

and

$$LCL = \bar{x} - 3 \cdot s - r \tag{9}$$

where

$$r = LCL - \bar{x} \tag{10}$$

4 Experimental Simulation of the Process Control with Associated Characteristics

In the experimental research conducted with the use of MATLAB software we used practical data from real control process and simulated a situation where the external diameter of a shaft and its concentricity were recorded during the production process (Fig. 5). The experiment involved a series of k = 12 measurements in one-element samples.

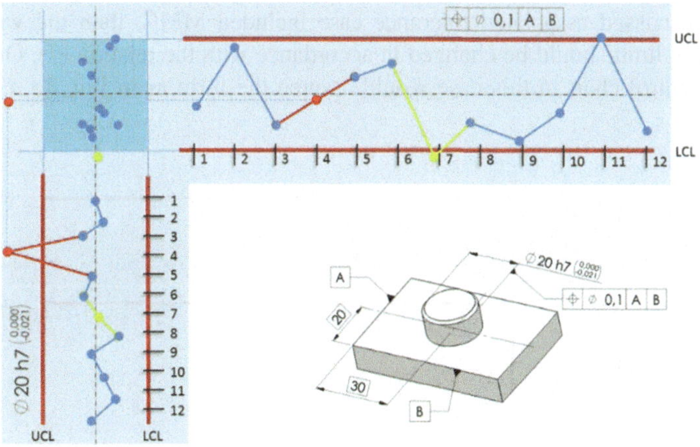

Fig. 5. Distribution of diameter sizes and positions recorded during experimental tests.

For the above situation, a design condition was simulated in which the measured diameter of the shaft is associated with its position by LMC - Fig. 6. The unused reserve of the shaft diameter is transferred to the position tolerance. The value of the lower control limit is determined by (9).

Figure 5 shows a violation of the lower control limit for shaft position tolerance. In this situation (not allowing for LMC) the operator would be forced to stop the process and introduce a technologically difficult adjustment to the shaft position. After the introduction of LMC and monitoring by means of the modified T2 Hotelling multi-dimensional control chart made, this violation is also visible (Fig. 6), but in this case, a point located close to the lower limit is not a cause for concern. We can also see that in order to improve the potential capability of the process, it is possible to adjust the diameter of the shaft, which is easy to implement. Thus, the proposed design of the multidimensional chart is very intuitive for the operator and should be a good form of visualization of the process.

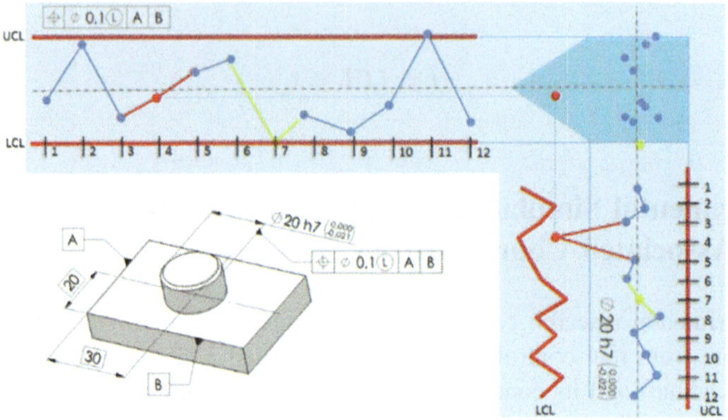

Fig. 6. Control chart of dimensions associated on the external feature with a specified LMC.

If the discussed associated tolerance case included MMC, then the value of the upper control limit should be changed in accordance with the relation (7). On the other hand, the control chart in this case would assume the form as in Fig. 7.

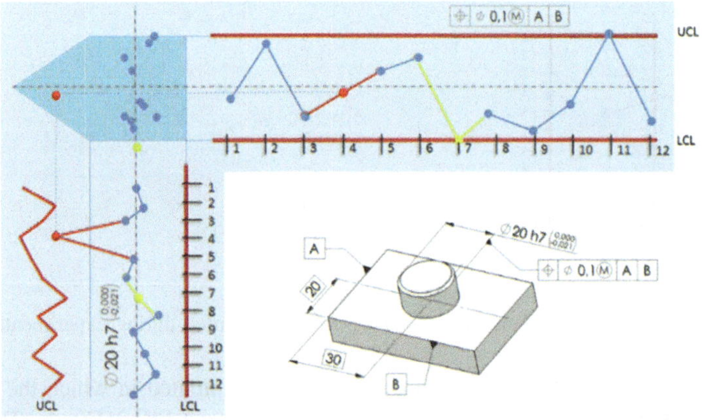

Fig. 7. Dimensional control chart for the external feature with a specified maximum material condition MMC.

In the situation shown in Fig. 7 there was no change in the decision where the shaft position exceeded the lower control limit. However, thanks to the application of the MMC modifier, it can be seen that the process of establishing the shaft position has been "moved away" from the upper control limit, at the same time giving the operator and the quality engineer a sense of stability and safety of the production process.

In addition to violations of control limits, the developed charts can also apply to other configuration tests defined in literature.

5 Conclusions

The application of LMC/MMC brings many benefits, including a reduction in the number of parts classified as non-compliant, without increasing the risk of loss of product functionality.

Their use allows a reduction in the amount of scrap and time associated with machine downtime resulting from corrections to the geometrical specifications of products, which include errors of circular runout, direction and position. These two effects can certainly lower the costs of the production process.

Until now, no statistical production control where the measurement characteristics were associated by these modifiers has been possible. On the market as well as in literature there has been a lack of dedicated statistical tools that could be used for this purpose.

The tool proposed in this paper fills this gap. Control charts are the focal point of all activities in the SPC system and the graphical representation of measurement data alone allows very good forecasting of the process status and its qualitative potential. Thanks to the developed chart it is possible to visualize the location of the process and dispersion of measurement results. The data collected on the chart make it possible to perform complementary analyses, allowing an increase in the quality potential of the process. Of course, this seemingly simple task, i.e. introducing minimum/maximum material conditions, results in complications, e.g. calculation of the capability indicators which depend on the range of tolerance, which in this case is a variable.

This research should be continued in the future in order to determine the specific economic benefits of the proposed SPC tool.

References

1. Hamrol, A.: Quality management at examples. PWN, Warsaw (2007). (in Polish)
2. Montgomery, D.C.: Introduction to Statistical Quality Control, 6th edn. Wiley, Hoboken (2009)
3. Dietrich, E., Schulze, A.: Statistical Procedures for Measuring Systems. Machines and Process Qualification, Notika System, Warsaw (2000). (in Polish)
4. Konczak, G.: Statistical Methods in Quality Control of Production. University Publishing - University of Economics in Katowice, Katowice (2007). (in Polish)
5. Myszewski, J.M.: Variability Management. A Systematic Look at Statistical Methods in Quality Management. Publishing by Institute of Organization and Management in Industry “ORGMASZ”, Warszawa (1998). (in Polish)
6. Vander Wiel, S.A., Tucker, W.T., Faltin, F.W., Doganaksoy, N.: Algorithmic statistical process control: concepts and an application. Technometrics **34**(3), 286–297 (1992)
7. Wheeler, D.J.: Advanced Topics in Statistical Process Control. SPC Press, Knoxville (1995)
8. Montgomery, D.C., Keats, J.B., Runger, G.C., Messina, W.S.: Integrating statistical process control and engineering process control. J. Qual. Technol. **26**(2), 79–87 (1994)
9. Woodall, W.H., Montgomery, D.C.: Research issues and ideas in statistical process control. J. Qual. Technol. **31**(4), 376–386 (1999)
10. Woodall, W.H.: Controversies and contradictions in statistical process control. J. Qual. Technol. **32**(4), 341–350 (2000)

11. Grochała, D., Berczyński, S., Grządziel, Z.: Modeling of burnishing thermally toughened X42CrMo4 steel with a ceramic ZrO_2 ball. Arch. Civ. Mech. Eng. **17**(4), 1011–1018 (2017)
12. Dudzińska, S., Bachtiak-Radka, E., Grochała, D., Berczyński, S.: A study on technological properties of metallic surfaces marked with Data Matrix codes with laser technology. J. Mach. Constr. Maint. **108**(1), 91–101 (2018)
13. Bachtiak-Radka, E., Dudzińska, S., Grochała, D., Berczyński, S., Olszak, W.: The influence of CNC milling and ball burnishing on shaping complex 3D surfaces. Surf. Topogr.: Metrol. Prop. **5**(1), 015001 (2017)
14. Chen, C.H., Shiou, F.-J.: Determination of optimal ball-burnishing parameters for plastic injection moulding steel. Int. J. Adv. Manuf. Technol. **21**(3), 177–185 (2003)
15. Shiou, F.-J., Chuang, C.-H.: Precision surface finish of the mold steel PDS5 using an innovative ball burnishing tool embedded with a load cell. Precis. Eng. **34**(1), 76–84 (2010)
16. Dzionk, S., Scibiorski, B., Przybylski, W.: Surface texture analysis of hardened shafts after ceramic ball burnishing. Materials **12**, 204 (2019)
17. Swirad, S.: High-precision finishing hard steel surfaces using hydrostatic burnishing tool. In: MATEC Web of Conferences, vol. 249, p. 03002. EDP Sciences (2018)
18. Görög, A., Görögová, I., Stančeková, D., Janota, M.: Influence of roller-burnishing on surface roughness parameters and roundness. In: MATEC Web of Conferences, vol. 244, p. 01021. EDP Sciences (2018)
19. Rodríguez, A., Fernández, A., López de Lacalle, L., Sastoque, P.L.: Flexible abrasive tools for the deburring and finishing of holes in superalloys. J. Manuf. Mater. Process. **2**, 82 (2018)
20. Tamilarasan, A., Renugambal, A., Mohan, T., Iyer Akshay, R., Krish, V., Pramoth, K., Rajkumar, A., Sakriya Himanshu, N.: Optimization of roller burnishing process parameters using lion optimization algorithm. In: IOP Conference Series, Materials Science and Engineering, vol. 390, no. 1, p. 012063. IOP Publishing (2018)
21. Patyk, R., Kukielka L., Kaldunski, P., Bohdal, L., Chodor, J., Kulakowska, A., Kukielka, K., Nagnajewicz, S.: Experimental and numerical researches of duplex burnishing process in aspect of achieved productive quality of the product. In: AIP Conference Proceedings, vol. 1960, no. 1, p. 070021. AIP Publishing (2018)
22. Dudzińska, S., Szydłowski, M., Grochała, D., Bachtiak-Radka, E.: Application of correlation function for analysis of surface structure shaping by hybrid manufacturing technology. In: Advances in Manufacturing. LNME, pp. 651–659. Springer (2018)
23. Nobuo, S.: Meteorological Handbook Guide on Measurements. Mitutoyo (UK) Ltd., Andover (2015). (in Polish)
24. Plowucha, W., Jakubiec, W., Humienny, Z., Hausotte, T., Savio, E., Dragomir, M., Bills, P. J., Marxer, M., Wisla, N., Mathieu, L.: Geometrical product specification and verification as toolbox to meet up-to-date technical requirements, pp. 131–139. CMT Bielsko-Biała, Bielsko-Biała (2014)
25. Płowucha, W., Jakubiec, W., Wojtyła, M.: Possibilities of CMM software to support proper geometrical product verification. Procedia CIRP **43**, 303–308 (2016)
26. Srinivasan, V.: Reflections on the role of science in the evolution of dimensioning and tolerancing standards. Proc. Inst. Mech. Eng. Part B: J. Eng. Manuf. **227**(1), 3–11 (2013)
27. PN-EN ISO 2692:2015-02: Geometrical product specifications (GPS)—Geometrical tolerancing—Maximum material requirement (MMR), Least material requirement (LMR) and Reciprocity requirement (RPR)
28. Ratajczak, E., Woźniak, A.: Coordinate Measuring Systems. University Publishing - Warsaw University of Technology, Warszawa (2016)
29. Henzold, G.: Geometrical Dimensioning and Tolerancing for Design. Manufacturing and Inspection. Butterworth-Heinemann, Oxford (2006). (in Polish)

30. Humienny, Z.: Generalization by the requirements of the material maximum and material minimum of the geometric tolerance. Arch. Mach. Technol. Autom. **30**(2), 79–85 (2010)
31. Ngoi, B., Lim, L., Ong, A.S., Lim, B.H.: Applying the coordinate tolerance system to tolerance stack analysis involving position tolerance. Int. J. Adv. Manuf. Technol. **15**(6), 404–408 (1999)
32. Jiang, G., Cheraghi, S.H.: Evaluation of 3-D feature relating positional error. Precis. Eng. **25** (4), 284–292 (2001)
33. Dantan, J.-Y., Ballu, A., Mathieu, L.: Geometrical product specifications model for product life cycle. Comput.-Aided Des. **40**(4), 493–501 (2008)
34. ISO 7870-2:2013 Control charts - Part 2: Shewhart control charts
35. Koutras, M.V., Bersimis, S., Maravelakis, P.E.: Statistical process control using Shewhart control charts with supplementary runs rules. Methodol. Comput. Appl. Probab. **9**(2), 207–224 (2007)
36. Montgomery, D.C., Mastrangelo, C.: Some statistical process control methods for autocorrelated data. J. Qual. Technol. **23**(3), 179–193 (1991)
37. Runger, G.C., Alt, F.B., Montgomery, D.C.: Contributors to a multivariate statistical process control chart signal. Commun. Stat. – Theory Methods **25**(10), 2203–2213 (1996)
38. Louwerse, D.J., Smilde, A.K.: Multivariate statistical process control of batch processes based on three-way models. Chem. Eng. Sci **55**(7), 1225–1235 (2000)
39. Lowry, C.A., Montgomery, D.C.: A review of multivariate control charts. IIE Trans. **27**(6), 800–810 (1995)
40. MacGregor, J.F., Kourti, T.: Statistical process control of multivariate processes. Control Eng. Pract. **3**(3), 403–414 (1995)

The Improvement of Sustainability with Reference to the Printing Industry – Case Study

Jan Lipiak(✉) and Mariusz Salwin

Faculty of Production Engineering, Institute of Production Systems Organization, Warsaw University of Technology, Warsaw, Poland
janlipiak@etigraf.pl

Abstract. The aim of the article is to present the manufacturing innovations as a tool supporting the implementation of the concept of sustainability. The improvement of sustainability will be monitored as the consumption of main media in the modern printing house. The impact for sustainability will be shown as a result of decrease media consumption. The innovations that are considered as one of the most significant subject matters with reference to sustainability have been widely discussed. The importance and the effectiveness of sustainability has been presented with the special attention paid to the social, economic and environment aspects. Furthermore, the significance and the efficiency of the innovative solutions aiming at the least possible impact on the environment has been identified. The theoretical part has been supplemented with the experience of the analyzed enterprise. On the basis of the analyses, it was demonstrated that the implemented manufacturing innovation (in terms of flexographic printing technology) contributed to the decrease of electricity consumption, production materials by 10–12% on a monthly basis and also to the increase of the employment by two employees. Innovations contribute to the decrease of inconvenience of production as far as the natural environment is concerned. Furthermore, the costs of production have been reduced and due to the increased employment, those have an impact on the social field of environment. Therefore, it can be stated that innovations can be considered as an indicator of efficient implementation of the sustainability with reference to eco-innovations in Poland.

Keywords: Innovations · Sustainability · Production improvement · Organization effectiveness · Printing industry

1 Introduction

Continuous climate changes, the increase of the temperature, the lowering ground water level, food and energy supplies that are progressively depleted are considered as ones of the most significant problems of the modern world. The environment in which humans live have been a subject matter of many researchers over several years, including environmentalists, natural science researchers, philosophers and finally, philosophy researchers. A quick advancement of science as well as a technical progress

A. Hamrol et al. (Eds.): *Advances in Manufacturing II* - Volume 3, LNME, pp. 254–266, 2019.
https://doi.org/10.1007/978-3-030-17269-5_19

contributed to a considerable deterioration of the natural environment what consequently threatens the life of humans, plants as well as animals. Concerning health care and environmental protection, as well as due to own international obligations, governments of numerous countries deal with environmental policy basing on the rules provided by sustainability. Sustainability is documented by both integration and cohesion of social, economic and ecological aspects in the country development process. Therefore, it can be stated that sustainability is one of the strategic area of economic activity as far as government administration is concerned.

Sustainability is characterized by the implementation of innovative solutions that allow to achieve a better quality and standards of living of a large number of people living all over the world, taking advantage of resources necessary for the production process in an efficient way. An enterprise, by providing novel solutions for sustainability should take a responsible implementation of its strategy in each area of activity into consideration as well as cover the entire activities chain up to the final product, its usage and last but not least, waste management.

Innovations can be considered as a determinant of the effectiveness as far as implementation of sustainability is concerned, not only on the macro-economic level – across the entire economy but primarily, on the micro-economic level – across one enterprise. The analyses provided in the following article are based on the innovation in terms of letterpress printing, implemented in the discussed company. The results of the research relate to the consequences of the indicated innovation in terms of enterprise functioning, with reference to the assumptions of sustainability concept.

The main aim of the present article is to discuss the good practices strictly connected with improvement of sustainability, basing on the example of the printing company.

The impact for sustainability will be shown as a result of decrease media consumption. The innovations that are considered as one of the most significant subject matters with reference to sustainability have been widely discussed.

The subject of the research is a printing company located in Masovian Voivodship, which produce labels and packaging.

2 Innovations as an Implication of Sustainability

2.1 The Notion and the Scope of Innovation

The concept of innovation was introduced by Schumpeter [1] and can be understood as launching either a novel product consumers are not familiar with yet, a novel production method, which was not tested before in any industry, a novel market, in which a certain industry of a certain country was not active before, regardless the previous existence or non-existence of that market. Furthermore, it can be defined as achieving a novel source of materials, regardless of the fact that it was already existing or it had to be created and last but not least, as implementing a novel organization of a certain industry [2–5].

It should be mentioned that innovations are observed to contribute to the technical development [6]. Innovations should be understood as a wide spectrum of all scientific, technological, organizational, financial and commercial activities that actually lead or aim at leading to the implementation [7–11].

The main purposes of innovation implementation include improving the quality of products and services and retaining or enhancing one's market position. Another significant factor is the need of customer as well as lowering costs [12].

The significant obstacles for entrepreneurs, as far as innovation implementation is concerned, include costs, difficulties in terms of funding and last but not least, bureaucracy.

There is a wide range of potential effects of innovation implementation. Nevertheless, the following benefits are mostly observed: increase of the functionalities, increase of the utility of products and services, modernization of obsolete systems, technology improvement, enhancement of human communication, optimization of working time and last but not least, environmental protection. It can be concluded that those effects correspond with the objectives of sustainability and therefore, a relation between innovations and this approach can be observed. The concept of sustainability is described in the following sub-chapter.

2.2 The Concept of Sustainability

For the first time, the notion of sustainability was formulated and adapted during the UN Conference on Environment and Development (UNCED) organized by the United Nations in 1992.

Sustainability is a socio-economic development in which a process of integration of political, economic and social actions can be observed while maintaining environmental balance and the stability of basic natural processes in order to guarantee the possibility of satisfying the needs of particular societies or citizens of current and further generations [6, 13–20]. Meeting the sustainability challenges is possible due to the integration of environment, economic and social policy. It requires treating natural resources as limited economic resources as well as using natural capital in a way which allows for marinating the functions of ecosystems in long-term perspective. Sustainability should be connected with both formulation and implementation of the strategy which concerns providing novel work places and economic development, supporting and conducting a business by appropriate subjects [21, 22].

Nevertheless, it should be mentioned that the progress in the areas such as air and water pollutions was observed. However, the impact on the environment is still unsustainable. Therefore, looking for additional factors of sustainability is necessary. Innovations may be considered as one of them. This concept is described in the following sub-chapter.

2.3 Innovation as a Determinant of Sustainability

Any innovation is connected with sustainability, especially innovations regarding the changes in terms of the way of organization and management of the enterprise of the economy of a country, technological innovations and last but not least, ecological innovations. The companies that decrease the usage of resources and components in its operations are perceived as "environmental-friendly" what consequently enables them to achieve additional incomes and to lead novel projects [23].

Innovative solutions are sometimes observed as serious threats for the environment due to the fact most of the problems regarding environment protection are difficult to be solved without implementing any innovation. The undertaken actions of an innovative character are considered as novel development opportunities and contribute to a considerable growth of attractiveness and market competitiveness [23].

It can be concluded that innovations and sustainability are connected with the cause-and-effect relationship. The wider analysis of this phenomenon is discussed in the research part.

3 Methodology of the Study-Case

3.1 Characteristics of the Analyzed Company

The analyzed company is a printing house with a long tradition, specialized in labels and laminate production. It is specialized in the production of labels and laminate. It is possible due to possessing a professional graphic studio, modern machinery, creative team and own digital production preparation. Thanks to it, the organization is able to face numerous tasks, starting from a graphic project, through printing, finishing and finally, delivering a finished product to the customer. Over a number of years, the mentioned company have been enjoying the confidence of numerous customers presenting the meat industry, food industry, pharmaceutical industry, cosmetics industry and last but not least, the chemical industry.

Among numerous types of labels produced by the described company, five of the main product-types can be listed:

- Self-adhesive labels,
- Heat shrinkable labels,
- Tea tags,
- OPP films,
- Laminates.

Basing on the observations of the pro-environmental tendencies of development and functioning of modern world economy, the company is aiming at being of top of such trends and take in into consideration while developing.

The assumptions of sustainability can be widely observed within the actions undertaken by the described organization. One of the areas which has an impact on adjusting to this phenomenon include technological innovations. It's assumed that all technological innovations taken over the years have the positive impact on the sustainability aspects as lower media consumption, and it will be commissioned in this article.

3.2 Innovation in Terms of Letterpress Printing Technology

The ability to adapt oneself to the quickly changing situation on the market, searching for novel outlets and novel customers, the development of printing techniques and technologies as well as using novel media, especially the Internet are observed to be the major factors of a proper development of a printing enterprise [12, 24–26].

The innovation was implemented in terms of letterpress printing technology, which is one of the basic graphic techniques (besides offset and intaglio), in which the print is made by the rebound of the ink placed on the protuberant parts of the printing form. It is the oldest graphic technique.

Currently, as far as the letterpress printing technology is concerned, the industry widely uses flexography and especially for special usage, typography. The flexographic units are also used for varnishing. Typography used to be a widely used printing technique before dissemination of exposure devices which enables to use the offset print in an easy way. It enhances the quality of prints and it is otherwise more economic.

The innovation is strictly connected with the technique of flexographic printing, used in the production of flexible packaging and labels on the non-absorbing surface (ABL and PBL barrier laminates). It allows to achieve the effects of overprint with relief through local application of a considerably thicker varnish layer what was previously used only in the technique of screen print. The described company owns a patent of letterpress printing no. 332/2016/PAT title: The way of creating flexographic prints.

One of the elements of the solution is to increase the capacity of the anilox which is a raster roller, one of the main elements of a printing press. Therefore, the quality of a raster roller has an enormous impact on the quality of print. The capacity of anilox is adjusted to the amount of ink that has to be placed on the material. Depending on the working ruling, the anilox ruling has to be adjusted.

The comparison of the transferring the varnish between a traditional printing plate and using the method implemented by the described company is presented in the (Fig. 1).

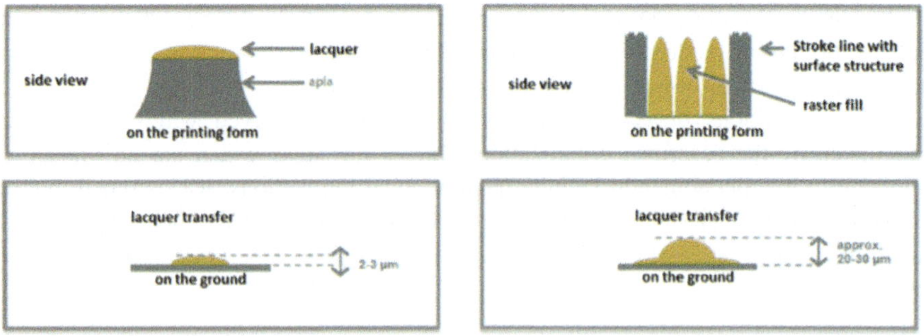

Fig. 1. The comparison of the type of printing plate and the letterpress printing technology.

Using this innovation enables to replace the screen printing varnishing unit with a properly modified varnishing unit in a flexographic technique. It eliminates the necessity of using screen printing varnishing and allows to print using low-migration inks and varnishes adapted to the current legal requirements of the food industry. The print of a considerable thickness is created in accordance to the description of the innovation allows to make the packaging with an increased value in use for recipients (producers) and customers. It is possible to make a printed inscription with a palpable thickness.

4 Results and Discussion

This chapter provides an analysis of the impact of the described innovation of the functioning of the analysed enterprise.

4.1 Analysis of the Impact of Innovation on the Functioning of the Enterprise with Reference to Sustainability

The concept of sustainability puts an emphasis on three areas, including economic, social and environmental areas. In order to consider all aspects of sustainability, the analysis connected with the impact of the described innovation on the functioning of the described organization, the following variables were selected: electricity consumption, consumption of film, consumption of paper, consumption of UV inks, consumption of UV varnish, photo polymers forms and employment. The data from the period before innovation implementation are compared to those after its implementation. The analyzed period of time amounts to 20 months (10 months before implementation of innovation and 10 months after it) – March 2016 – October 2017. The innovation was implemented at the beginning of January 2017.

The total electricity consumption in the period of time from March to December 2016 amounted to 898970 kWh. The average monthly consumption amounted to 89897 kWh. The comparison of electricity consumption before and after innovation implementation on a monthly basis is shown in the (Fig. 2).

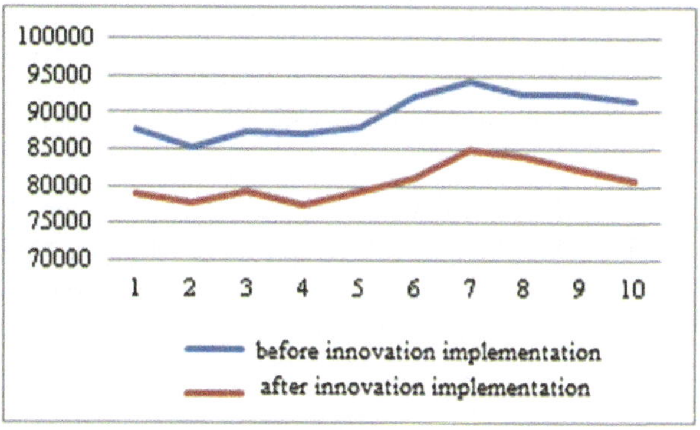

Fig. 2. The comparison of electricity consumption [kWh].

During the analyzed period of time the electricity consumption was observed to decrease by 10.31%. The average monthly electricity consumption decreased from 89897 kWh to 80625 kWh. Furthermore, the median before innovation implementation amounted to 89897 kWh, and after its implementation – 80062 kWh (Table 1).

Table 1. Comparison of electricity consumption [kWh].

Specification	Before	After
Mean	89897	80625,02342
Standard error	972,4770689	806,589151
Median	89897	80062,71769
Standard deviation	3075,24251	2550,658853
Scope	8989,7	7344,5849
Minimum	85402,15	77608,0801
Maximum	94391,85	84952,665

The total usage of film and paper from March to December 2016 amounted to 2923653 m^2. The average consumption of film and paper amounted to 2923650 m^2. The comparison of the consumption of film and paper before and after innovation implementation on a monthly basis is presented in the (Fig. 3).

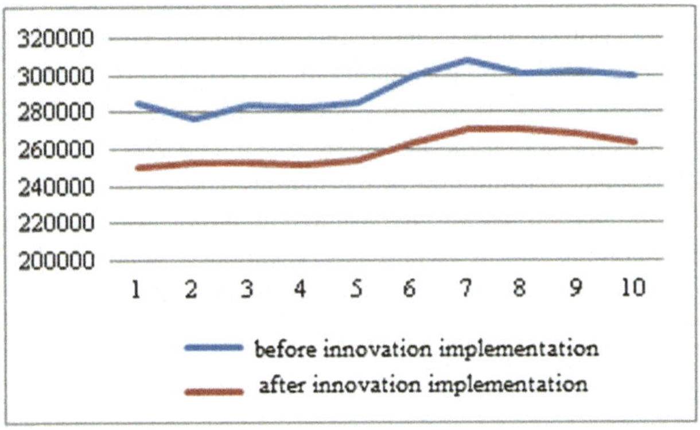

Fig. 3. Comparison of the consumption of film and paper [m^2].

During the analyzed period of time the decrease of consumption of film and paper by 9,81% can be observed. The average monthly usage of film and paper decreased from 292365 m^2 to 259868 m^2. What is more, before the implementation of innovation the median amounted to 292365 m^2, and after it - 258706,4794 m^2 (Table 2).

Table 2. Comparison of electricity consumption [kWh].

Specification	Before	After
Mean	292365	259868,0455
Standard error	3339,167	2667,051033
Median	292365	258706,4794
Standard deviation	10559,37	8433,9559
Scope	30405,96	20167,3377
Minimum	277162	250591,8888
Maximum	307568	270759,2265

The total consumption of UV inks from March to December 2016 amounted to 6050 kg. The average monthly consumption of UV paints amounted to 605 kg. The comparison of the consumption of UV inks before and after innovation implementation on the monthly basis is depicted in the (Fig. 4).

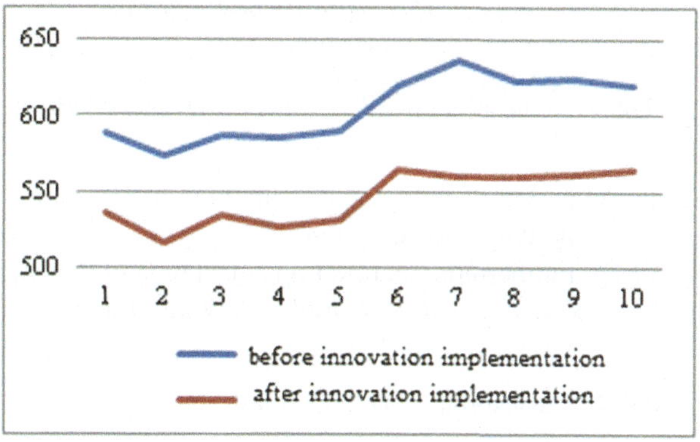

Fig. 4. The comparison of UV inks consumption [kg].

During the analyzed period of time the consumption of UV inks decreased by 11,12%. The average monthly consumption of UV inks decreased from 605 kg to 545,64 kg. Moreover, the median before implementation of the innovation amounted to 605 kg, whereas after it - 548,16 kg (Table 3).

Table 3. Comparison of electricity consumption [kWh].

Specification	Before	After
Mean	605	545,6447
Standard error	6,909842819	5,818755
Median	605	548,1603
Standard deviation	21,85084158	18,40052
Scope	62,92	48,6783
Minimum	573,54	516,186
Maximum	636,46	564,8643

The total consumption of UV varnish during the period of time from March to December 2016 amounted to 6100 kg. The average monthly consumption of UV varnish amounted to 611 kg. The comparison of the consumption of UV varnish before and after implementation of innovation on the monthly basis is shown in the (Fig. 5).

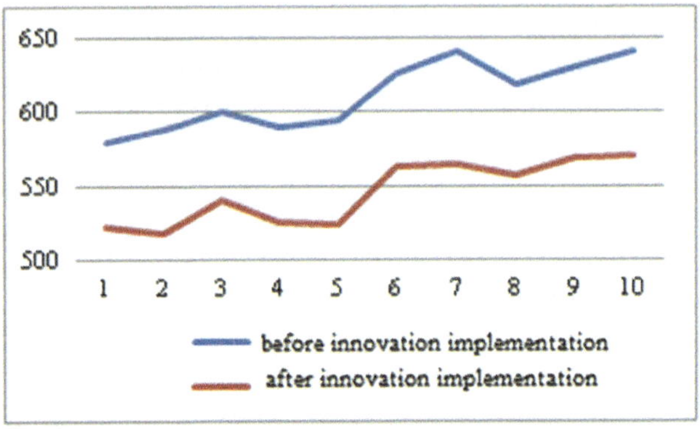

Fig. 5. Comparison of UV varnish consumption [kg].

During the analyzed period of time, the consumption of UV varnish decreased by 10.80%. The average monthly consumption of the UV varnish decreased from 611,28 kg to 545,27 kg. Furthermore, before implementation of innovation the median amounted to 610 kg, whereas after it decreased to 549 kg (Table 4).

Table 4. Comparison of the UV varnish consumption [kg].

Specification	Before	After
Mean	611,281	545,27168
Standard error	7,298760389	6,815435655
Median	610	549
Standard deviation	23,08070693	21,55229991
Scope	62,22	52,033
Minimum	579,5	518,012
Maximum	641,72	570,045

The total consumption of photo polymers form from March to December 2016 amounted to 1051,2 m^2. The average consumption of photo polymers form amounted to 105,32 m^2. The comparison of the consumption of photo polymers before and after the implementation of innovation during a month is shown in the (Fig. 6).

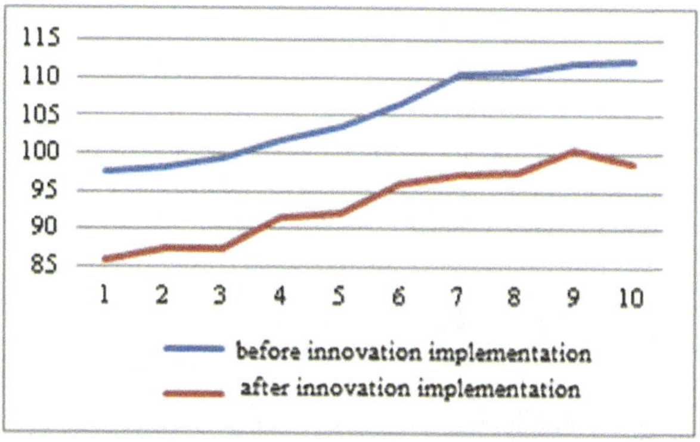

Fig. 6. The comparison of the consumption of photo polymers forms [m^2].

During the analyzed period of time the consumption of the photo polymers forms decreased by 11,20%. The average monthly consumption of them decreased from 105,32 m^2 to 93,52 m^2. What is more, the median before innovation implementation amounted to 105,15 m^2, whereas after its implementation it amounted to 94,12 m^2 (Table 5).

Table 5. Comparison of the consumption of photo polymers forms [m^2].

Specification	Before	After
Mean	105,32071	93,52492
Standard error	1,869116161	1,684471
Median	105,15255	94,11968
Standard deviation	5,91066428	5,326764
Scope	14,714	14,72451
Minimum	97,743	86,01384
Maximum	112,457	100,7384

The final analyzed element is employment which during the period of time from March to December 2016 amounted to 48 people. After the implementation of innovation, the company had to employ 2 more people. The decrease of electricity and production materials contributed to generating the savings. Consequently, the volume of production increased in order to satisfy the growing needs of recipients and that required the increase of the number of employed people.

Taking the effects of the discussed innovation implemented in the company into consideration, the decrease of consumption of both resources and energy contributed to generating savings (economic dimension), as well as the decrease of the negative impact on the environment (environmental dimension). Additionally, the increase of the employment has an impact on the social sphere of the enterprise surrounding, for example by decreasing the unemployment (Table 6).

Table 6. Effects of innovation implementation.

Specification	Effect
Electricity	−10,31%
Film and paper	−11,12%
UV inks	−9,81%
UV varnish	−10,80
Photo polymers forms	−11,20
Employment	4,17%

On the basis of the conducted analyses it can be stated that innovations can be considered as a catalysator for the sustainability from a micro-economic point of view. What is more, due to undertaking such actions by entrepreneurs, those can have a positive impact on wider implementation of sustainability of overall economy.

Furthermore, it can be stated that the sustainability with reference to the printing industry can be enhanced by the following aspects:

- sustainable forest management and the management of paper,
- effective management of materials and recycling,
- processes optimization,
- sustainable management of a company.

The described company is observed to introduce waste management in as sustainable way, minimize the amounts of pollutions emitted into the air and constantly monitor the used machinery and equipment in order to avoid the possibility of indicating the emergency occurrence reg. environment. What is more, the presented company used the materials and technologies which are examined in terms of safety for both people and environment. It is also constantly enhancing the technological processes in order to minimize the impact on the environment.

5 Conclusions

The assumption of technological innovation implementation with reference to sustainability is decreasing the negative impact on the natural environment. It is connected with almost each area of activity of modern enterprises as well as other market participants. In a sense, innovations are directed into the sustainability. They are beneficial for producer, consumers as well as other market participants.

Therefore, innovations can be considered as a solution to the modern global challenges due to the fact that the problems in terms of sustainability implementation occur, and in the longer term, it can cause permanent negative changes of widely understood environment.

Innovations can contribute the standard of living of current and future generations. Therefore, it can be stated that innovations constitute a tool, a catalysator used for sustainability.

The policy regarding environmental protection allows to avoid the problems connected with it. In a long term, sustainability gives a chance for the survival of civilization.

To conclude, it can be stated that the policy and good practices of the analyzed company are observed to improve sustainability.

This research was conducted only in one printing industry, therefore conclusions and findings from this study-case are not common for every printing industry. So the impact on the sustainability can be only set as a goal in narrow field. However in mentioned printing company, all implemented technological developments had significant impact – measured as consumption of media – on the sustainability. Further research with extended area of printing industries and impact on sustainability measured as a consumption of media will be needed.

References

1. Schumpeter, J.A.: Capitalism, Socialism and Democracy. Harper and Brothers, New York (1942)
2. Schumpeter, J.A.: The Theory of Economic Development. An Inquiry into Profits, Capital, Credit, Interest, and the Business Cycle. Harvard University Press, Cambridge (1934)
3. Carayannis, E.G.: Encyclopedia of Creativity, Invention, Innovation and Entrepreneurship. Springer, New York (2013)
4. McCraw, T.K.: Prophet of Innovation: Joseph Schumpeter and Creative Destruction. Belknap Press of Harvard University Press, Cambridge (2007)
5. Solow, R.M.: Heavy thinker. (Prophet of innovation: Joseph Schumpeter and creative destruction). New Repub. **236**(16), 48–50 (2007)
6. Smith, A., Stirling, A., Berkhout, F.: The governance of sustainable socio-technical transitions. Res. Policy **34**, 1491–1510 (2005)
7. Baregheh, A., Rowley, J., Sambrook, S.: Towards a multidisciplinary definition of innovation. Manag. Decis. **47**, 1323–1339 (2009)
8. Bessant, J., Lamming, R., Noke, H., Phillips, W.: Managing innovation beyond the steady state. Technovation **25**(12), 1366–1376 (2005)
9. Bledow, R., Frese, M., Anderson, N., Erez, M., Farr, J.: A dialectic perspective on innovation: Conflicting demands, multiple pathways, and ambidexterity. Ind. Organ. Psychol. **2**(3), 305–337 (2009)
10. Boer, H., During, W.E.: Innovation, what innovation? A comparison between product, process and organizational innovation. Int. J. Technol. Manag. **22**, 83–109 (2001)
11. Crossan, M.M., Apaydin, M.: A multi-dimensional framework of organizational innovation: a systematic review of the literature. J. Manag. Stud. **47**, 1154–1191 (2010)
12. Carrillo-Hermosilla, J., del Río, P., Könnölä, T.: Diversity of eco-innovations: reflections from selected case studies. J. Cleaner Prod. **18**, 1073–1083 (2010)
13. Gladwin, T.N., Kenelly, J.J., Krause, T.-S.: Shifting paradigms for sustainable development: implications for management theory and research. Acad. Manag. Rev. **20**(4), 874–907 (1995)
14. Pratima, B., Song, H.-C.: Similar but not the same: differentiating corporate sustainability from corporate responsibility. Acad. Manag. Ann. **11**(1), 105–149 (2017)
15. Geels, F., Schot, J.: Typology of sociotechnical transition pathways. Res. Policy **36**, 399–417 (2007)

16. Jordan, A.: Governance for sustainable development: taking stock and looking forwards. Environ. Plann. C: Polit. Space **26**, 17–33 (2008)
17. Kemp, R., Loorbach, D., Rotmans, J.: Transition management as a model for managing processes of co-evolution towards sustainable development. Int. J. Sustain. Dev. World Ecol. **14**, 78–91 (2007)
18. Kern, F., Smith, A.: Restructuring energy systems for sustainability? Energy transition policy in the Netherlands. Energy Policy **36**, 4093–4103 (2008)
19. Meadowcroft, J.: National sustainable development strategies: a contribution to reflexive governance? Eur. Environ. **17**, 152–163 (2007)
20. Voß, J.P., Bauknecht, D., Kemp, R.: Reflexive Governance for Sustainable Development, pp. 3–28. Edward Elgar, Cheltenham (2006)
21. Bäckstrand, K.: Civic science for sustainability: reframing the role of experts, policy maker and citizens in environmental governance. Glob. Environ. Polit. **3**, 24–41 (2003)
22. Folke, C., Hahn, T., Olsson, P., Norberg, J.: Adaptive governance of social ecological systems. Annu. Rev. Environ. Resour. **30**, 441–473 (2005)
23. Bossink, B.: Eco-Innovation and Sustainability Management. Routledge Taylor & Francis Group, New York (2012)
24. Lipiak, J., Ejsmont, K.: Implementation of the SMED method in a printing company. In: Economics and Business Organization, no. 4(783), pp. 104–116, Published by the Institute of Organization and Management in Industry ORGMASZ, Warsaw (2015)
25. Lipiak, J., Werpachowski, W.: Use of the Kaizen philosophy in packaging companies. In: Economics and Business Organization, no. 8(787), pp. 43–53, Published by the Institute of Organization and Management in Industry ORGMASZ, Warsaw (2015)
26. Ejsmont, K., Krystosiak, K., Lipiak, J.: Application of the selected data mining technique to the printing industry. In: Knosala, R. (ed.) Innovations in Management and Production Engineering, vol. 2, pp. 75–86. OW PTZP, Opole (2015)

Improvements in the Production Environment Made Using Quality Management Tools

Adam Górny(✉)

Faculty of Management Engineering, Poznan University of Technology, Poznań, Poland
adam.gorny@put.poznan.pl

Abstract. Improvements are the outcomes of management activities aimed at ensuring that the performance of tasks is increasingly enhanced, often beyond expectations. Improvements can be defined as radical modifications that help achieve expected results. They can also be viewed as efforts to help create an environment in which activities, including manufacturing tasks, are executed with efficiency and consistency. Success in achieving the desired efficiency depends on employing appropriate tools that support efforts to boost the ability to achieve desired outcomes. Under the seventh principle of quality management, effective quality management can be assumed to require solid data and facts rather than speculation and ungrounded opinions. This principle also applies to improvements made to the manufacturing environment, which are only successful if based on robust data and information on the working environment, with all of its hazards and strains, as well as any other irregularities that compromise the ability to perform work safely and effectively. Thus, the manufacturing environment can only be improved once one has assessed the nature of specified hazards as well as the extent and nature of their impacts.

The article outlines some of the key (traditional and new) quality management tools available for collecting and processing data and information on the manufacturing environment. While the article's primary focus is on a literature review, it presents sample applications of quality-assurance tools based on real-life cases taken from industry. These come predominantly from small and medium-sized enterprises engaged in various types of production.

Keywords: Improvement · Quality management tools · Work environment · Safety · Elimination of hazards and nuisances

1 Introduction

High-quality efficient production processes are commonly considered to be vital for an enterprise's survival. This is especially important when faced with growing competition. It should also be noted that these aspects are strongly related to the development of SME's enterprises [1]. To achieve such quality, it is essential to identify the factors that affect process efficiency and secure opportunities and conditions for the adequate completion of tasks. To make this possible, it is commonly necessary to identify the root causes of any potential non-conformities and irregularities that may impair the ability to achieve desired outcomes.

A. Hamrol et al. (Eds.): *Advances in Manufacturing II* - Volume 3, LNME, pp. 267–276, 2019.
https://doi.org/10.1007/978-3-030-17269-5_20

One of the factors for the ability to conduct production processes to a satisfactory quality standard is the working environment defined as a setting, occupied by workers, in which process are conducted [2–4]. Every time an improvement is attempted, one should take proper notice of the people it is presumed to address. The key target group for improvements are internal process customers, i.e. persons employed in the concerned company who are responsible for the performance of manufacturing tasks. Their performance depends on the extent to which the environment in which processes are conducted can be improved by measures that modify the work setting and respond to worker needs [5, 6]. Concern with adequate working conditions needs to be integrated into the overall improvement process [5–7].

If working environment improvements are recognized as falling within the scope of management, then such improvements should be considered part and parcel of a management process and undertaken to facilitate the completion of tasks to an ever higher quality standard [8]. This requires creating the right conditions for the efficient management of the organization and proper supervision over whether and how tasks are performed. Efficient working environment is predicated upon management execution with proper account taken of the specific needs of the entities in which improvements are undertaken [5]. The nature of the related activities and adequate resources are also of great importance [9, 10]. The end results are successful improvements that meet expectations and they affect the improvement of production results and development of enterprises [5].

Early in the process, one should eliminate any barriers to improvements [11]. While this approach will enable the organization to exceed any originally presumed outcomes, it requires that the issues at hand be identified accurately and that any decisions be based on fact. Quality improvement tools are of great help in correctly identifying nonconformities, streamlining cause identification and motivating the persons involved to work effectively as a team [11, 12]. It is essential that the most appropriate quality-management tools be selected for given circumstances [7, 13]. The aim of this study is to specify the basic features of quality-management tools and the potential for their deployment in specified conditions with a view to improving the working environment.

2 Selected Aspects of Working Environment Improvements

2.1 Nature of Systemic Improvements

To employ improvement measures systemically, it is crucial to meet certain conditions. First and foremost, one needs to structure one's efforts [11]. Regardless of the course of action taken, improvement measures must enable the organization to achieve specific presumed effects [14].

Improvement efforts commonly aim to enhance the capability to meet requirements. The aim can be described as the creation of an environment that fosters an effective and stable implementation of specific projects and, more specifically, helps perform production processes in a stable manner [14]. When adopting a systemic approach, one should ensure that the improvement measures that are undertaken reflect actual capabilities and needs and are adequate to the circumstances. Actions taken in different areas of business management must take into account the safety and the work environment aspects [15].

In keeping with the seventh principle of quality management, the management of any function in a company can be presumed to be effective and efficient if it is based on actual and verified data and factual descriptions rather than mere suppositions and/or opinions [16]. It is therefore essential to systematically collect information, properly process data by tried, tested and reliable methods, and submit the findings to the right people who may then reach appropriate decisions [12, 17].

The same principle can be applied to indicate the need for an opportunity to improve the working environment in a manufacturing setting. One can presume that such improvements are successful if they are based on data and information that describe actual process environments. Information on the state of the working environment defines the feasibility of performing production tasks. This applies in particular to [2, 18]:

- Risk and strain factors,
- Other irregularities that adversely affect the ability to work safely, effectively and efficiently.

The ability to improve an existing environment is contingent upon:

- Obtaining real data through observation, control, bench-mark tests, measurements, etc.,
- Processing such data to produce an accurate picture of the circumstances,
- Choosing further feasible actions based on current data and information on areas of improvement,
- Selecting a course of action that will help achieve desired improvement outcomes,
- Taking actions that will modify the working environment.

By completing the above actions, an organization will modify its working environment and boost its capacities to perform work processes [2, 7]. The flow of the above process is shown in Fig. 1. Each stage of the process shown below requires the use of dedicated quality-management tools that improve the performance of tasks.

An analysis of the potential benefits of employing quality-management tools to improve the working environment will not be complete without examining the very nature of such benefits closely. Measures that alter the status quo lead to the creation of what can be described as a "new reality". To improve the new reality, one needs to re-apply the improvement methodology. Such steps lead to a gradual improvement of the working environment through resolving identified issues. These include discrepancies between the desired and existing state as well as any identified positive impacts of the improvement process on work efficiency, an example of which are contributions to process performance efficiency.

The ability to use quality improvement tools correctly is frequently central to effectively solving problems [2, 12, 19]. In the event the desired effects are not achieved, an organization has to either re-use quality-management tools, once it makes sure the tools are being applied in a correct manner, or employ alternative tools that are tailored to its specific needs and that will help it obtain the expected benefits. It is also in situations where performance efficiency improvements are sought that a decision is

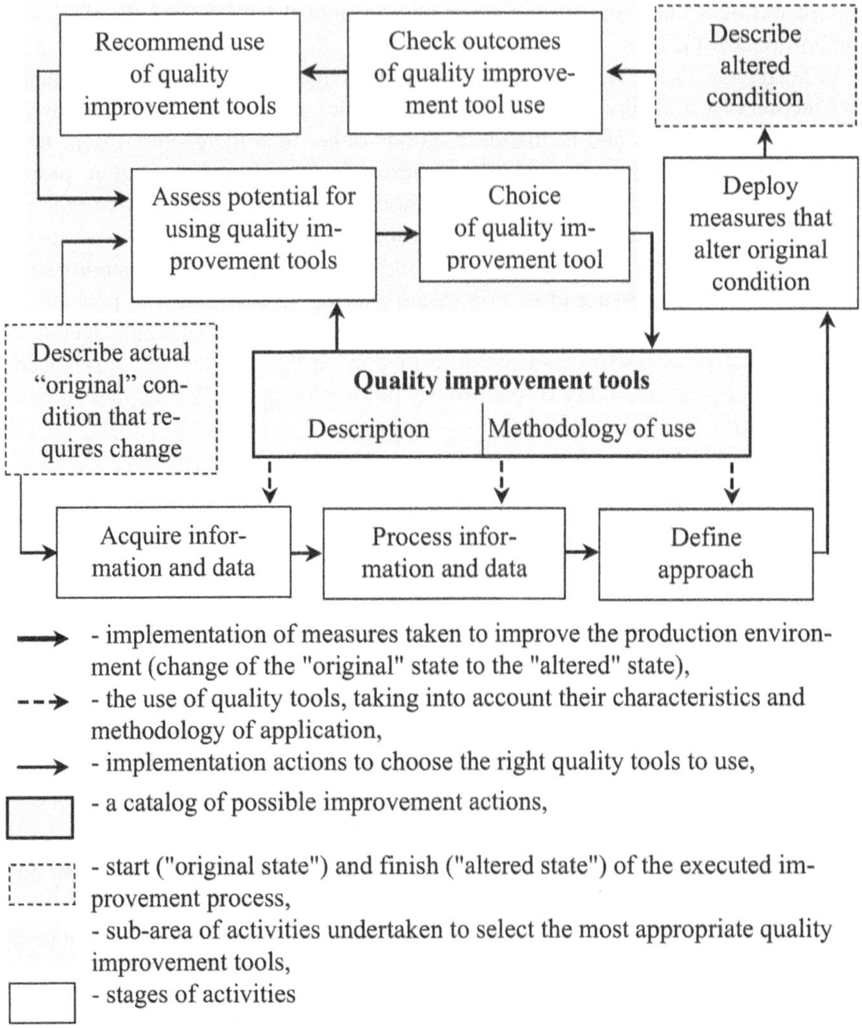

Fig. 1. Process sequence designed to improve working environment by means of quality improvement tools. Source: Own work.

needed on whether to continue to use quality improvement tools. Here again one needs to check whether proper tools have been selected and if they are used in a correct manner.

2.2 Application of Quality Improvement Tools to Enhance Working Environment

A wide range of instruments are available for improving the quality of the environment in which processes are performed. These include an array of improvement tools and methodologies. Quality-management tools, which tend to be simple and easy to use, can be employed to capture and process data associated with various aspects of quality management and define the course of action that will lead to a favorable modification of the existing state [12, 20, 21]. Their design usually reflects the specific needs of the areas in which they can be used.

Quality-management tools help gather information and data that are real at the time of their acquisition [22]. It is therefore necessary to give proper consideration to the choice of quality-management tools as such a choice will affect the capture of relevant information that will be instrumental in advancing process improvements. The benefits of using such tools also depend on the skill level of their users [22].

Two basic types of quality management tools can be distinguished [12, 17, 21]:

- Traditional (basic) quality-management tools, which include tools used to collect and analyze data,
- New-generation quality-management tools, derived from organizational techniques, which support problem analysis, suggest sequences of measures and help choose the right improvement actions.

The purpose behind using quality-management tools is to gather the necessary data, ensure that improvement measures are properly executed and that quality-management-support methods are applied. In particular, quality-management tools help [17, 21]:

- Group existing situations by appropriate grouping criteria,
- Monitor changes as they unfold,
- Identify relationships between analyzed factors,
- Rank risks and strain factors and their underlying causes,
- Determine the capacity to undertake improvements and identify the required pre-conditions that need to be met to achieve desired benefits,
- Visualize the course of improvement activities to support a more accurate evaluation.

The potential for employing quality-management tools to assess and improve the working environment and, ultimately, increase occupational safety, is presented in Tables 1 and 2. The tools help boost work safety and improve the working environment to meet the required standard.

The above list leaves out a third category of tools that are statistical in nature and that describe a quantitative relationship and take due account of the uncertainty factor. These are omitted in line with the intended focus on easy-to-use working-environment improvement tools. In order to use statistical tools appropriately, one needs to master and apply a set of complex mathematical instruments. Insufficient command of statistical knowledge on the part of quality-management-tool users may impair the effectiveness and efficiency of their application [5].

An additional prerequisite for the use of statistical tools is to ensure that the source data at hand lend themselves to processing by means of mathematical statistics and whether such processing is advisable. The author realizes that statistical tools often produce reliable information on the target population, headcounts, risk factor categories, etc. They are frequently indispensable, especially in identifying the stochastic changes that take place in an organization.

Table 1. Options to use traditional quality improvement tools to enhance working environment.

Quality improvement tool	Sample applications of quality improvement tools
Check sheet	– Groups risks by possible impacts of their occurrence – Groups risks by the severity of impacts on workers – Monitors headcounts in view of possible risks and strenuousness – Monitors variations in the impact of harmful and onerous factors on labor – Monitors the rates and impacts of occupational accidents – Identifies causal relationships between working environment parameters and occupational disease prevalence (unfitness for work) & accident rates – Determines links between technological process parameters and their impact on workplace safety
Check sheet	– Monitors the causes and occurrence of working environment non-conformities with legal requirements and applicable standards – Monitors safety (accident rates) and prevalence of occupational diseases (caused by working environment factors) – Monitors occupational safety improvement spending and related effects – Assesses working environment in terms of workers' ability to operate and perform duties safely – Assesses effectiveness of improvement measures in view of the nature and severity of non-conformities (risk and strain factors)
Ishikawa diagram	– Groups causes of irregularities in view of their potential impacts – Associates specific event causes with irregularities – Determines chances of irregularities having adverse impacts – Identifies circumstances in which working environment irregularities lead to losses (e.g. those resulting from occupational accidents and diseases)
Pareto diagram	– Groups factors into categories in view of their contributions to the occurrence of risks and strenuousness having adverse effect on workers – Identifies prime causes of accidents and diseases – Identifies working environment factors that need to be improved to increase safety – Identifies recommended improvement measures to be taken by companies operating under resource/capacity constrains (e.g. financial) – Assesses accident rates in view of their causes (such as worker age, years served at workstation, and education attainment)
Flow chart	– Defines the timelines of working environment improvement measures – Identifies factors that disrupt the launch and execution of improvement measures – Unambiguously specifies the nature and scope of ongoing improvement measures (including the start and conclusion of the ongoing improvement processes) – Identifies benefits to be gained from improvement measures designed to address existing non-conformities – Specifies the order of improvement measures and areas of potential disruptions that may impair the ability to achieve intended results
Distraction diagram	– Identifies predominant non-conformities – Identifies causes of irregularities, such as non-conformities resulting from deviations from desired state (accepted as normal)

Source: Own work based on [11, 12, 23, 24].

Table 2. Applications of new quality improvement tools to enhance working environment.

Quality improvement tool	Sample applications of quality improvement tools
Relationship diagram	– Groups irregularities by cause – Defines the link between irregularities and their causes – Improves identification of a large number of problems caused by a multitude of factors and identifies reasonable improvements – Structures working environment improvement solutions in view of parallels between measures (action strategies, target groups, costs, etc.)
Subjection diagram	– Defines links between working environment parameters and any arising risk and strain factors – Defines relationship between events (such as occupational accidents) and irregularities potentially triggering events during process performance – Establishes causal links between existing irregularities and situations and conditions that may occur in the working environment and working conditions – Identifies key causes of accidents
Matrix diagram	– Assigns identified risk factors to risk groups (categories) – Assesses the strength of relationship between right causes and impacts – Improves the identification of factors that require interventions to alleviate risks
Schematic diagram	– Groups factors into categories in view of their contributions to the occurrence - Arranges causes of problems on the assumption that many analyses will be conducted simultaneously – Identifies (determines) predominant issues that contribute to irregularities – Identifies actions necessary to improve current status (e.g. reduction in the number of risk factors or the severity of their impacts) – Suggests the most appropriate course of action to assuage the impact of irregularities or eliminate their causes – Presents an in-depth structure of issues (i.e. issues underlying the occurrence of risk factors in working environment)
Matrix data analysis	– Identifies relationships between existing risks and their impacts – Ranks causes of irregularities by frequency of occurrence, among other criteria – Ranks effects of irregularities by severity of impact on worker safety and health, among other criteria – Identifies relationships between the causes of events, such as accidents, and their impacts, including those previously unrealized by workers – Improves evaluation (choice) of improvements selected for implementation by a range of previously unrelated criteria
Arrow diagram	– Describes links between the follow-up measures to be adopted and their impact – Defines the course of recommended improvement measures – Defines the start (preparation) and end (outcomes) of tasks – Specifies the sequence of measures that ensure best results – Defines the stages (sequence) to be followed to complete planned activities – Presents parallel steps needed to complete specific tasks
Action plan	– Streamlines the procedure for selecting the most appropriate course of action as needed to achieve specified benefits (such as reduced accident rates) – Logically arranges recommended actions (including intermediate goals) – Outlines all possible actions that lead to the achievement of comparable outcomes – Identifies other viable follow-up measures that lead to comparable outcomes

Source: Own work based on [11, 12, 23, 24].

3 Benefits of Employing Quality-Management Tools to Improve Working Environment

Quality-management tools are used to capture and process data on various aspects of an organization's operations and performance. Their advantage lies in the ability to provide real data and information about the subject matter. Reliable and complete information is a sine qua non condition for effectively and systematically improving any company function [25], including the working environment [9].

To ensure that improvement efforts attain the desired level of effectiveness, it is critical to properly identify issues and challenges [26–28]. Information on the above is a prerequisite for the effective implementation of activities that will increase a company's capacity to meet its goals. Quality improvement tools will help identify irregularities in the areas that need improvement. Once the nature of irregularities and the severity and nature of their impacts is known, it becomes possible to improve the environment in which manufacturing processes are conducted. Such knowledge may be viewed as a determinant of the successful use of quality-management tools.

The available quality-management tools include both qualitative and quantitative types. The tools provide detailed information that is relevant for their specific applications. They benefit both individual employees as well as the business organization as a whole. Specifically, the tools help effectively perform processes, improve performance and cut costs [9, 10]. They can be applied to the benefits traditionally obtained as a result of the use of quality tools in order to improve the systemic management and functioning of the organization. Organizations that use them to assess the working environment and adopt improvement measures are capable to:

- Acquire intimate knowledge on the state of the working environment and any existing or potential irregularities, which are viewed as risk/strain factors,
- Provide excellent insights into problems and thus apply improvements that best reflect operating conditions in enterprises,
- Identify information (data) of particular relevance for the decision on whether to undertake improvement measures,
- Provide all stakeholders with data and information on the improvement measures proposed to facilitate the completion of specific tasks,
- Outline a range of practicable actions and help select those that ensure the greatest benefits within the shortest time,
- Identify solutions that meet the adopted optimization criteria,
- Provide concerned parties with data, information and knowledge on possible solutions to problems, including the problem of the incapacity to perform production tasks,
- Bring order to actions and eliminate chaos from solutions aimed at boosting the organization's ability to operate and carry out tasks that increase its market value,
- Improve the organization's perception by external clients as a company that cares about its working environment and seeks to increase its potential to meet the expectations of such clients,

- Improve working comfort and worker efficiency with an eye to achieving job satisfaction and increasing interest in solutions that enhance work performance,
- Meet all needs and expectations of workers seen as internal clients of processes.

A wide range of benefits can be identified by referring to the guidelines of systemic safety management [9, 10] and corporate social responsibility [6, 26, 27].

4 Summary

Working environment improvements achieved with the use of quality improvement tools should be viewed as a way to make far-reaching changes that will ultimately increase company value. Such tools can be employed to both collect and process data as well as define measures designed to bring manufacturing processes to a high quality standard. The need to include such resources in improvement efforts is stressed, inter alia, in working environment management standards [9, 28]. The resulting approach guarantees compliance with mandatory standards, which suggest that the data used in process implementation must be relevant to outcomes, complete and unambiguous [9, 25].

The simplest quality improvement tools also tend to be the easiest to use. However, the effectiveness of their application and the success in obtaining critical information depends on the proper selection of tools, and especially on their relevance in view of the organization's approach to improving the working environment. It should also be noted that the applicability of such tools depends crucially on the type of business which the organization is pursuing and its ability to obtain vital information. Also of significance is company size, its organizational structure and how well the organization has trained its workers in the use of quality-management tools.

References

1. Goerziga, D., Luckeb, D., Lenza, J., Dennerb, T., Lickefetta, M., Bauernhansl, T.: Engineering environment for production system planning in small and medium enterprises. Procedia CIRP **33**, 111–114 (2015)
2. Górny, A.: The role of safety in ensuring efficient working conditions. In: Fertsch, M., et al. (eds.) 24th International Conference on Production Research (ICPR 2017). Engineering and Technology Research, pp. 348–353. DEStech Publications, Inc., Lancaster (2017)
3. Górny, A.: The use of working environment factors as criteria in assessing the capacity to carry out processes. In: MATEC Web of Conferences, vol. 94, no. 04011 (2017)
4. Rosness, R., Blakstad, H.C., Forseth, U., Dahle, I., Siri, B.W.: Environmental conditions for safety work - theoretical foundations. Saf. Sci. **50**(10), 1967–1976 (2012)
5. Boer, J., Petruţa, B.: A more efficient production using quality tools and human resources management. Procedia Econ. Finan. **3**, 681–689 (2012)
6. Škulj, G., Vrabič, R., Butala, P.: Experimental study of work system networking in production environment. CIRP Ann. **63**(1), 401–404 (2014)
7. Górny, A.: Man as internal customer for working environment improvements. Procedia Manuf. **3**, 4700–4707 (2015)

8. Sinay, J.: Safety Management in a Competitive Business Environment. CRC Press, Taylor & Francis Group, Boca Raton (2017)
9. ISO 45001:2018: Occupational health and safety management systems - requirements with guidance for use. ISO, Geneva
10. Górny, A.: Occupational health and safety management in the international condition (consistent with objectives the ISO 45001 standard). Mod. Manag. Rev. **XX(22)**, 73–88 (2015)
11. Aqlan, F., Al-Fandi, L.: Prioritizing process improvement initiatives in manufacturing environments. Int. J. Prod. Econ. **196**, 261–268 (2018)
12. Hamrol, A.: Quality Management with Examples. PWN, Warsaw (2005). (in Polish)
13. Dahlgaard, J.J., Khanji, G.K., Kristensen, K.: Fundamentals of Total Quality Management. Taylor & Francis, London, New York (2008)
14. Starzyńska, B., Hamrol, A.: Excellence toolbox: decision support system for quality tools and techniques selection and application. Total Qual. Manag. Bus. Excellence **24**(5–6), 577–595 (2013)
15. Sanz-Calcedo, J.G., González, A.G., López, O., Salgado, D.R., Herrera, J.M.: Analysis on integrated management of the quality, environment and safety on the industrial projects. Procedia Eng. **132**, 140–145 (2015)
16. Konarzewska-Gubała, E. (ed.): Management Through Quality. Concepts, Methods, Case Studies. Publisher House of the University of Economics, Wrocław (2006). (in Polish)
17. Starzyńska, B.: Practical applications of quality tools in Polish manufacturing companies. Organizacija **47**(3), 153–164 (2014)
18. Papazoglou, I.A., Aneziris, O.N., Bellamy, L.J., Ale, B.J.M., Oh, J.: Multi-hazard multi-person quantitative occupational risk model and risk management. Reliab. Eng. Syst. Saf. **67**, 310–326 (2017)
19. Jeppe, A., Dastjerdi, E.L., Dyreborg, J., Kines, P., Jeschke, KCh., Sundstrup, E., Jakobsen, D.M., Fallentin, N., Andersen, L.L.: Safety climate and accidents at work: cross-sectional study among 15,000 workers of the general working population. Saf. Sci. **91**, 320–325 (2017)
20. Tarí, J.J., Sabater, V.: Quality tools and techniques: are they necessary for quality management? Int. J. Prod. Econ. **92**(3), 267–280 (2004)
21. McCormick, K.: Quality. Butterworth-Heinemann, Oxford (2002)
22. Górny, A.: Use of quality management principles in the shaping of work environment. Commun. Comput. Inf. Sci. **529**, 136–142 (2015)
23. Blaga, P., Boer, J.: The influence of quality tools in human resources management. Procedia Econ. Finan. **3**, 672–680 (2012)
24. Asaka, T., Ozeki, K.: Handbook of Quality Tools: The Japanese Approach. Taylor & Francis, Milton Park (1996)
25. Woźniak, K. (ed.): Contemporary Tools for Improving Organization Management Systems. Mfiles.pl, Cracow (2012). (in Polish)
26. Kolus, A., Wells, R., Neuman, P.: Production quality and human factors engineering: a systematic review and theoretical framework. Appl. Ergon. **73**, 55–89 (2018)
27. Górny, A.: The OHS management in a development of small enterprises (for example of welding factory) In: Rebelo, F., Soares, M. (eds.) Advances in Ergonomics in Design. Advances in Intelligent Systems and Computing, vol. 841, pp. 161–171. Springer, Heidelberg (2016)
28. Sousaa, S., Rodriguesb, N., Nunes, E.: Application of SPC and quality tools for process improvement. Procedia Manuf. **11**, 1215–1222 (2017)

The Analysis of the Occurrence of Faults in Passenger Cars as an Element of Improving the Management of the Production Process

Piotr Sliż(✉) and Elżbieta Wojnicka-Sycz(✉)

Institute of Organization and Management, Gdańsk University, Gdańsk, Poland
{piotr.sliz, elzbieta.wojnicka-sycz}@ug.edu.pl

Abstract. The main goal of the article was to present the results of the fault analysis on the example of one model of a passenger car "X" during the two-year warranty period. The partial goal was to present the possibilities of using data generated in the mega-process of after-sales service and warranty multi-process in order to streamline production processes in the automotive sector. On the basis of the bibliometric analysis, a cognitive gap was identified, consisting in the lack of publications regarding car failures based on data obtained from warranty repairs in the perspective of improving the car production process. The quantitative study was carried out in 2015–2018 in Poland. The subject of the study identified in the article as intervention was car repairs carried out in the Polish network of authorized services stations (ASO). As a result, repairs analyses, based on the statistical and econometric calculations, the probability of the occurrence of faults was presented, broken down into nine construction groups of passenger cars. Moreover, the possibilities of applying the results of the analysis in the improvement of the production process with the use of data generated in the after-sales process and the warranty multi-process. The following methods have been selected in the study: bibliometric analysis, observation, opinion poll, and statistical and econometric methods.

Keywords: Automotive · Production process · After-sales process · Risk management · Warranty

1 Introduction

The turbulent environment of contemporary organizations operating in the automotive sector [1] is determined by the dynamic development of car production technology [2, 3], implementation of innovations and implementation of methodologies and tools aimed at improving the after-sales [4–9] and production processes. In the discussed issue, this concerns the implementation of tools and methods of data analysis in the assessment and improvement of vehicle repair processes, while assisting in the forecasting of faults in new products (cars and components from which they are built) [10–12] and dynamic modification of the production process by changing production technology, selecting suppliers, modifying technical documentation and repair technologies. This creates new challenges for the creation of management systems. This means that both contemporary researchers and managers should manage the diffusion

A. Hamrol et al. (Eds.): *Advances in Manufacturing II* - Volume 3, LNME, pp. 277–289, 2019.
https://doi.org/10.1007/978-3-030-17269-5_21

of knowledge and information, using modern solutions in the field of technology enabling its use [13, 14]. For this purpose, it is possible to use data sets generated by users stored in the memory of computers installed in cars, as well as data on the level of customer satisfaction, spare parts rotation, data on vehicle operation acquired during periodic inspections as well as failures and warranty events. The analysis described includes data generated during the implementation of warranty repairs in the Polish dealer network. As a result of literature analysis, a cognitive gap was identified, consisting in the lack of publications on car failure analyses based on data obtained from warranty repairs in the perspective of improving the production process and after-sales processes of cars. This article focuses on the analysis of data sets obtained by authorized service stations in the implementation of after-sales service processes. The following processes were qualified in them: after-sales service, diagnosis, verification and repair process and warranty multi-process [9], defined as a process with inter-organizational features [15–17], constituting a new generation of business processes [18]. The research hypothesis was that certain types of faults appear more frequently in given intervals of the warranty period, and also that they are associated with a specific type of car engine (petrol or diesel). Thus, their diagnosis should enable preventive repairs to be carried out during car inspections before the fault occurs, and thus reduce repair costs during the warranty period.

2 Research Problem

2.1 Bibliometric Analysis

The bibliometric analysis was the motive to undertake the issues described in this article. The analysis was based on two scientific databases: Web of Science and Google Scholar. Two criteria were adopted in the conducted analysis, formulated on the basis of [19, 20]. The first criterion is a criterion that includes only reviewed scientific articles, monographs and post-conference materials into the analysis. They have been divided into publications of: type A – warranty management and after-sales management in the automotive, agricultural and aviation sectors, type B – preventive repairs in the automotive sector, type C – literature reviews and comments to articles A and B. In addition, an elimination criterion has been formulated for type D of publications – regarding after-sales and warranty services outside the areas indicated in type A of publications.

As a result of the literature analysis, the relationships between organizations in the after-sales area of the automotive sector were verified, taking into account the course of the warranty multi-process [21] (Fig. 1).

On the example of a simplified model shown in Fig. 1, it has to be clarified that the manufacturer (NSC) provides a warranty for the final product, built of components manufactured by NSC, as well as by the component manufacturer (OEM). NSC distributes new cars through the manufacturer's local branches (NSC) or private importers (IMP), who are also the grantors on local markets. The authorized service stations (ASO) are the grantors. NSC/IMP by selling a car, ASO is a guarantor on the local market of the product sold by the operator and an intermediary in payments for repairs on product warranties belonging to an external customer who is also an external customer to the grantor (NSC) and operator (ASO). The warranty multi-process or its selected actions can be carried out by external companies (EXT) (Fig. 1).

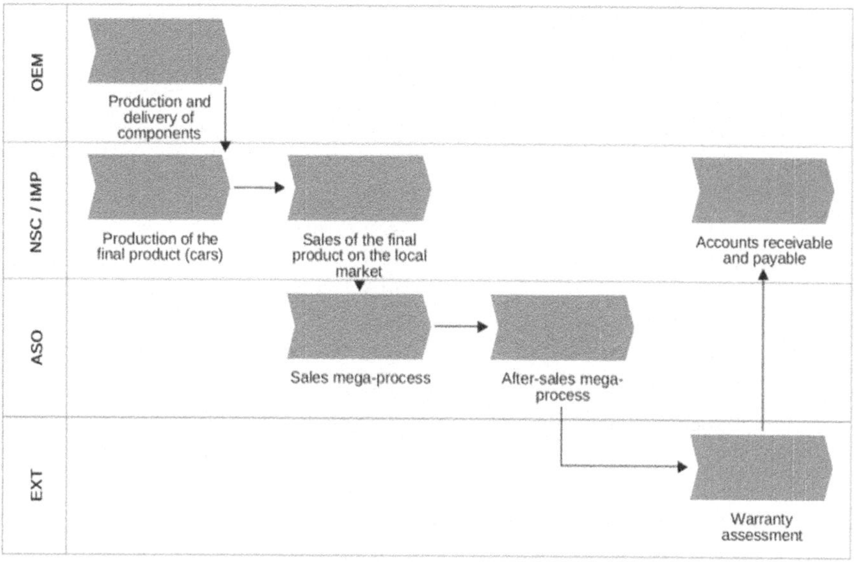

NSC – manufacturer, OEM – component manufacturer, IMP – private importer, ASO – authorized service station, EXT – external companies.
Source: own study using the Adonis NP.

Fig. 1. Model of warranty multi—process in the automotive sector.

2.2 Selection of the Subject of the Study

Based on the literature analysis, it was decided that among the identified organizations in the automotive sector, the study will be carried out at authorized service stations (ASO) (Fig. 1). Next, grouping of the car repair orders in the examined units was made. As a result, paid, complaint, body and paint, internal and warranty orders were identified. At the stage of preliminary verification of data in the documentation, enabling the computational analysis, it was found that the largest amount of data is contained in warranty orders. In addition, the quality of collected data in warranty orders is controlled by the importer (NSC), which positively affects the results of the computational analysis. Moreover, in order to determine the share of warranty repairs in relation to all repairs, a survey was conducted on a sample of 163 authorized service stations in Poland. The empirical study was conducted in Poland in 2015. The study used the surveys used in the process maturity studies of the authorized service stations [22]. The results are shown in Table 1.

Based on the results obtained (Table 1), it was considered that the share of warranty repairs in all types of repairs is significant. As a consequence, a decision was made regarding the selection of the subject of the studies, i.e. the warranty repairs. At this point, it should be emphasized that the analysis of warranty repairs reduced the interventions resulting from perforation, corrosion, varnish defects, service actions and good-will.

Table 1. Percentage share of warranty orders in all repair orders

Country class	Poland n = 163
0–10%	27,61%
11–20%	41,72%
21–30%	9,82%
31–40%	17,79%
41–50%	1,23%
51–60%	1,23%
61–70%	0,61%
71–80%	0,00%
81–90%	0,00%
91–100%	0,00%

Source: own study based on own research.

2.3 Structure of Empirical Proceedings

The empirical proceedings were carried out in 2015–2018 on the example of after-sales service in 30 authorized service stations in Poland. The subject of the study were the completed warranty repairs of one car manufacturer. At this point, it should be emphasized that the results of the computational analysis published in this article concern one model of passenger car marked as "X".

The conclusions presented in this article were formulated on the basis of quantitative data of the entire population of the "X" model repairs (1651 repairs) and based on a random sample. The selection of the research sample was carried out using a probabilistic technique with a simple individual draw. The population in the draw was 1651 of completed repairs. The study assumed a maximum error not exceeding ±5%, with a confidence level = 0,95. The fraction p = 0,5 was adopted as the maximum value of the product (1) [23]:

$$\hat{p} \cdot \hat{q} = \hat{p} \cdot (1 - \hat{q}) \tag{1}$$

On this basis, the sample size was n $\geq$ 312.

2.4 Selection of Variables for the Model

The model of variables for the assessment was based on a review of the analysis of defects in passenger cars, trucks [24, 25], tractors and agricultural machines [26, 27], as well as aircrafts [28] (Table 2).

Table 2. List of reference explanatory variables in the analysis of defects of passenger cars

Variable	Characteristics of the variable
Brand_type	Type of brand of studied vehicles
Model_type	Type of model of studied vehicles
Engine_type	Type of engine installed in the vehicle (combustion, hybrid or electric)
Repair_type	Type of repair carried out in the vehicle (mechanical, electrical, paint, body, service)
Gearbox_type	Type of gearbox installed in the vehicle (manual, automated or automatic)
Client_type	Type of vehicle (private, company or special)
Manufacture_type	Place of car manufacturing
Age_fabr	The number of days from the date of manufacturing of the vehicle to the date of reporting the failure
Age_warr	The number of days from the sale of the vehicle to the date of reporting the failure
Mileage	The number of kilometres of the vehicle from the date of manufacturing the vehicle to the date of reporting the failure
Total_cost	The total cost of the repair
Parts_number	The total number of parts used in the repair

In the next stages of the analysis, based on the databases prepared during the research, four variables were taken into account: engine_type, age_warr, age_fabr and mileage. First, the correlation between the age_fabr, age_warr variables was verified with the mileage variable. Strong positive correlations were obtained for the age_fabr and age_warr variables, 0,56 and 0,54 (p = 0,05) (Fig. 2).

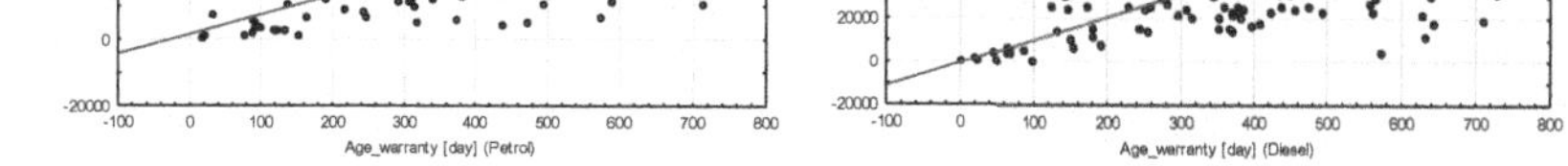

Fig. 2. Linear regression between variables afe_fabr and age_warr
Source: Own study using the Statistica 13 program.

Then, the number of interventions (completed repairs) was verified in the studied group of vehicles both with petrol and diesel engines (Table 3).

Table 3. List of vehicles in a random sample according to the engine_type variable

Parameter	The engine_type variable		Total
	Petrol	Diesel	
Number of vehicles	107	161	268
Number of repairs	114	198	312
Number of repairs per car	1,065	1,300	1,164

Summing up, the number of repairs was summarized for the age_warr variable and the intervals determined by the performance of service reviews after 12 and 24 months of operation were standardized (Table 4).

Table 4. Summary of the number of repairs according to the operating periods expressed in months

Engine	Parameter	(0, 3)	<3, 6)	<6, 9)	<9, 12)	<12, 15)	<15, 18)	<18, 21)	<21, 24)	<24, 27)
Petrol	Number	6	12	8	13	11	16	13	24	11
Share in the total number of repairs		5%	11%	7%	11%	10%	14%	11%	21%	10%
Diesel	Number	11	9	11	21	36	24	27	43	16
Share in the total number of repairs		6%	5%	6%	11%	18%	12%	14%	22%	8%

3 Results

In the further part of the study, a view was formulated that the identified and removed defects should be groups in accordance to the general structural layout of motorized road vehicles qualified to the group of passenger cars [24, 25] (Table 5). The aim of the grouping was: the ability to identify parts and assemblies, most often failures, and consequently: the possibility of implementing preventive repairs, expanding the range of periodic reviews depending on the age_warr and mileage variable, providing information about defects to streamline the production process and identification of sub-contractors.

Using the logit regression, the factors determining the appearance of faults of each of these types were checked. The purpose of the analysis using the logit regression was to check which factors in terms of the engine type (Diesel or Petrol), mileage in terms of exact kilometres travelled or in interval terms or age on a monthly or interval terms – in 3-month intervals, as well as the repair time in hours increased the probability of occurrence/denoted the given type of defect.

Table 5. Characteristics of groups of the passenger car's structural system

Group	Name of the group
G1	Engine
G2	Gearbox
G3	Brake system
G4	Steering
G5	Suspension system
G6	Ventilation and air conditioning system
G7	Electric system
G8	Audio system and communication
G9	Others

The logit models take the following form (2) [28]:

$$P(Y_n = 1|X) = \Lambda(X\beta) = \frac{\exp(X\beta)}{1 + \exp(X\beta)} \tag{2}$$

where, $\Lambda(X\beta)$ is a logistic cumulative distribution, Y_n the variables reflecting a specific result variable, e.g. a kind of a defect at n = 1,…,N cases expressed in binary terms, X is a vector containing a set of determinants of the occurrence of this defect (in the form of a discrete and/or continuous variable).

Tables 6 and 7 present the results of the logit regression calculations with regard to the division into structural groups, presented in Table 5.

Table 6. Results of I logit regression

N = 1649	Group 8	Group 9	Group 1		Group 2
Constant	−2,68***	−2,10***	−1,16***	1,12***	−3,21***
Diesel	−0,57** (−0,03)	−0,48*** (−0,4)	−0,28** (−0,06)		0,48*** (0,04)
age_warr < 3, 6)	0,71** (0,04)				
age_war < 6, 9)		0,78*** (0,08)			0,66* (0,08)
age_warr < 15, 18)					0,82*** (0.10)
age_warr < 18, 21)					1,13*** (0,14)
age_warr21, 24)				0,4*** (0,08)	0,68*** (0,07)

(*continued*)

Table 6. *(continued)*

N = 1649	Group 8	Group 9	Group 1		Group 2
Mileage			8,9e−06*** (1,8e−06)		
Total repair time				0,0004* (7,6e−05)	0,001***
m(0,15000)					
Pseudo R^2	0,02	0,01	0,01	0,01	0,06
Computational R^2	95%	91%	72%	72,2%	87,8%

Source: Own calculations in Gretl, ***- significance level 0,01, **- significance level 0,05, *- significance level 0,1, final margins are given in brackets.

Table 7. Results of II logit regression

N = 1649	Group 7		Group 3	Group 4	Group 6	Group 5
Constant	−1,29***	−1,82***	−4,55***	−5,05***	−2,45***	−1,22***
Diesel		0,41** (0,02)			0,55*** (0,05)	
Age_warr [months]				0,10*** (0,002)		
age_warr (0, 3)			1,45** (0,04)		−0,99* (−0,06)	
age_warr < 9, 12)			1,42*** (0,03)			0,41** (0,08)
age_warr < 12, 15)						0,51*** (0,10)
age_war < 18, 21)	−0,52** (0,03)					
age_warr < 21, 24)					0,37* (0,03)	
Mileage				−1,2e−05* (−2,2e−07)	−3,6e −06	−1,06e−05 *** (−1,7e −06)
Total repair time	−0,01*** (−0,0005)	−0,008*** (−0,0005)				0,001*** (0,0002)
m(0,15000)		0,84*** (0,06)				
Pseudo R^2	0,07	0,08	0,04	0,03	0,01	0,04
Computational R^2	89,8%	89,8%	98,4%	97,9%	90,1%	76,4%

Source: Own calculations in Gretl, ***- significance level 0,01, **- significance level 0,05, *- significance level 0,1, final margins are given in brackets.

The performed analysis showed that:

- the chance of occurrence of a fault in the audio system and communication type was increased by the age of the car of 3–6 months, while it was decreased by the Diesel engine. The age of the car of 3–6 months increased the probability of this problem twice, i.e. by exp(0,71) = 2,03 that is by 103%. In the case of Diesel cars, the probability of this fault was by exp(−0,57) = 0,56 i.e. 43% lower than for petrol cars;
- the chance of occurrence of other defects was increased by the age of the care of 6–9 months, while was decreased by the Diesel type of engine. The age of car of 6–9 months increased the probability of occurrence of this fault almost 2 times, i.e. by exp(0,78) = 2,18 that is by 118%. In the case of Diesel cars, the probability of this defect was by exp(−0,48) = 0,62 i.e. 38% lower than in petrol cars;
- each 10 000 km of mileage increased the probability of occurrence of engine fault by $e^{10000*\beta}$ = exp(0,089) = 1,093, that is by 9,3%, while in the case of Diesel engines, the probability of occurrence of engine defect was by exp(0,28) = 0,76 i.e. by 24% smaller than in the case of petrol engines;
- in the case of car age of 21–24 months, the probability of occurrence of engine fault was by exp(0,4) = 1,49 i.e. by 49% higher than in other age ranges. One-hour longer repair time increased the chance that it was an engine fault by 0,04%;
- the car age of 6–9 months (while in this case the relationship was not very important, because at the level of 0,1), 15–18 months, 18–21 months and 21–24 months, as well the Diesel engine increased the chance of occurrence of a fault of gearbox. The car age of 15–18 months increased the chance of occurrence of a fault of gearbox by 127%, the car age of 18–21 months meant the probability of the fault of gearbox bigger by 209% than in all other time intervals, while the age of 21–24 months by 97%. In the case of Diesel engines, the probability of a gearbox defect was higher by 62% than in the case of petrol engines;
- a greater total repair time reduced the chance of an electrical system failure, as did the age of the car in the range of 18–21 months. The chance for an electrical system fault was however increased by 130% by the mileage of the car in the range up to 15 thousand km;
- the probability of occurrence of a defect of a braking system was increased by 326% by the age of the care up to 3 months, while the age of the car of 9–12 months increased the probability of occurrence of a defect by 313% in relation to the other age ranges of the car;
- each month more of car use increased the probability of a steering system failure by 10,5%, while the higher mileage in kilometres reduced the probability, but in this case the relationship was not very important;
- the probability of a fault in the air-conditioning and ventilation system was 45% higher in the car's age range of 21–24 months, and by 169% lower than in other age ranges of the car up to 3 months (but here the relationship was not very important). In addition, the probability of occurrence of a defect of the ventilation and air-conditioning system was higher by 73% in the case of cars with a Diesel engine;
- the probability of occurrence of a defect of the suspension system was 51% higher at the car's age of 9–12 months than in other ranges, while with the car's age of 12–

15 months it was 67% higher. Therefore, it can be assumed that the greatest chance of occurrence of this defects takes place in the case of vehicles 9–15 months old. Each 10 thousand km of mileage increased the chance of occurrence of the defect of the suspension system by 10%. The time repair longer by one hour meant a 0,1% greater probability that it was a defect of the suspension system.

4 Discussion and Conclusions

The conducted analysis made it possible to obtain original results that may contribute to the implementation of the presented results in business practice, but also to broader research in the area of management and facilitation of car manufacturing processes. As shown in Table 8, it can be indicated that attention should be paid to gearbox, ventilation and air-conditioning and electrical system defects in the case of Diesel engines of the "X" car that occur relatively frequently. In the case of petrol engines, it is necessary to counteract the defects of the engine, audio and communication system, and others which appeared relatively frequently during the warranty period. In the first year of vehicle use, there were frequent defects of the gearbox (6–9 month), braking system (0–3 and 9–12 month) and others (6–9 months). The greatest probability of failures of the suspension system occurred in the 9–15 months of the car's use. The gearbox defects occurred mainly in the 15–25 months of the car's use. In the 21–24 months of the warranty period, there was a higher probability of occurrence of engine, ventilation and air-conditioning system than in other analysed periods.

Table 8. High probability of failure of a given type

Type of defect	Diesel	Petrol	Months							
			0–3	6–9	9–12	12–15	15–18	18–21	21–24	Age-warr
Engine		X							X	
Gearbox	X			X			X	X	X	
Braking system			X		X					
Steering system										X
Suspension system					X	X				
Ventilation and air-conditioning system	X								X	
Electrical system	X									
Audio and communication system		X								
Others		X		X						

Five summary conclusions were formed based on the completed analyses:

- firstly, the approach presented in the article may be an incentive to improve the quality of products as a result of a dynamic analysis of their use (in the discussed issue – cars), selection of sub-contractors of subassemblies and spare parts in the process of car design and manufacture, analysis and prediction of costs connected to the warranty service and forecasting, as well as a detailed verification of the impact of implementation of individual parts or subassemblies on the occurrence of undesirable events during the operation of passenger cars;
- secondly, as a result of the analysis of repair order documentation, a model of data transfer between a mega-process of after-sales service and the warranty multi-process and the processes of designing and testing new products and manufacture has been presented (Fig. 3).

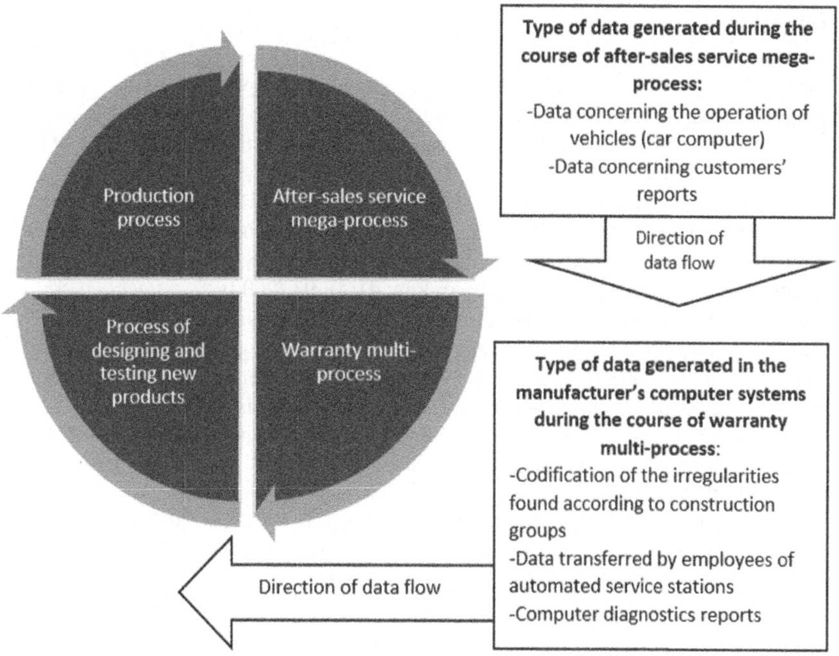

Fig. 3. The model of data flow concerning the failure of car processes based on data generated in the mega-process of after-sales service of passenger cars.

- another conclusions, the implementation of the presented method of defect analysis may have a positive impact on the improvement of the after-sales system measures at the level of organization, implementation and business processes. This mainly concerns the after-sales customer service process, including the improvement of such indicators as: the time of the diagnosis, verification and repair process, the degree of retention and satisfaction of the external customer;

- the fourth conclusion, the extension and implementation of the computational analysis presented in the article may be a contribution to the design and use of dynamic operation inspections, understood as periodic inspections, the implementation scope of which would be determined by each verification of variables: engine_type, mileage and age_warr.
- the final conclusion, additional models that could be used could be multilevel modelling which would allow to check if control variables connected with a specific dealer and derived from this fact probable conditions of the usage of a car (like climate) could have impact on the occurrence of the car defects. Moreover, general linear modelling with multiple dependent variables might be used. Also probit, instead of logit transformation could be used in the modelling, which belongs to the same as logit regression type of microeconometric models.

References

1. Gonzalez, R.V., Melo, T.M.: Innovation by knowledge exploration and exploitation: an empirical study of the automotive industry. Gestão Produção **25**(1), 1–15 (2017)
2. Ili, S., Albers, A., Miller, S.: Open innovation in the automotive industry. R&d Manag. **40** (3), 246–255 (2010)
3. Cavalieri, S., Gaiardelli, P., Ierace, S.: Aligning strategic profiles with operational metrics in after-sales service. Int. J. Prod. Perform. Manag. **56**(5/6), 436–455 (2007)
4. Cohen, M.A., Cull, C., Lee, H.L., Willen, D.: Saturn's supply-chain innovation: high value in after-sales service. MIT Sloan Manag. Rev. **41**(4), 93 (2000)
5. Gaiardelli, P., Saccani, N., Songini, L.: Performance measurement systems in after-sales service: an integrated framework. Int. J. Bus. Perform. Manag. **9**(2), 145–171 (2006)
6. Gaiardelli, P., Saccani, N., Songini, L.: Performance measurement of the after-sales service network—Evidence from the automotive industry. Comput. Ind. **58**(7), 698–708 (2007)
7. Saccani, N., Songini, L., Gaiardelli, P.: The role and performance measurement of after-sales in the durable consumer goods industries: an empirical study. Int. J. Prod. Perform. Manag. **55**(3/4), 259–283 (2006)
8. Verstrepen, S., Deschoolmeester, D., Van den Berg, R.J.: Servitization in the automotive sector: creating value and competitive advantage through service after sales. In: Global Production Management, pp. 538–545. Springer (1999)
9. Bakker, A., Kent, P., Noss, R., Hoyles, C.: Alternative representations of statistical measures in computer tools to promote communication between employees in automotive manufacturing. Technol. Innov. Stat. Educ. **3**(2), 1–31 (2009)
10. Dale, B.G., Shaw, P.: The application of statistical process control in UK automotive manufacture: some research findings. Qual. Reliab. Eng. Int. **5**(1), 5–15 (1989)
11. Oakland, J.S.: Statistical Process Control. Routledge, Abingdon (2007)
12. Sliż, P.: Statistical analysis of the process of repair of automotive vehicles using methods of CPM and PERT network analysis. Sci. Pap. Silesian Univ. Technol. Organ. Manag. Ser. **18**, 515–524 (2018)
13. Grudzewski, W.M., Hejduk, I.K.: The enterprise of the future. Changes in management paradigms. (Przedsiębiorstwo przyszłości. Zmiany paradygmatów zarządzania). Master Bus. Adm **116**, 95–111 (2011)
14. Bouchbout, K., Alimazighi, Z.: Inter-organizational business processes modelling framework. In: Proceeding of the ADBIS, vol. 2, pp. 45–54 (2011)

15. Garmann-Johnsen, N.F., Eikebrokk, T.R.: Critical success factors for inter-organizational process collaboration in eHealth. In: Proceeding of the 6th International Conference on eHealth, Telemedicine, and Social Medicine, pp. 217–223 (2014)
16. Legner, Ch., Wende, K.: The challenges of inter-organizational business process design – a research agenda. In: Proceeding of European Conference on Information Systems, ECIS, pp. 1643–1654 (2007)
17. Breu, R., Dustdar, S., Eder, J., Huemer, C., Kappel, G., Köpke, J., Langer, P., Mangler, J., Mendling, J., Neumann, G., Rinderle-Ma, S.: Towards living inter-organizational processes. In: Business Informatics (CBI), pp. 363–366 (2013)
18. Jones, M.V., Coviello, N., Tang, Y.K.: International entrepreneurship research (1989–2009): a domain ontology and thematic analysis. J. Bus. Ventur. **26**(6), 632–659 (2011)
19. Servantie, V., Cabrol, M., Guieu, G., Boissin, J.P.: Is international entrepreneurship a field? A bibliometric analysis of the literature (1989–2015). J. Int. Entrepreneurship **14**(2), 168–212 (2016)
20. Sliż, P.: The role of multiprocess maturity in establishing interorganizational relations - report of empirical studies. (Rola dojrzałości multiprocesowej sieci w kształtowaniu relacji międzyorganizacyjnych-wyniki badań empirycznych). J. Manag. Finance **14**, 325–336 (2016)
21. Sliż, P.: Business process maturity – report of empirical research (Dojrzałość procesowa organizacji – wyniki badań empirycznych). Res. Pap. Wroclaw Univ. Econ. **421**, 530–532 (2016)
22. Szreder, M.: Methods and Techniques of Public Opinion Polls. (Metody i techniki sondażowych badań opinii). Polskie Wydawnictwo Ekonomiczne, Warszawa (2010)
23. Prochowski, L.: Traffic Mechanics. (Mechanika ruchu) (2005)
24. Prochowski, L., Unarski, J., Wach, W., Wicher, J.: Motor Vehicles Traffic Mechanics. (Pojazdy samochodowe. Mechanika ruchu) (2008)
25. Juściński, S., Piekarski, W.: Logistic management of an authorized service for agricultural tractors and machines. (Zarządzanie logistyczne autoryzowanym serwisem ciągników i maszyn rolniczych). Eksploatacja i Niezawodnosc – Maint Reliab **2**(38), 25–33 (2008)
26. Rotz, C.A.: A standard model for repair costs of agricultural machinery. Appl. Eng. Agric. **3** (1), 3–9 (1987)
27. Pogačnik, B., Duhovnik, J., TAvčAr, J.: Aircraft fault forecasting at maintenance service on the basis of historic data and aircraft parameters. Eksploatacja i Niezawodnosc – Maint. Reliab. **19**(4), 624–633 (2017)
28. Greene, W.H.: Econometric Analysis. Prentice Hall, Upper Saddle River (1993)

Author Index

A. Hamrol et al. (Eds.): *Advances in Manufacturing II* - Volume 3, LNME, pp. 291–292, 2019.
https://doi.org/10.1007/978-3-030-17269-5

The manufacturer's authorised representative in the EU is Springer Nature Customer Service Centre GmbH, Europaplatz 3, 69115 Heidelberg, Germany. If you have any concerns regarding our products, please contact ProductSafety@springernature.com

Printed and bound by CPI Group (UK) Ltd, Croydon, CR0 4YY
30/09/2026
02225230-0006